LOUIS AGASSIZ: A LIFE IN SCIENCE

LOUIS AGASSIZ

A LIFE IN SCIENCE

BY EDWARD LURIE

THE JOHNS HOPKINS UNIVERSITY PRESS
BALTIMORE and LONDON

© 1988 by Edward Lurie
All rights reserved
Printed in the United States of America

Originally published, 1960, by the University of Chicago Press
Johns Hopkins Paperbacks edition, 1988

The Johns Hopkins University Press
701 West 40th Street, Baltimore, Maryland 21211
The Johns Hopkins Press Ltd., London

Library of Congress Cataloging-in-Publication Data

Lurie, Edward, 1927–
  Louis Agassiz, a life in science.

  Originally published: Chicago : University of Chicago Press, 1960.
  Bibliography: p.
  Includes index.
  1. Agassiz, Louis, 1807–1873. 2. Naturalists—Switzerland—Biography.
3. Naturalists—Switzerland—Biography. I. Title.
QH31.A2L8 1988   508'.092'4 [B]   88-45392
ISBN 0-8018-3743-X (pbk.)

Dedicated to the Memory of

ALEXANDER LURIE

who loved men, ideas, books, and nature

PREFACE

Louis Agassiz was a student of nature. This biography seeks to report the conditions and passions of that existence.

To Agassiz, the meaning of the creative process was spiritual, and understanding came to him through emotional involvement in intellectual effort. The study of nature was for Agassiz the study of the universe, and so he thought of himself as mirroring the grandeur of natural history through his perceptions. The inner strength that came from this conception enabled Agassiz to think of himself as a unique and singular individual.

Agassiz was perpetually youthful in outlook because he was always optimistic, always looking toward the future. His romantic soul was forever thrilled by the potentialities of a new venture. People were rarely neutral in their reactions to this vibrant, dedicated man. A giant of the nineteenth century, he strode through the world determined to succeed. Success came to Agassiz because he was a person deeply involved in his surroundings, a man who understood the possibilities of life with an uncommon awareness, who knew the uses of power and the techniques by which society could be shaped to provide for the wants of his intellect and the sustenance of his romance with nature.

A man of science, Agassiz was no less a man of the world. He mastered the social environment of Europe and then demonstrated an even greater ability to dominate the culture of the United

States, his second and permanent home. This talent, enriched by the captivating, almost magical charm by which he influenced men of position and ordinary folk, meant that he always received social support for his ambitions. Kings and commoners alike were somehow uplifted from participating in the realization of an Agassiz dream.

Agassiz was a person of contrasts; he loved nature and man in general, but he was determined to let no one stand in the way of his ambition. Ambition for him was always a selfless aim; personal welfare and the progress of science were one and the same. He exemplified a belief that the life of the mind was a noble impulse in and for itself, yet he always acted to demonstrate that knowledge, to be culturally valuable, had to be shared with others. In America, the culture of democracy responded with high enthusiasm to Agassiz, whose public ambitions seemed perfectly in tune with his times and who only wanted to share the excitement that was his when he reconstructed a fossil fish, scaled a mountain peak, discovered signs of ancient glaciers, or collected turtle eggs.

Standing between two diverging intellectual climates—the idealism of an older world view and the empiricism of modern times—Agassiz's mind reflected the larger philosophical contrasts of his age. Possessed by universal compulsions, he taught men to appreciate specialized knowledge and impressed society with the need to support science and advance the professional status of its practitioners. In approaching nature with a complete identity of mind and spirit, his subjective assessment of creation was in sharp contrast to the objectivity of a developing modern attitude, but he had done much to inspire that attitude by his lasting contributions to ichthyology, paleontology, and geology. In an important sense, Agassiz helped create cultural conditions that made his cosmic world view less and less meaningful with the passing of time.

Surrounded by multitudes of admirers, Agassiz was nevertheless a man who lived in personal loneliness and in growing intellectual isolation. These were the penalties resulting from years of authority and from the urge to domineer society and science. Yet he never knew failure, and he left the world richer for his presence.

The writing of this book was facilitated by many people and institutions. It is a distinct pleasure to acknowledge the deep indebtedness I feel for such aid. The volume began as a doctoral dissertation dealing with Agassiz's American career, written at Northwestern University. Professor Ray Allen Billington directed this study, providing unfailing encouragement at every stage of writing and research. Initial research was made easier and my understanding of American science increased, by the award of an Advanced Graduate Fellowship from the American Council of Learned Societies. While working under this grant I benefited from the advice and kindness of Professors I. Bernard Cohen and Richard H. Shryock. Subsequent research was made possible in part by a grant-in-aid from the Horace H. Rackham Faculty Research Fund of the University of Michigan, and a Faculty Research Fellowship from the Graduate School of Wayne State University.

I should like to thank Harvard University for permission to consult and use manuscripts on deposit at the Houghton Library, the University Archives, the Gray Herbarium, and the Museum of Comparative Zoology. My work in Cambridge was made rewarding by the kindness of many people. I am particularly grateful for the assistance of Professor William A. Jackson and Miss Carolyn Jakeman of the Houghton Library, Mr. Kimball C. Elkins of the University Archives, and the staff of the Gray Herbarium. I am indebted to Professor A. Hunter Dupree, who shared with me his extensive knowledge of manuscripts relating to Gray and who made copies of letters available to me. My manuscript was completed before I could benefit from his recently published biography of Gray. I owe a special debt of gratitude to the staff of the Museum of Comparative Zoology at Harvard College. Without the interest, advice, and unfailing assistance of people who work in the spirit of Agassiz and in the building he began, this book could not have been written. Of these, I am appreciative of the help of Miss Jessie B. MacKenzie, librarian, and for the many kindnesses of Miss Ruth C. Norton. Dr. Tilly Edinger of the museum was of direct assistance in research, and was a welcome auditor, hearing my opinions on Agassiz and his work with understanding and sympathy. Professor Alfred S. Romer, who as director of the Museum of

Comparative Zoology is Agassiz's modern successor, enhanced my understanding of Agassiz as a man and a naturalist. His advice and assistance were invaluable. The illustrations in the volume are all reproduced through the courtesy of the museum, unless otherwise acknowledged.

I also wish to acknowledge the kind co-operation of the staffs of the following archives and institutions: the American Philosophical Society, the Academy of Natural Sciences of Philadelphia, the Historical Society of Pennsylvania, the Library Company of Philadelphia, the Manuscripts Division of the Library of Congress, the Smithsonian Institution Archives, the United States National Archives and Records Service, the Yale University Library, the Collection of Regional History and the University Archives of Cornell University, and the New York State Museum at Albany. I am also grateful to the following people who helped in special ways: George W. White, Claude W. Hibbard, Ernst Mayr, William J. Clench, James H. Zumberge, Albert Hazen Wright, Donald F. Zinn, Nathan Reingold, Karl Guthe, Francis C. Haber, and W. B. McDaniel, II. Mrs. Louise Hall Tharp was very kind in giving me information on Elizabeth Cary Agassiz and providing me with copies of manuscript material.

My greatest intellectual obligation is to Professor Bert James Loewenberg of Sarah Lawrence College. He first stimulated my interest in Agassiz and American science and culture, advised me in the revision of the manuscript, and was a source of constant support. This book reflects his friendship and also, I hope, his teaching. My wife, Dr. Nancy Oestreich Lurie, was a hard taskmaster and a wonderful critic who took valuable time from her own scholarly labors to help me in mine. This book is the product of her encouragement, and I hope she likes it.

CONTENTS

ILLUSTRATIONS
(Following page 84)

LOUIS AGASSIZ: A LIFE IN SCIENCE

1

THE FORMATIVE YEARS

1807–1827

Louis Agassiz was almost fifteen years old when he outlined his first program for future accomplishment. Agassiz would always live with mind and spirit keyed to the grand promise of tomorrow. Visions born from imagination and ambition would become realities through the application of intense energy, the enthusiasm of a ruling passion, and the help of others who were somehow uplifted through participation in an Agassiz venture. The youth of fifteen and the mature man of fifty both looked toward the future with the same optimistic self-confidence. Projects and dreams that might seem impossible or unrealistic to ordinary men were pursued and fulfilled because Agassiz and those who helped him believed that large plans are made real through the faith of large-minded men. For Agassiz, the future would always be the spiritual resting place for his searching intellect; the promise of things to come would always be more attractive than the realities of life in the present. The ideals born of imagination were what made the present meaningful. They comprised the inner personal resources that made life bearable while anticipating grand accomplishments to come.

1

The first Agassiz plan for the future took form as the young man was completing his final year of preparatory study at the Collège de Bienne in his native Switzerland. Having received a good classical education that stressed languages, history, mathematics, and the natural and physical sciences, Agassiz set about recording, in a private memorandum, a series of personal and intellectual goals.

"I wish to advance in the sciences," were the first words of this memorandum, reflecting a determination that would never yield to circumstance. "I have resolved, as far as I am allowed to do so, to become a man of letters, and at present I can go no further." Intellectual tools and the assistance of others were primary requirements, Agassiz knew, in the furtherance of this aim. Teachers at Bienne had been helpful in the past, and if more assistance was forthcoming, success would be possible. Books were of first importance:

In ancient geography, for I already know all my note-books, and I have only such books as Mr. Rickly can lend me, I must have d'Anville or Mannert; 2nd, in modern geography, also, I have only such books as Mr. Rickly can lend me, and the Osterwald geography, which does not accord with the new divisions; I must have Ritter or Malte-Brun; 3d, for Greek I need a new grammar, and I shall choose Thiersch; 4th, I I have no Italian dictionary, except one lent me by Mr. Moltz. I must have one; 5th, for Latin I need a larger grammar than the one I have, and I should like Seyfert; 6th, Mr. Rickly tells me that as I have a taste for geography he will give me a lesson in Greek (gratis), in which we would translate Strabo, provided I can find one. For all this I ought to have twelve *Louis.*

With these intellectual and financial necessities set forth, there was the future to consider, and the future required serious planning for a youth so determined to "advance in the sciences":

I should like to stay at Bienne till the month of July, and afterward serve my apprenticeship in commerce at Neuchâtel for a year and a half. Then I should like to pass four years at a University in Germany, and finally finish my studies at Paris, where I would stay about five years. Then, at the age of twenty-five, I could begin to write.[1]

For more than fifty years, the world of science and letters would feel the impact of the dedication and ambition revealed by this youthful conception. For an adolescent to dream of the next ten years of his life was not, in itself, unusual. What was uncommon was that young Agassiz fully expected to realize these ends, even though the material means for accomplishing them were not at all predictable or certain. A practical and prosaic acceptance of the world would never be Agassiz's habit of mind. He had to act upon it, to transform it and make it sympathetic to his own images and purposes. And the world would always meet this singular individual on his own terms. Environment and circumstance would yield to will and imagination, because Agassiz's conception of the world comprised a personal faith in his own ability that would and could move mountains. This talent for building from inner resources to make dreams come true was proved at an early age. The essential goal young Agassiz set for himself in his ten-year plan was fulfilled in ample measure. At twenty-five he had profited from an excellent education in natural science and had indeed begun to write. He had become a man of letters of considerable promise who could envision new achievements on the strength of past accomplishments.

The education of Louis Agassiz is a story of determination, resolve, and personal transformation. Born a Swiss, Agassiz grew to manhood within an intellectual tradition that boasted of such figures in the history of European thought and culture as Ulrich Zwingli, Heinrich Johannes von Müller, Albrecht von Haller, Charles Bonnet, Horace Bénédict de Saussure, Jean Jacques Rousseau, and Johann Heinrich Pestalozzi. Agassiz was fortunate in his place of birth, but this single fact cannot account for his early and unique achievement. The youth who thrilled to the natural grandeur of the lakes and ice formations of his homeland, who attempted to appease parental desires but also to create opportunities for an independent life, was indelibly impressed by his environment and at the same time was able to transcend its limitations when greater opportunities, real or imagined, rose to inspire him. He could find strength and support in being a good and devoted son, in absorbing knowledge so that he could add to the distinction of the Agassiz name, yet the respectful son and bright student who grew up honoring his native traditions and surroundings was, during his mature life, a Frenchman by friendship and association, a German by intellectual aspiration, and an American by adoption.

But such a status in the world had to be worked for. The days of youth and early fame were marked by restless dissatisfaction with things as they were. Young Agassiz strove to emancipate himself from the confining bonds of a middle-class, mercantile society, where success and achievement were measured by propriety, by behavior within an established framework. Conscious of the virtues of stability and tradition, Agassiz nevertheless rejected these values and strove to identify himself with the larger world of cosmopolitan academic and intellectual life, with the free spirit of science and learning. His youthful outlook was thus radical and libertarian, set as it was against the milieu of Swiss middle-class respectability.

Though christened Jean Louis Rodolphe, Agassiz was always called Louis by family and friends of boyhood days. In later years, most intimates would address him as Agassiz, or "Agass," the identifying symbol of so many animal forms first described by him. Born on May 28, 1807, in the Motier parsonage, he was his parents' fifth child but the first to survive infancy. Motier, a grouping of four small villages, was a beautiful place. Surrounded on three sides by water—the lakes of Morat and Neuchâtel and the Broyé River—the Agassiz birthplace was situated at the eastern foot of Mount Vuly and looked out upon the majestic Bernese Alps. To the north and west stretched the Jura Mountains. Encompassed thus by mountains, valleys, lakes, grassy fields, and wandering streams and rivers, Motier was an appropriate boyhood home for one who would spend his mature years in the study of both marine life and glacial forms.[2]

The Agassiz home in Motier was a place of some distinction, since Louis's father, Rodolphe, was the assistant pastor to the Protestant congregation. Rodolphe was thirty-one years old when his son Louis was born, and the year 1807 was an important one in his life. He had arrived in Motier the year before, having previously served in the parish of St. Imier in the Jura Mountains. Motier was much more congenial and friendly to Rodolphe and his family, and in moving there the elder Agassiz had returned to a region of west central Switzerland that had long been the home of his relatives. Motier was an island of devout Protestantism in the midst of the Catholic canton of Fribourg. There a pastor had to be a strong and resourceful spiritual and intellectual leader of his congregation. That Rodolphe Agassiz was successful is attested to by the fact that by 1810 he had become the head pastor and had developed a not insignificant reputation as an inspiring teacher in secular as well as religious subjects.

Louis's father was a man of strong will and abiding purpose. He was conscious of the importance of his position in the community as a member of the middle-class elite of professional people who were looked upon as arbiters between peasants and artisans and the landed aristocracy. The Agassizes had long been noteworthy figures in the cultural life of their region. The family traced its Swiss origins as far back as the thirteenth century to the villages of Orbe and Bavois located in the canton of Vaud adjacent to Fribourg. The Protestant Reformation had figured significantly in family history. Since the mid-seventeenth century there had been an unbroken succession of Protestant ministers in the Agassiz line. Rodolphe was the sixth of such pastors.

Rodolphe's position in the community was enhanced by the fact that his wife and her family also represented an influential segment of Swiss society. Rodolphe had married well. At Constantine, a village near his father's parsonage, he had met and won Rose Mayor, the daughter of a physician from neighboring Cudrefin, a village not far from Motier. Her family had long been shopkeepers, commission merchants, and physicians, occupations that were eminently respectable and influential in the social and economic structure of Switzerland. Rose Mayor Agassiz was a woman of strong convictions, remarkable intelligence, a keen wit, and energetic disposition. She was a delightful and charming woman, able to engage in spirited and informed conversation on a wide range of subjects. After Louis's birth in her twenty-fourth year, she presented her husband with three more children, a son, Auguste, and two daughters, Olympe and Cécile. Louis was clearly her favorite child, the source of her highest hopes and aspirations.

A sound education would be followed, she hoped, by some respectable career—perhaps the ministry, although she would prefer an occupation such as merchant or businessman where Louis could be helped by her successful brothers. The long-established roots of both families in the region permitted a certain amount of mutual aid in economic and social occupations. It also meant that young Louis had many cousins to share boyhood amusements at Motier and Cudrefin.

The prosperous cantons of Fribourg, Vaud, and Neuchâtel comprised one of the most important food-producing and small manufacturing regions of French-speaking Switzerland. Vineyards covered the shores of Lake Neuchâtel, and the Agassiz home in nearby Motier was typical in having both a vineyard and an orchard. These cantons are still a primary source of Switzerland's home-produced food; dairy farming, truck gardening, fruit-growing, and the manufacture of wine were all familiar occupations among the people Agassiz knew as a boy. Watchmaking and domestic lace manufacture, profitable and highly respected skills, occupied the native artisans. Horology in particular was a principal livelihood for many people of the region, and Switzerland has derived a large share of her clock-making fame from the work of this area. The region was also the most important transportation and communications route through the western Alps; generations of travelers thrilled to the awesome beauty of the surrounding mountain peaks and glaciers. Young Agassiz himself learned of such wonders at an early age by exploring them on foot.

Agassiz's native canton of Fribourg, part of the rolling country typical of the region between the Rhine River to the north and Lake Geneva to the south, was peopled by a mixture of Alamannians, Burgundians, and French Huguenots. The last group was probably represented in the Agassiz family.

The social divisions of French Switzerland in the early nineteenth century consisted of a landed aristocracy and officialdom at the top, a substantial mercantile and professional class in the middle, and peasants and artisans at the bottom. The predominant middle-class mores of the period played a significant part in Agassiz's development. What was the natural and proper way of life for a Swiss of this background? First, and perhaps most important, there was the driving ambition for material success exemplified by Rose Mayor's family. There were, of course, exceptions to this social compulsion, and these were to be found in the professional classes, the teachers and clergy, who, while not enjoying the affluence of the merchant, nevertheless knew a social status of equal rank. Rodolphe Agassiz and his wife's physician-father belonged in this group, a social entity or considerable significance in the political and intellectual history of Switzerland. The Agassiz family, then, was notable for its secure and significant social and economic role in a region that was typified by a prosperous, stable society with a firmly established class structure.

In this society, whatever avenue toward the fulfilment of family social patterns one decided to follow, education was a primary requirement, a psychological as well as a practical necessity. It provided the middle class with that superior command of history, political theory, geography, languages, and literature that gave impetus to the effort to achieve a distinct Swiss nationality. It supplied a mark of quality and distinction that set the middle class apart from people who tilled the fields or worked the vineyards. Understandably, then, even before Louis's formal schooling began, his father took care to instruct him in languages and literature in order that he might follow the family pattern of achievement.[3]

Formal studies were not, however, the most important aspect of Louis's early life. Of much greater interest were the concerns and activities of a rural boyhood. He was an energetic youth, constantly organizing jaunts through the countryside with friends, cousins, sisters, and his brother. Among his favorite diversions, quite understandable in a boy reared in his surroundings, was collecting specimens of natural history. Louis, however, showed a more than usual interest in animate creation. Fish held a special fascination for him, and he found such a variety of them in nearby lakes and streams that he converted the stone catch basin of the spring behind his home into an aquarium. This was but the first in an increasingly larger number of depositories destined to house the objects of Agassiz's searching curiosity about the wonders of nature.

With Auguste as an eager helper, Louis learned the crafts of fisherman, hunter, trapper, and field naturalist. Swimming in the lake, the two boys discovered that they could attract fish by remaining very still; a sudden grab with their bare hands and they had another specimen for study in the catch-basin aquarium at home. A variety of animals took up residence in Louis's room, which in his own words became "a little menagerie." Raising caterpillars until they became butterflies, studying the habits of fish and birds—these were the happy diversions of a pastoral youth that passed, as such days must, much too quickly.

In 1817, at the age of ten, Louis was sent twenty miles away to the Collège de Bienne to begin his formal education; here he spent almost five years in grade-school studies typical for a boy of his background. While the physical sciences and mathematics were not particularly appealing to him, Agassiz was attracted to the study of languages, especially Greek, Latin, German, and Italian. Showing a marked proficiency, he became effectively multilingual, a skill which was to be an asset throughout his life. Intellectual curiosity was already crowding into his mind in a manner more than usual for a schoolboy. As the private memorandum written a few years later demonstrated, the study of ancient and modern geography fascinated him. He tried to learn all he could about the history of the ancient world and about the discoveries of geographers and explorers of his own day.

Vacations at home were spent adding to collections of insects, birds, fish, and small land animals. By the time of Agassiz's adolescence, the study of nature had become something more than the accumulation of curious bits and pieces of interesting data. He wanted to know the underlying reasons for the phenomena he observed, to discover relationships, to understand general concepts. In this quest, young Agassiz proceeded to educate himself. He learned the linguistic and technical classifications of nature study. If one caught an unusual fish, what was one to do with it? First, he realized, it was necessary to know the exact identity of the creature. Did it have a name? Did it bear a relationship to other fishes it seemed to resemble? What were its habits of life? These were some of the basic questions which, according to Agassiz's own account of his youth, he asked himself.

I picked up whatever I could lay my hands on, and tried, by such books and authorities as I had at my command, to find the names of these objects. My highest ambition was to be able to designate the plants and animals of my native country correctly by a Latin name, and to extend gradually a similar knowledge in its application to the productions of other countries.[4]

Nomenclature and classification were obviously interesting and important, but what fascinated Louis much more was the study of the thing-in-itself, the raw material of nature. It seemed regrettable that a fifteen-year-old had to spend time studying mathematics and the classics. "I spent most of the time I could spare . . . in hunting the neighboring woods and meadows for birds, insects, and land and fresh water shells," he confessed at a later date.[5] Books were important, but he thought it more educational to study living things in their natural habitat. Because of this intense interest Agassiz could claim in later years that what he knew of the habits of central European fresh-water fishes he had learned from his youthful observations of them. Nevertheless, he realized that accuracy in recording data was also necessary. From the age of eleven until he was nineteen, Agassiz kept minute and detailed accounts of his natural history observations, set down in fine script in a series of notebooks, with subjects classified and divided carefully under proper subject headings. The importance with which he regarded this activity is revealed by the fact that Agassiz carefully preserved these notebooks, treasuring them as intellectual landmarks of his first scholarly efforts in natural history.

The days of youth were not so entirely given over to nature study that Agassiz neglected other pleasures. Holidays were always a festive and welcome interlude, especially the Easter season, when he would visit the Mayor home at Cudrefin and join in the gaiety. The pastor's son home from Bienne was attractive to the girls of the village, who gathered at the Mayor household for dances, captivated by the handsome lad with deep-set eyes and laughing countenance. Mayor cousins and their village girl friends counted themselves fortunate if they could engage the attention of the energetic, gay youth who could provide such fascinating stories and amusing diversions. Louis was exceptionally strong, and he delighted in opportunities to demonstrate his physical prowess. He loved to swim and hike as well as to join in the singing and wine-drinking fetes typical of the autumn grape harvest season in Swiss villages around Lake Neuchâtel.

During the final year of his Bienne schooldays, Louis was conscious of the need to plan his future life. Parents and uncles had their ideas for him, all of which were predictable. Now that the boy had a good education, he could begin to earn his way in the world. In the town of Neuchâtel, Uncle François Mayor directed a prosperous trading establishment, a kind of commercial clearinghouse for financial transactions. The perfect course presented itself. Louis would go to Neuchâtel, live in his uncle's house, and put his education to use by serving an apprenticeship. The time was at hand to end the frolicking of youth, to pass beyond the dancing parties, swimming, and the gathering of strange bugs and fish. Four or five years of practical experience in business would be the right sort of higher education needed to make Louis a success in the world. The arena of commerce and material accomplishment, in the eyes of realistic parents and relatives, was certainly a proper and gainful place for a good son.

His parents could not have understood their son less. Despite the careful recording of notations regarding Latin names and the coloration of fish, Agassiz had shown no interest in the facts of business or in matters pertaining to finance. A career in the ministry would have been an understandable and perfectly traditional alternative. Yet what his parents never suspected, and Louis himself realized only dimly, was that while at Bienne he had decided to free himself entirely from the demands and expectations of his middle-class origins. He would not be a clergyman, that much was plain, for a rural Swiss pastor would have little opportunity to study nature professionally. No matter how much Louis respected his father, a parochial clergyman's life in French Switzerland was much too humdrum an existence for a boy filled with the energy and ideals of young Agassiz. He understood full well that status and social position were important and that these could be gained through a career in business or the ministry. Yet commerce and theology represented the binding forces of localism and village conformity. Suppose one followed a career in business at Neuchâtel? What could the future hold for a youth who took up life in this environment?

Since 1707, both the canton and the town of Neuchâtel had been the domain of the King of Prussia, an authority that was broken but for a brief period of French ascendancy under Napoleonic rule. Prussian dominance was restored in 1806 by Napoleon and strengthened in 1814 by a constitution granted the Neuchâtelois by Frederick William III. This document perpetuated the power and privileges of the aristocracy and landed nobility who filled all the high offices of civil power except that of the governor, appointed by the Prussian monarch. This native aristocracy ruled the canton with a firm hand; if anything, the Prussian monarch was often regarded by the bourgeoisie as a countervailing power to set against the native nobility. When Neuchâtel joined the Swiss confederation and Federal Pact in 1815, the dominance of Prussia, while not removed, went unrecognized by the federal government. It was nevertheless a potent factor in the social structure of Neuchâtel, a canton that represented a curious mixture of aristocracy, foreign monarchy, and republican elements. The town of Neuchâtel, as the capital of the canton, reflected the entire social and political structure of the whole principality. The town was the seat of cantonal government, its politics and culture dominated by the ruling Prussian officialdom and native aristocracy. It was difficult for a rural Swiss youth of Agassiz's mental outlook to imagine how he might achieve distinction in such an environment. Identification with the radical republican faction of Neuchâtel represented a possible route to political prominence, but such a role was not usual for a businessman. Wealthy and successful businessmen enjoyed close ties with the native aristocracy, but it was hard to envision immediate success of this sort, and Agassiz wanted to succeed quickly. Besides, a business occupation would leave little time for the study of nature.[6]

This environment was clearly not attractive to a romantic, independent youth of Agassiz's temperament. How could he resist the demands of parents and relatives and strike out on his own? This was the problem that faced Agassiz in 1822. He understood that some alternative was necessary if he were to oppose familial pressures, and that success of the type he sought was not to be found through the paths followed by his paternal and maternal relatives. Like Stendhal's Julien Sorel, Agassiz would discover that success could be achieved by breaking the ties of environment and doing the unusual; but such feats must be accomplished by personally distinctive yet traditionally acceptable modes of behavior and social relationships. Julien chose theology as his entrance into a new world. Agassiz chose science, a career that was respectable in that such distinguished names in the intellectual history of Switzerland as Bonnet, Haller, De Saussure, and Augustin de Candolle had followed it with marked success.

At fifteen, Agassiz had three essential attributes that would help him in his ambition: an insatiable curiosity about nature, a good knowledge of languages, and an intense desire to succeed. These were the qualities that enabled him to avoid the quiet, predictable, and dull life of a Neuchâtel apprentice in commerce. When he finally did go to Neuchâtel many years later, it was not as a businessman, but as a professor of natural history. Neuchâtel would know his talents, but it would be the Neuchâtel of the nobility, the foreign aristocracy, and the academic community.

At this juncture in his life, Agassiz wrote the determined memorandum to himself. He would dedicate his energy to the world of science and letters. To do this he would attend a German university, study in Paris, and then "begin to write" as a professional scholar. Plans of this sort required personal resolution, self-discipline, and sustained energy, as well as assistance from others. Almost by an act of self-will Agassiz determined to transform himself from a conventional rural youth into a dedicated scholar. Had he taken religious orders, or decided to follow a military career, the transformation from youth to an almost frightening maturity could not have been more profound. His decision to attend a German university meant intellectual and personal liberation, challenge, excitement, and an identification with a world community of scholarship where individual ability and attainment would be the measure of success. He would then go to Paris, a cultural complex that signified political stability, refinement, urbane manners, and worldly attitudes. Paris also meant the Academy of Sciences, the university, and the National Museum of Natural History. Agassiz now looked toward the intellectual centers of Europe to supply him with both the moral and mental armament to resist and reject the demands of his cultural origins.

Strength of purpose would be of no avail without material support. But Rodolphe and Rose Mayor Agassiz had decided that Louis's next activity would be in the commercial establishment of François Mayor. Young Agassiz recognized the force of this parental compulsion by planning to compromise with family wishes and spend a brief period in business at Neuchâtel before going on to an academic career. Realizing that this was a meaningless temporary expedient, however, he tried to effect a complete acceptance of his personal goals by his parents. His admiring teacher at Bienne, Mr. Rickly, was chosen as the person who could best prevail upon his mother and father because a teacher's opinion would be respected. Louis wisely confided his innermost ambitions to him, and Mr. Rickly, clearly impressed with the intelligence and exceptional motivation of the student to whom he had lent so many books, interceded with the Agassiz family.

The immediate result was that Louis was allowed to spend two years at the Academy of Lausanne, in the canton of Vaud. The decision on Agassiz's commercial career was apparently suspended for this period, and the ostensible purpose of the Lausanne years was to round out studies begun at Bienne, particularly in literature and other humanistic studies. Faced with their son's determination, the Agassizes found much to recommend their choice of Lausanne. It assured that he would not be far from home and family influences. In 1821, Rodolphe had left the Motier parish and moved his family to the village of Orbe in the canton of Vaud. In so doing he had returned to the place of origin of the Agassiz family, serving here, as he had at Motier, as minister to the local congregation. Lausanne was relatively close to the new parish town, and Louis would be among people who had known his family long before he was born.

The Vaudois were warm, good-natured people. Located on Lake

Geneva in the heart of the rich food- and wine-producing country, the town of Lausanne itself had been hospitable to Voltaire and Rousseau. Edward Gibbon had spent some years in Lausanne, where he had completed *The Decline and Fall of the Roman Empire;* he described the environment as "a beauteous landscape, in a life of leisure . . . among a people of easy and elegant manners."[7] The largest urban center Agassiz had yet seen, Lausanne took pride in its academy. This institution was a college that prepared students for education at a university. The academy had a distinguished history, marked by an eminent faculty in the arts and sciences.

This was the social and intellectual climate into which the fifteen-year-old Agassiz moved in 1822. He had taken a first step, albeit a short one, on his journey into the wider world. But if Agassiz thought he was escaping his family entirely, he was mistaken. His uncle Mathias Mayor, one of the most prominent physicians of Lausanne, was to oversee his nephew's college education, making sure that family financial assistance was not spent frivolously. Rose Agassiz's brother was held in high respect by Louis's parents, since he was both eminent in his profession and prosperous. He was well qualified to be a mentor to their son.

Mathias Mayor figures prominently in Louis's early life, for the boy learned much from him during the next two years. They spent many happy hours together, the doctor imparting facts and concepts from his store of anatomical knowledge and teaching his nephew aspects of natural science that he could not learn at the academy. These discussions fascinated Agassiz. So did the cantonal museum of Lausanne, the first natural history museum he had ever visited. Humanistic studies at the academy were entirely secondary in interest, compared to experiences of this sort.

By 1824, Agassiz was even more determined to continue his education as a natural scientist. During the two years in Lausanne the world of nature study that Agassiz had earlier perceived but fleetingly from Mr. Rickly's few books had revealed to him a vista of profound intellectual significance. He had become acquainted with the contents of a museum, a small collection, to be sure, but one which held evidence of limitless horizons to the youth of seventeen. If the collections he had gathered at Motier and Bienne had excited his curiosity, here he saw representations of the natural history of the entire Vaudois region. Animals and plants he had never heard of, except perhaps through books, were available for close study.

Of greater importance was the fact that he had profited from the tutelage of a professional naturalist. D. A. Chavannes, director of the museum as well as professor of zoology at the academy, was a scholar who could explain zoological affinities, laws of classification, and the relationships of the animal kingdom. From his lectures Agassiz learned that the romantic pleasures of youthful collecting could become a precise science pursued by trained men adding to the world's fund of systematic knowledge. Finally, Agassiz's Lausanne natural history studies had made him acquainted with the writings of two great naturalists of the age, Jean Baptiste de Monet de Lamarck, and Georges Cuvier.

From Lamarck's *Animaux sans vertèbres* and Cuvier's *Règne animal* he discovered that "the learned differ in their classifications."[8] It required no great insight to understand that Lamarck's ordering of nature into a continuous series from higher to lower forms, judged by the presence or absence of particular organs, differed markedly from Cuvier's approach to the problem, which was based on a rigid division of the animal kingdom into four primary groupings separate and distinct from one another. The two systems derived from different conceptions of the meaning of nature. What did require perception on Agassiz's part was the realization that there was much to be done in developing reliable criteria of zoological classification. At this early stage of intellectual development Agassiz was struck by the magnitude of the problem of finding the process whereby an individual plant or animal might be readily identified in the scheme of nature. How did the naturalist know what to call something? How did he know when a particular specimen belonged to one or another unit of natural identity? How, exactly, did he know what constituted a species?[9]

These were among the questions that arose in Agassiz's mind during his Lausanne years, years that were supposedly to be devoted to humanistic studies. It was impossible to abandon such inquiries, to withdraw from the world of science that had opened so invitingly before him, and to be what his family expected and anticipated. Once more, the aid of elders was important for the accomplishment of life goals. Professor Chavannes reported to the Agassizes regarding the intellectual superiority of their son. It remained for the anatomy lessons with Dr. Mayor to have their effect. Louis, who had confided in his uncle as he had earlier confided in Mr. Rickly, now pleaded with him to make possible a greater opportunity to increase his knowledge. It was foolish to ask for support to become a professional naturalist just yet, because it would be difficult to demonstrate how a really significant career could result from such an occupation. But a convenient and acceptable alternative was at hand. If it were possible to study medicine at a university, such an opportunity would also provide new knowledge in natural history, since zoology, comparative anatomy, and kindred subjects were part of the usual training for physicians. Many distinguished naturalists, moreover, had received degrees in medicine and gone on to pursue studies in natural history, including Ignatius Döllinger, Johannes Peter Müller, Marie François Bichat, Johannes Purkinje, Richard Owen, and Theodor Schwann.

Dr. Mayor may or may not have been convinced that Louis really wanted to become a physician. His nephew quite probably understood that if he were to become a naturalist, his only possible course was to give evidence of a desire to study medicine. At seventeen, the important goal was to gain permission to enter a university. With the medical profession a respected family tradition through the careers of his grandfather and uncle, this was an acceptable compromise with family ambitions for him. Accordingly, Dr. Mayor became the instrument whereby Louis's parents were prevailed upon to give up their aspirations for a business career for their son and agreed to support a medical education instead.

In the light of limited family finances, this decision was not reached easily, but it did seem to promise excellent prospects for future success. The matter was settled; Louis would learn the physician's craft at the medical school of the University of Zurich. He would then return to Orbe or some other nearby village and begin his practice as a respected professional person, hopefully gaining the status and prestige which had come to other medical men in the family.

For Agassiz, Zurich was another forward step on the road to personal freedom. A center of Swiss learning, Zurich had been the home of such representatives of science, arts, and letters as Ulrich Zwingli, Johann Heinrich Pestalozzi, Johann Jakob Bodmer, Johann Kaspar Lavater, and Heinrich Füssli. A Protestant community, Zurich was a focal point for Swiss liberal, nationalist aspirations and a cosmopolitan cultural center. Here Agassiz would know that part of Switzerland dominated by Germanic influences. In language, in customs, and in tradition Zurich was closer to Germany than any region of Switzerland Agassiz had yet known. In many respects Zurich was a different culture for the young French Swiss, a window that looked out upon the larger world in which he hoped to achieve eminence. His years there proved to be wonderfully full of new experiences and associations.

When Louis's parents consented to send him to the university, they were apparently reluctant to favor one son over the other, and it was thus decided that Auguste would continue his education in company with his older brother. In common with many people closely associated with Agassiz in his lifetime, Auguste was irresistibly drawn by the powerful force of his brother's drives and ambitions. The two boys shared a room in a private home, and Auguste, it seems, was more Louis's secretary than fellow student. He helped his brother copy volumes of Lamarck's works, because limited means did not allow the purchase of such treasures. He cheerfully acquiesced in Louis's making their quarters serve as a combination zoo and museum, where a variety of preserved and living creatures were kept. Being Louis's creation, the depository was somewhat spectacular—forty birds, for example, flew about the study and roosted on a large pine tree which had been set up in the room. Visitors were admonished not to disturb the inhabitants of the miniature zoo around them.

At seventeen Agassiz had not entirely lost the charm of adolescent precociousness or the romantic tendencies of a schoolboy.

The student from Vaud cut a striking figure. Strong, of compact build and athletic, he also conveyed the impression of a thoughtful and poetic nature. His thick, chestnut-colored hair was worn shoulder length in the fashion admired in his era. He was a handsome young man with dark-brown eyes, aquiline nose, and full mouth. A large, well-shaped head and an alert sensitivity of expression seemed to betoken unusual intelligence and personality. His demonstrated ability fulfilled all that his appearance promised and made a clear and favorable impression on his fellow students at the medical school.

Agassiz made it plain to colleagues without giving offense that his studies took precedence over all else and that while he enjoyed the convivial pleasures of the table and the wine cup, these matters were secondary to the pursuit of knowledge. Social excursions to the town were infrequent, and his favorite recreation took the form of scholarly discussions with friends. "Agassiz knew everything," a fellow student recalled:

He was always ready to demonstrate and speak on any subject. If it was a subject he was not familiar with, he would study and rapidly master it; and on the next occasion he would speak in such brilliant terms and with such profound erudition that he was a constant source of wonder to us.[10]

He made friends easily and quickly but chose them with care. One such companion, Arnold Escher, was the son of the famous Swiss engineer and geologist, Hans Conrad Escher. The older Escher was famous for his engineering work on the Linth River, but his avocational interests ranged over a wide field of natural history. The Eschers belonged to an old and aristocratic family. Agassiz enjoyed his friendship with both father and son, the association quickening his appetite for more knowledge and the opportunity to begin a professional career as a naturalist. The bright young man with the striking gift for verbalization impressed his professors no less than he did personal friends. He displayed an unlimited curiosity about every aspect of the natural world, and he applied to his studies remarkable resources of energy and patience for painstaking investigation. Anatomy, zoology, and physiology were Agassiz's formal subjects of instruction at Zurich, but in an important sense he went beyond the experience of the class-room and laboratory and educated himself. When Professor Heinrich R. Schinz, his instructor in zoology, found that he had described every bird in his collection, he gave Agassiz the key to his private library so that he might read as he pleased in natural history.

There was something deeply appealing about Agassiz that in-spired contemporaries like Auguste or elders like Professor Schinz to extend him aid in his endeavors. Intellectual promise, the charm of personality, clarity and ease in manner of speech, and above all an intense dedication to learning were traits that attracted and inspired sympathy. Once, when Louis and Auguste were walking to Orbe from Zurich because they could not afford to pay for transportation, they were stopped by a wealthy gentleman who took them home in his carriage. This man was so impressed by Louis, by his plans and his obvious talent, that soon thereafter he wrote to Rodolphe and Rose Agassiz offering to adopt their older son and to assume all costs of his education and professional train-ing. The offer was rejected, but it is indicative of the charismatic force of Agassiz's personality. The gentleman of the Swiss highway was only the first in a long and impressive series of people who, during the course of more than fifty years, were won in an almost magical way to the support of projects, causes, and endeavors in-spired and captained by Louis Agassiz.

The Agassizes did not need the incident of the affluent gentle-man to convince them that their son was no ordinary student. Evaluations from professors at Lausanne and Zurich, and excellent reports from Dr. Mayor, testified to his exceptional quality. His parents knew he was energetic, tenacious of purpose, and even precocious, but now they began to believe that he was truly bril-liant, as he continued to attract favorable notice from persons more qualified than they to judge his talents objectively. Thus, when Louis told his parents that Zurich was not quite good enough an institution to advance his medical studies, they listened sympa-thetically to his complaint. Zurich could not compare with the great universities of Germany, where a student could enjoy the instruction of distinguished scholars and a spirit of free critical inquiry. The natural sciences in particular were beginning to assume a high rank in the intellectual structure of German higher education. The University of Heidelberg, for example, had an excellent medical faculty, a fine anatomical laboratory, and such celebrated scientists as the paleontologist Heinrich Georg Bronn, the embryologist Friedrich Tiedemann, and the zoologist Friedrich S. Leuckart. These were the arguments Agassiz advanced to his parents in 1826 as reasons for leaving Switzerland and entering the larger world of German scholarship. He assured them that it was quite usual for a medical student to attend two or three schools before getting his degree. He would attend church services regularly. He would exercise regularly and not jeopardize his health by remaining too much indoors with his books. After Heidelberg, he would return home and begin medical practice.[11]

What could parents do in the face of such entreaties from a brilliant nineteen-year-old son who seemed to feel himself destined for great accomplishments? Permission was granted. Auguste, the faithful copier of Louis's books, the good-hearted lad who shared his room with birds and animals, was instructed to' return to Neuchâtel and begin a career in business with Uncle François Mayor. At least one Agassiz son would be close to home. As for Louis, the Agassizes could hope that ultimately his sign would be nailed beside a door in Orbe announcing that here Dr. Louis Agassiz practiced.

Having achieved one of the great ambitions envisioned four years before, Agassiz arrived in Heidelberg to begin his German education as a professional naturalist. He found quarters in the comfortable and pleasant home of a tobacco merchant, where he had ample opportunity for uninterrupted study. Almost everything that Agassiz read was new to him, and he was inspired by a firm sense of personal and intellectual independence. In 1826, the year Agassiz came to Heidelberg, G. A. Goldfuss' first volume describing the fossils of Germany was published, and the student read it in utter fascination. Before this he had had no insight regarding the relationship between paleontology and geology. Study of the still young science of fossil remains revealed to him that there was a challenging connection between animals that had lived in the past, the ancient structure of the earth, and the present character of animals and earth strata. Agassiz now realized that he had "never thought of the larger and more philosophical view of nature as one great world." Modern life forms bore some relationship, as yet only partly understood by investigators, to life in former epochs. No more would a solitary fish seem to Agassiz an interesting specimen to be described and catalogued. This one animal represented an entire history of nature. This single creature symbolized the entire sweep of creation, past and present. By comparison, the study of medicine was hardly challenging. How could one spend a life healing only a single species while the entire natural world awaited examination?[12]

Agassiz's interest in paleontology led him to attend the lectures of Heinrich Bronn, who took a personal interest in him and showed him how to study collections of fossils illustrating the history of the earth and its extinct species. The Heidelberg museum was richer in its materials than that of Zurich, and, inspired by Bronn's teaching, Agassiz wasted no time in becoming familiar with its contents. Nor did he abandon his interest in living animals. The instruction of the zoologist Friedrich Leuckart was of deep interest to Agassiz. Leuckart appreciated the unusual curiosity of the Swiss student. Like Bronn, this professor gave Agassiz special instruction, guiding him in an understanding of the habits of life of particular groups of animals. The same consideration was accorded by the botanist and embryologist Theodor Bischoff, who introduced Agassiz to the science of botany. Bischoff, moreover, performed a singular service when he taught Agassiz the use of the microscope, showed him how to collect and preserve plants, and took him on long excursions through the surrounding countryside.[13]

The promising student met with more than cordiality from Friedrich Tiedemann, the distinguished embryologist. Tiedemann had been a student of Heinrich Schinz, Agassiz's zoology teacher at Zurich. Schinz wrote a letter of introduction for Agassiz to Tiedemann, who promised the delighted student that "he would be for me here as Professor Schinz . . . had been for me in

Zurich."[14] Like his predecessors, Tiedemann offered to lend Agassiz the contents of his library. Chancellor of Heidelberg and an impressive lecturer, Tiedemann was among those scientists who contributed to the foundation of the theory of recapitulation, the concept that held that the fetal life of the individual repeated the embryonic history of the race. He had published his findings and doubtless spoke of his research in the classroom; as far as can be determined, it was Tiedemann's influence that led Agassiz at a later date to carry the idea of recapitulation further in his research on fossil fish.[15]

Tiedemann performed another, more personal, service for Agassiz. He introduced his Swiss student to another outstanding pupil, Alexander Braun, the son of the postmaster-general of Baden. The two young men felt an intellectual kinship from their first meeting, and the friendship that resulted had a lasting influence on Agassiz. Both young men were highly intelligent, energetic, and enthusiastic. Both were determined to learn all they could about natural history. Braun, like so many others, was impressed by the force of Agassiz's personality and struck by his grasp of a wide range of scientific knowledge. He wrote to his father:

I sometimes go with this naturalist so recently arrived among us (his name is Agassiz and he is from Orbe) on a hunt after animals and plants. Not only do we collect and learn to observe all manner of things, but we have also an opportunity of exchanging our views on scientific matters in general. I learn a great deal from him, for he is more at home in zoology than I am. He is familiar with almost all the known mammalia, recognizes birds from far off by their song, and can give a name to every fish in the water. In the morning we often stroll together through the fish market, where he explains to me all the different species. He is going to teach me how to stuff fishes. . . . Many other useful things he knows; speaks German and French equally well, and English and Italian fairly. . . . He is well acquainted with ancient languages also, and studies medicine, besides. . . .[16]

Agassiz had found his first student and his first disciple.

If Braun had much to learn from his Swiss friend, Agassiz also benefited much from the association. Another German student, Karl Schimper, was a close companion of Braun, both of them attracted to the science of botany. These two friends were fascinated by that aspect of botanical science of which Johann Wolfgang von Goethe was the major philosophical exponent. Plants, Goethe held, developed through a process of metamorphosis whereby the leaves grew through successive stages, one form followed by another. Such transformation of growth phases represented an effort of the individual form to approximate an idea of higher order and reality. Agassiz was tremendously impressed by the intellectual attitude such speculation represented. Although he was not particularly interested in botanical studies as such, the ideas Braun and Schimper expounded gave him a great appreciation for the brilliance of German philosophy and science. Intellectual influences of this sort convinced him that "grand generalizations, such as were opening upon botanists, must be preparing for zoologists also."[17]

Life with Braun and Schimper was indeed full. Days, weeks, and months passed in exciting activity shared by the three young men. Collecting trips to nearby fields and streams were an almost daily occupation. Agassiz and Braun decided to share their living quarters and arranged to divide their labors as novice naturalists. While one of them prepared specimens for mounting or dissection, the other read aloud from books on botany, embryology, or ichthyology. In the evenings they held long discussions and reviewed facts and theories gleaned from the day's lectures. Was there, perhaps, a mathematical order and statistical precision by which the leaves of plants appeared on the stem, no matter what the plant or how many leaves it had? If true, this was an amazing discovery, and they spent long hours discussing the concept of phyllotaxy, an idea inspired by Braun's reading in Goethe's botanical works. The impetus that had led Agassiz to Heidelberg was strengthened and reinforced by all such exposures to German philosophical discussions.

It would be difficult to imagine even so dedicated a student as Agassiz spending all his time in intellectual labor. Physical exercise came easily enough in the walking trips in search of specimens, but attractive diversions were offered in the society of fellow students. Braun and Schimper were not the only young men of Heidelberg to be drawn to Agassiz. Recognizing a leader in the

boy from Vaud, his fellow Swiss students at the university elected
him president of their *Burschenschaft,* or club of fellow countrymen. This meant beer-drinking sessions and an hour's practice in
fencing every day. Agassiz soon became an accomplished swords-
man and took to the sport so enthusiastically that he once acci-
dentally wounded his dear friend Braun. He put his skill to better
use when he accepted a challenge from the German student club
to meet one of their best fencers and, not satisfied with a single
victorious encounter, insisted on dueling with as many Germans
as wished to try their skill with him. After four such matches, the
president of the Swiss club had clearly upheld the honor of his
fellows. Whether fencing or fish-collecting, any activity that engaged Agassiz's interest had to be pursued with energy, enthusiasm,
and high competence.

Agassiz's amazing store of physical energy was wisely channeled
and directed according to a rigorous schedule. Rising at six, he
spent the morning attending lectures and working in the anatomical laboratory. Fencing took up an hour before lunch, which was
followed by a walk and then reading until five, when he again
attended lectures. After that, a swim in the Neckar River was a
refreshing diversion before dinner. Studies took up one or two
hours in the early evening, and the day ended with philosophical
debates or an evening at the Swiss club. Prayers closed this typical
day of combined physical and intellectual activity. Occasionally
the pattern was varied by a visit to the opera or a day spent camp-
ing out in the country and cooking lunch over an open fire.

Agassiz's activities were all reported faithfully to his parents
back home in Switzerland. Rodolphe and his wife could only marvel at their son's diligence and take pleasure in his happiness.
They were impressed that famous professors had shown Louis such
special attention and kindness. Paleontology, botany, and zoology
were certainly interesting subjects, it seemed, but the elder Agassizes wondered at their importance for a doctor's education. Louis's
attention to them must have quelled their doubts, for Rodolphe,
reflecting on his son's letters, wrote to a friend: "We have the
best possible news of Louis. Courageous, industrious and discreet,
he pursues honorably and vigorously his aim, namely the degree
of Doctor of Medicine and Surgery."[18]

"He . . . studies medicine, besides . . . ," Alexander Braun had written of the varied characteristics of his Swiss friend. Yet there was hardly any mention of formal medical subjects in Agassiz's letters home, except for anatomy instruction under Tiedemann. The same thing had occurred at Zurich. Professors were sought out not for what they might offer in the way of a medical education but because they excelled in knowledge of natural history. Fish, rocks, minerals, fossils, and plants—these were the aspects of nature that fascinated Agassiz. Louis knew he was a good son: frugal, devout, devoted to his parents, and even lonely for them while away from home. If he could succeed as a naturalist, they would then surely be proud of his ability to make his way in the world. Having become aware of such grand generalizations as the metamorphosis of plants and phyllotaxy, having examined Professor Bronn's collection of fossils and talked with Braun and Schimper about the exciting vistas open to naturalists, he found it impossible to contemplate a return to Orbe as a country physician. He might still return to Switzerland, but he would be undistinguished there unless he came back as a scholar with a full German education. With this ambition, there was still one great problem. After eight months at Heidelberg, with so many fascinating intellectual challenges before him, Agassiz did not have sufficient funds to continue.

Rather than add to the burden of his parents, he proposed that his father make an "arrangement" with the prosperous François Mayor, who could well afford to advance money for further education. "I am confident that when I have finished my studies," he declared to his father, "I could easily make enough to repay him."[19] If Agassiz was determined to be a naturalist, then it would be many years before he could repay his uncle. If he really intended to practice medicine, the extension of his studies was not warranted. Nevertheless, father and uncle were persuaded to undertake joint sponsorship of Louis's further education.

The Heidelberg education was not confined to the university. Alexander Braun's home was at nearby Carlsruhe, and it was much more sensible to spend short vacations there than make the longer journey to Orbe. The Braun home offered a new social experience for Agassiz. The spacious house, surrounded by attractive gardens, stood at the edge of a large oak forest. The home itself abounded in treasures to delight a student of natural history; Carl Braun, Alexander's father, an ardent collector of minerals and plants, had one of the finest private collections to be found in the German states. The work tables, microscopes, plant presses, books, dissecting tools, and other scientific instruments, all there in the house, impressed Louis deeply. He noted, too, the comfort of the home, the delights of music sung and played by the family in congenial sessions of four-part singing, the friendliness and cordiality of Carl Braun, the intelligence and worldly charm of Mrs. Braun, and the distinction added to the family circle by Arnold Guyot, a brilliant Swiss theology student who boarded at the house. Most pleasant was the unquestioning family acceptance that Agassiz's future career would be that of a naturalist.

The Brauns belonged to a world heretofore unknown to Agassiz. Here were people who appreciated art, music, and history and who, as a matter of course, set aside space in their home for a small scientific laboratory. In fact, all the money Carl Braun could spare was devoted to the purchase of scientific books, materials, instruments, and collections. The ease of manner, worldly attitudes, and informed outlook of this German official were all attributes that impressed the rural Swiss youth. Agassiz was also attracted by the two beautiful Braun daughters, Emily and Cécile. Cécile, Louis's favorite, was an appealing girl with deep-set eyes, long brown hair, and a small, winsome mouth. Of slight build, Cécile was shy and reserved, expressing her individuality and creativity in skilful painting and sketching. The Christmas vacation of 1826 gave Cécile the opportunity to have Louis sit for her. She was quite plainly interested in this brilliant, expansive, charming, talkative young man, and she was flattered that he appreciated and praised her talents. By his encouragement he persuaded her to sketch the specimens that he and Alexander had collected. She proved particularly adept at drawing fish, and Agassiz expressed enthusiastic admiration for each line and shading. It was difficult to leave this happy atmosphere when vacations came to an end.

Agassiz's studies at Heidelberg were interrupted early in 1827 when he was stricken by typhoid fever, then rampant in the city. As soon as he could be moved, Alexander Braun accompanied him to Carlsruhe, where the Braun family tended him during his recovery. Cécile was his constant nurse, and a strong bond of affection developed between the young artist and the young naturalist. When Agassiz, at the request of his parents, returned to Orbe to complete his convalescence, thoughts of Cécile and the hospitable Braun family went with him.

Back in Switzerland in May, 1827, Agassiz was much too restless to lead the life of an invalid. His Heidelberg companions were soon reminded of their Swiss friend when a shipment of specimens of an interesting species of toads arrived from Orbe, the result of an investigation Agassiz had made of their embryology, habits of life, and mode of reproduction. When Alexander Braun showed Professor Leuckart these specimens, the zoologist was delighted and urged his Swiss student to continue observing the reproductive habits of this species, since such data were unknown. Also, Pastor Fivas, Rodolphe Agassiz's assistant, was pressed into service to help the twenty-year-old naturalist make a catalogue of all the plants growing nearby in the valley of the Jura Vaudois. Another clergyman, Pastor Mellet, showed Agassiz his large collection of indigenous insects, but the university student thought it a pity that the older man "knew nothing of distribution, classification, or the general relations" of fauna and flora, for there was a more philosophical aspect of nature than mere collection could reveal.[20]

In addition to studying his specimens Agassiz spent a good part of the summer reading the *Lehrbuch der Naturphilosophie* by the German naturalist and philosopher Lorenz Oken. These volumes gave Agassiz "the greatest pleasure," as well they might after Tiedemann's lectures and Braun's speculations concerning the significance of the metamorphosis of plants and the mysteries of phyllotaxy.[21] Familiar with the interpretations of nature expounded by Lamarck and Cuvier, Agassiz had long since realized that "the learned differ in their classifications." He was hardly surprised to find, therefore, that Oken approached the problem from yet another point of view. Oken affirmed that man was the key to understanding the anatomy, physiology, and order of the entire animal kingdom. Nothing was present in the lower animals that was not somehow represented in the structure of man himself. The body of man, in segmented and incomplete form, could be seen represented in the bodies of animals. The animal kingdom was symbolic of man, or, as Oken wrote, "Animals are only the persistent foetal stages or conditions of man." Under this assumption, nature could be viewed in a cosmic sense; all creatures aspired in their course of development to the fulfilment of an ideal type, a final and perfected product.[22] Grand and inspiring generalizations such as these made Agassiz even more enthusiastic over the prospect of a naturalist's life. Made confident by the success of his Heidelbreg disputations and his ability to understand Oken, Agassiz yearned to contribute his own original share to the refinement of the concepts of *Naturphilosophie*.

During the same summer of 1827 he journeyed to Neuchâtel to reassure his uncle François about financial arrangements and also to study the fishes of lakes Neuchâtel and Morat. This undertaking was quite different from the amateurish collecting he had done during his adolescence. He now set about making a catalogue of fish species on the order of his catalogue of plants, bringing to this task his new knowledge of classification, anatomy, vertebrate zoology, and comparative ichthyology. He sought an example of every species indigenous to the area in order to make dissections for the purpose of exact description. Louis made collections for Professor Leuckart once again, and again the zoologist admired the work of his outstanding student. The summer thus passed quickly, and strengthened by his outdoor activity and the lavish care of loving parents, Agassiz made ready to return to Heidelberg.

Then a letter came from Braun in Heidelberg with a proposal that had to be accepted. In 1826, the Bavarian government, anxious to match the expanding educational achievements of other German states, had established a university in the city of Munich and had gathered there a group of outstanding intellectual figures. Braun had inquired of one of these men, the embryologist Ignatius Döllinger, whether two serious students of natural history would be welcome. The professor had answered that instruction in this and allied subjects at Munich was second to none and that they would certainly be welcome; he would even try to arrange lodgings for them. Braun went on, tempting Agassiz with the names of Lorenz Oken, his colleague in *Naturphilosophie* Friedrich Wilhelm Joseph von Schelling, the botanists C. F. P. von Martius and Joseph Gerhard Zuccarini, the zoologist T. D. Schubert, and equally outstanding scholars in the fields of mineralogy, mathematics, physics, and astronomy. The library was excellent, there were fine collections of zoological and botanical specimens, most of the lectures were free, lodging would be no more expensive than at Heidelberg, and the beer and schnapps would be plenty and good. "Let all this persuade you," Braun urged. "We shall hear Oken in everything. . . . We shall soon be friends with all the professors." They could share a room together. Wonderful vacation trips to Salzburg and the Carinthian Alps awaited them.[23]

Such an opportunity had to be seized, its potentialities exploited, its challenges accepted. Study with Döllinger, Oken, and Schelling was too fascinating a prospect to be ignored. Agassiz was reminded that, until his discussions with Braun and Schimper, his philosophical education had been neglected. German students were clearly his masters in generalizations about the meaning of nature. He would go to Munich with Braun and Schimper "and drink new draughts of knowledge," in an intellectual environment that would broaden and inspire his developing awareness that philosophy and nature should be approached as a composite unit of mental experience.[24] Besides, Braun had written that at Munich "the clinical instruction will be good," and no doubt medicine could be studied much more fruitfully in Munich than in Heidelberg.

Once more Agassiz's parents were subjected to a barrage of appeals. Costs could be no more than at Heidelberg. This would be the last absence from home and family. Only two years at Munich, he promised, and he would secure his medical degree, return home, and begin his practice. Attendance at the University of Munich was positively the best way for him to achieve a proper medical education. His guiding principles would be economy, devotion, and intellectual application. He would write home often. Reluctantly, but resigned to the idea that Louis probably knew what was best for him, his parents consented to the new venture and, aided by Uncle François, provided an allowance of $250 a year for all expenses. Certainly, education at three universities would make their son the most learned of local physicians.[25]

Thus provided for and sternly admonished to complete his education as soon as possible, Agassiz joined Alexander Braun in October, 1827, at Carlsruhe before going on to Munich. Welcomed by the elder Brauns almost as a son and by Cécile as much more than a brother, Agassiz, with his companion, prepared for the journey to Bavaria. Since they were not due in Munich for a month, they decided to take a leisurely trip, visiting points of interest on the way. Leaving Carlsruhe for Stuttgart, their first stopping place, Agassiz could feel confident in having accomplished an important part of his youthful ambitions. He had spent nearly two years at one German university and enjoyed the prospect of at least two years at another. He had a good knowledge of zoology, anatomy, embryology, classification, and languages. He was an expert on the natural history of his native region. The wider world of philosophy had beckoned to him and found him a receptive intellect. Professors had volunteered their friendship and special aid to help him along his way. He rejoiced in an understanding family, a true friend, and the admiration of a particular young lady. Alive to the promise of the world, expecting to meet its challenges with energy and enthusiasm, Agassiz walked the road to Munich with a happy heart and confident spirit.

2

THE MAKING OF A NATURALIST

1827–1832

I wish it may be said of Louis Agassiz that he was the first naturalist of his time, a good citizen and a good son, beloved of those who knew him. I feel within myself the strength of a whole generation to work toward this end, and I will reach it if the means are not wanting.[1]

WRITTEN a little over a year after Agassiz arrived in Munich, these words reflect a personal and intellectual self-assurance derived from experience, from the practical results of earlier ambition and determination. The Munich years supplied him with further special knowledge of the sciences of anatomy, embryology, ichthyology, and paleontology, together with a philosophical outlook that in one form or another would remain with him his entire life. With professors extolling his talents, treating him as an unusual student, paving his way into the professional world of science, his Munich successes meant that Agassiz had conquered a complex and demanding environment and could go on to seek grander and more demanding ambitions and goals. He returned to Switzerland a scholar, a professor, a man of science and the world—no longer a rural youth attracted to the study of nature but an established naturalist bringing the glories of German accomplishment to the villages of his early days. He could be, truly, a good citizen and a good son, not because he had followed parental wishes, but because he had triumphed over them.

The achievements of these years were not attained easily. Intense physical and mental exertion, much independent research, long hours of attending classes and reading books—all these strengthened habits of work learned at Zurich and Heidelberg. Never content to be just an ordinary diligent student, Agassiz stood out from his fellows and was admired by his teachers. It was Agassiz who had prepared the fish skeletons that Döllinger sent the distinguished anatomist Johann Meckel to inspect, Agassiz who led student field trips to the countryside along the banks of the Isar River and to the Tyrolean Alps, Agassiz to whom Braun and Schimper turned as a constant source of inspiration and instruction. All who came to know him at Munich were left with an impression of purpose, of self-confidence, and all had a desire to do everything they could to aid him. Döllinger, Von Martius, and Oken, these men joined Schinz and Tiedemann to form the front ranks of a coterie of individuals, some great, some small, that from the days of Zurich and German universities to Vienna, Paris, and London would grow larger and larger, united in an effort to help Agassiz conquer his world. Munich witnessed the maturing of Agassiz's talent—talent not only in the sense of energy and dedication to purpose, but talent to convince people that what he did and planned was vitally important. Actually, Agassiz discovered the feeling of success and distinction before he had fully accomplished the objectives leading to real success in his chosen field.

One person attracted by the spell of Agassiz's zealous pursuit of knowledge at Munich was Joseph Dinkel, a young artist who had come to the city to work at the Bavarian Academy of Fine Arts. Dinkel became Agassiz's personal artist, his salary paid out of the small allowance Louis received from home. For Agassiz, the services of an artist were essential, since accurate drawings of living and fossil animals were of primary importance to the naturalist's craft. Dinkel remained associated with Agassiz for nearly twenty years, so great was his admiration for his Swiss employer and friend. The artist's word picture of Agassiz at Munich is revealing:

He never lost his temper. . . . He remained self possessed and did everything calmly; he had a friendly smile for everyone. . . . [He] was at that
time scarcely twenty years old, and was already the most prominent
amongst the students. . . . They loved him. . . . I had seen him at the . . .
Student Clubroom several times and had observed him amongst the
jolly students. . . . He picked out the highly gifted and learned students
and would not waste his time with ordinary conversation. . . . Agassiz
often times when he saw a number of students going for an empty
pleasure trip, made the observation to me, saying "there they go with
the other fellows because none of them can go his own way. Their motto
is, *'Ich gehe mit den andern.'* " He said to me, "Mr. Dinkel, I will go
my own way, and even not alone, but I will be a leader of others."[2]

To aspire to be a leader of his peers at twenty, Agassiz had to master his new environment. His first impression of Munich, therefore, was important for his purpose. Arriving with Braun in November of 1827, Agassiz encountered a society distinguished in arts and letters, one purposefully fostered by the Bavarian monarchy under Ludwig I. In the effort to make Munich and Catholic Bavaria the rival of Protestant Prussia in intellectual achievement, the university, formerly located at Landshut, had been moved to the capital city. There, distinguished scholars from all over Germany found one of the greatest libraries of Europe, the royal cabinets of botany and natural history, the Royal Botanic Garden, and the academies of fine arts and science. With the university, such institutions made Munich a center of southern European culture, a cosmopolitan community that welcomed scholars and travelers from all over the world. The museum, library, hospital, and university halls all beckoning to him, Agassiz was impressed by the importance of wealth and power for the support of intellectual activity. The Bavarian monarchy became his instructive model for the non-intellectual requirements useful for the advance of scholarship.

The city teemed with resources for the students in arts, letters, philosophy, and science. . . . The King seemed liberal; he was the friend of poets
and artists, and aimed at concentrating all the glories of Germany in
his new university. I thus enjoyed . . . the example of brilliant intellects,

and that stimulus which is given by competition. . . . Under such circumstances a man either subsides into the position of a follower in the ranks that gather around a master, or he aspires to be a master himself.[3]

In the effort to become a master, it was instructive to have a master as a model. Ignatius Döllinger was such a person. Döllinger lived in a large house overlooking the city, and fortunately for Agassiz and Braun he had been in the habit of renting rooms to students to provide money for increasing his library. The two friends, like many transient scholars in the past, became part of the Döllinger household. Among their predecessors who had lodged with Döllinger and benefited from his personal instruction they could number Karl Ernst von Baer, who in this very year 1827 was publishing his epoch-making treatise on the nature and development of the mammalian egg.[4] Döllinger's talents as an embryologist and anatomist were of primary significance for the development of Agassiz's mind. Initially, however, the fact that Döllinger provided the Heidelberg friends with wonderful working and living quarters was more important than intellectual stimulation.

Agassiz had his own room, where chairs and work tables were plentiful, that looked out over the Alps stretching toward his homeland. The studio had wide windows, a bed in an alcove, and was, in Dinkel's words, "a perfect German student's room." Agassiz's room connected with those of Braun and Schimper, who had joined them soon after their arrival. The three rooms were converted into one common studio apartment. Chairs and tables had to be littered with books from the start; microscopes were set up; a machine of Agassiz's invention made coffee in the morning, was used to prepare small animals for skeletal reconstructions during the day, and in the evening brewed tea. Visitors were impressed by the clutter of books, by the tobacco smoke, and by the once white walls now covered with anatomical sketches, some technical and others understandable to anyone. These comrades-in-science gave each other names to denote the character of their interests: Braun was "Molluscus," after a current fascination with shells, Schimper was "Rhubarb" for a botanical flair, and Agassiz was "Cyprinus," the carp, a fish notable for its fecundity, rapid growth, size, and great appetite.

In such an atmosphere the education of the young man who aspired to be the first naturalist of his time progressed with the same order and discipline that had marked Agassiz's prior search for knowledge. C. F. P. von Martius, a most instructive teacher who was fond of taking his students on botanical excursions through the countryside, initiated Agassiz into the study of botany, an interest he had already begun to develop through friendship with Braun and Schimper. Soon realizing that both Agassiz and Braun were exceptional students, Von Martius invited them to his home weekly for tea. At these gatherings the professor told the young men tales of his recent exploration of Brazil with the zoologist Johann B. Spix. Such accounts were marvelous vicarious experiences for Agassiz. While at Heidelberg, he had made an excursion to the museum at Frankfurt am Main to talk with its director, Edward Rüppel, just returned from Africa. Inspired by Rüppel's stories, he had then visited the museum at Stuttgart, where he saw elephants from Africa, a bison from North America, and the bones of a mammoth recovered from Siberian ice. Now, as Von Martius talked of Brazil, the same excitement came over him and he longed for the day when he could undertake a great exploration to a distant, exotic land. One could not, of course, show the emotion one really felt as such adventures were described. Agassiz and Braun smoked their pipes and sipped tea, all the while "discussing with all our might and main subjects of which we often knew nothing."[5]

If the wide world of nature opened up in grand display through such experiences, another equally compelling attraction resulted from study with Döllinger. Agassiz came to know the value of personal instruction from a master in his special subject:

With Döllinger I learned to value the accuracy of observation. . . . He gave me personal instruction in the use of the microscope, and showed me his own methods of embryological investigation. Döllinger was a careful, minute, persevering observer. . . . He gave his intellectual capital to his students without stint, and nothing delighted him more than to sit down for a quiet talk on scientific matters. If he found himself understood . . . he was satisfied. . . .[6]

Agassiz, like the embryologists Von Baer and Christian Pander, publicly acknowledged his debt to Döllinger. Thirty years after Munich, when he commenced publishing a series of volumes he hoped would serve as models of the kind of precise analysis he had learned from his teacher, he dedicated them "to the memory of Ignatius Döllinger, the founder of Embryology, who first taught me how to trace the development of animals."[7]

For the student who had been studying botany and zoology and imagining how he would undertake a great exploration, the science of embryology unfolded a vast new world under the microscope. Agassiz had realized the importance of exact investigation in his earlier studies but had not known how exact a scientific investigation could be. He had used the microscope in study with Bischoff but never with the precision and intensity that Döllinger impressed upon him. Döllinger was a pioneer in exploring the uses of this instrument, having procured one of the earliest achromatic microscopes, an innovation pioneered by that highly talented craftsman, Professor Joseph von Fraunhofer of the Physical Academy of Munich. Though far from perfect, this microscope and its uses signaled a notable improvement in embryological investigation because of the superior flat plane of vision it produced. Since Döllinger used this instrument with such success, Agassiz had to own such a microscope himself. By practicing strict personal economy, he had saved enough from his small allowance by 1828 to purchase a Fraunhofer achromatic microscope from the Munich firm of Utzschneider und Reichenbach. Now Agassiz could enjoy a private dominance over this aspect of the organic world. As nature revealed her secrets under the glass of Agassiz's microscope, he gained a private insight into the processes of the development of life; as the egg grew and changed through the various and predictable stages of its development, Agassiz's perception of the creative process increased in proportion. He was so impressed by the facts and generalizations learned in this way that he engaged in a lifelong search for ever more precise instruments with which to study the data of embryology.[8]

Through long evenings of work with Döllinger in his home laboratory, Agassiz became acquainted with an aspect of nature study as close to an inner knowledge of reality as a man can have: knowledge of the process by which life develops. The laboratory became as wide a theater of nature as the lakes and mountains of Switzerland. He discovered for himself what Döllinger and Von Baer already knew, that certain forms of animals resemble one another in their embryonic state. Observing the growth of the egg of various kinds of mammals, he also learned to compare different stages of embryonic growth. His experiences here had a lasting effect upon Agassiz's conceptualization of the method and purpose of natural history. Although at different phases in his career he was paleontologist, ichthyologist, or geologist, the science of embryology held such fascination for him that he returned time and again to it. So firmly convinced was he of the ultimate knowledge to be derived from this kind of investigation that he even proposed an entire system of classification based upon "embryological principles." The growth of animals from eggs, their rise to maturity, and the comparative study of embryos were the essential basis for any understanding of organic creation. Later in life, under the spell of the religion of embryology learned from Döllinger, he wrote:

A superficial familiarity with the microscope gives no idea of the exhausting kind of labor which the naturalist must undergo who would make an intimate microscopic study of these minute living spheres. . . . The deeper insight he has gained by long training . . . as well as the concentration of intellect that makes the brain work harmoniously with . . . [hand and eye] he cannot communicate.[9]

Embryology provided Agassiz with an understanding of nature in general, but the study of fishes promised the opportunity to gain a special competence. "It will interest you to know that I am working with a young Dr. Born upon an anatomy and natural history of the fresh-water fishes of central Europe," Agassiz wrote his brother in December of 1827. Agassiz was happy because he had a large project before him, one which he had outlined in a memorandum book at Heidelberg but which became more meaningful at Munich, where he found the security and distinction he needed.[10] Of all the eminent naturalists at Munich, none was a specialist in ichthyology, and the chance to continue a youthful interest and at the same time to establish a claim to particular excellence was not to be ignored. The young "Dr. Born" disappeared from Agassiz's correspondence as quickly as he had emerged. The project became Agassiz's alone. He visited the Munich market place every week, discovering many fishes previously unknown to him, and one entirely new species, a carp, appropriately enough. Dinkel was kept occupied drawing all the new discoveries. The faithful Auguste, hard at work at the Mayor business house in Neuchâtel, was asked to scour the Swiss lakes and send a collection of desired fishes to Munich. By the spring of 1828, the first effort at intellectual distinction was well under way.

If the plan to publish a comprehensive account of European ichthyology was a presumptuous effort for a twenty-one-year-old student, Professor von Martius did not think his student unprepared for a professional enterprise. The journey of Von Martius and Spix to Brazil, undertaken during the years 1817–20, had yielded a great deal of zoological data, some of which had already been described and published.[11] One of the most important results of the exploration—the fishes collected from the Amazon River system in Brazil—remained undescribed because Spix, whose task this was, had died in 1826. The following year, Von Martius, knowing no one at Munich competent to undertake the task, had sent some plates and descriptions that Spix had completed to Cuvier for his evaluation. The French naturalist was struck by the importance of these materials. He made brief observations on them and urged that the entire collection be described by a capable ichthyologist.[12]

In the early spring of 1828, Von Martius asked Agassiz to perform this task. "The opportunity for laying the foundation of a reputation by a large undertaking seemed too favorable to be refused," Agassiz told his sister Cécile, impressing on her the importance of the venture.[13] One came to be a master by working like one, and with this prospect before him, Agassiz set to work, despite the obligations of daily lectures and his research on European fishes. The real task involved was the description of about ninety species, some of which had already been described and sketched by Spix, although Agassiz found that "several of these . . . were not very exact" and had to be done over.

Agassiz felt he had entered the world of professional scholarship. He was at work on materials that the great Cuvier had examined. He had planned a comprehensive treatise on ichthyology, and only Cuvier, who had begun work on a monumental study which would depict all known fishes in the world, seemed to dwarf his effort. Cuvier had asked his fellow naturalists to send him descriptions of species known to them, and young Agassiz, who felt he had an exact knowledge of Bavarian and Swiss fishes, hastened to provide the master with information because "I could not better launch myself in the scientific world than by sending Cuvier my fishes."[14]

All this activity made the present seem rewarding and the future optimistic. Agassiz did not draw back from reporting his successes. Thus Rodolphe Agassiz learned of his son's progress: "I can tell you from a good source (I have it from one of the professors himself) that the professors whose lectures I have attended have mentioned me more than once, as one of the most assiduous and best informed students of the university; saying also that I deserved distinction."[15] These professors, it seemed, were anxious that their superior student and his talents be recognized in the professional arena. Von Martius and Oken, attending the convention of all German naturalists in Berlin in the summer of 1828, announced Agassiz's forthcoming volume on Brazilian fishes. Oken, moreover, presented a paper describing Agassiz's discovery of a new carp species, a description that was published soon thereafter, representing Agassiz's first scientific publication.[16] Scholars at this conclave who knew him spoke of him "in very favorable terms." In October, the first part of the Brazil volume had been sent to the publisher. "Will it not seem strange," Agassiz playfully asked his sister, "when the largest and finest book in Papa's library is one written by his Louis?"[17]

In May of 1829 this aspiration was realized when *Brazilian Fishes* appeared in print. Agassiz was three years ahead of schedule. The work of the twenty-two-year-old naturalist was a moderately distinguished, precise description of fishes indigenous to Brazil. He was proud of what he had done. The book was a step on the right road, he confidently told his father, because it would make him favorably known and give him "a European name." The road to success yielded pleasant rewards. There was the opportunity for the author of a book to dedicate it to someone, parents, perhaps, or an old teacher. Agassiz, with his face turned to the future, dedicated the volume to Cuvier. "I beg that you will judge this work . . . as the first literary essay of a young man. . . . I had as my guide the observations you had kindly made . . . on the plates of Spix." Thus Agassiz wrote to the dean of French natural science.

The dedication and letter to Cuvier provided Agassiz with the opportunity to tell his hopes and ambitions to the man he had come to admire so much. Agassiz reviewed for Cuvier the character of his education since Zurich. While he had studied for the ostensible purpose of learning medicine, every scientific subject only increased his desire for a naturalist's career. Work on the Brazilian fishes had intensified this ambition. "Allow me to ask some advice from you, whom I revere as a father, and whose works have been till now my only guide." He had no fortune, to be sure, but would "gladly sacrifice my life if, by so doing, I could serve the cause of science." He thought himself suited for the life of an explorer; he could skin animals, knew how to use a hammer and ax, was proficient in the use of the sword. "I am strong and robust, know how to swim, and do not fear forced marches. . . . I seem to myself made to be a traveling naturalist. I only need to regulate the impetuosity which carries me away. I beg you, then, to be my guide."[18]

Reading these words, and the volume that accompanied them, Cuvier was impressed. He wrote Agassiz how much he admired the Brazil volume, that he had already incorporated its contributions into the second edition of his *Règne animal,* and that it would be useful for the forthcoming volumes of his natural history of fishes. As for Agassiz's view of himself and his role in science, Cuvier did what his young admirer asked. Plans for the future were all worthwhile and noble aims, but life should be lived one step at a time. He observed, for example, that the scheme to write a natural history of the fresh-water fishes of central Europe was an important one, but too ambitious. Steady application to the problem of describing one family, first, was a more reasonable approach.[19]

Agassiz was flattered by Cuvier's praise and attention, but his impetuosity was not to be regulated. Why retreat, when the world of nature seemed his personal intellectual domain? The unbridled enthusiasm of a romantic youth would not be curbed, even by entrance into the serious arena of scholarship. Thus, a short time before the Brazil volume appeared, Agassiz and Braun visited Oken's home and learned from him that Baron Alexander von Humboldt was planning another expedition of the kind that had won him renown, this time to the Ural Mountains and the deep interior of Siberia. Oken wondered if the two friends would be interested in accompanying the Prussian scientist. "Yes," was Agassiz's immediate reply, "and if there were any hope that he would take us, a word from you would have more weight than anything." Going home that evening, Agassiz rolled in the snow in joyous anticipation of such a possibility. Even if Humboldt did not take them, he would at least have learned their names. The decision went against them, even though Agassiz, not content with Oken's support, had François Mayor arrange for influential Swiss and Prussians to intercede for him with Humboldt.[20]

These real and imagined conquests left still unresolved the essential dilemma Agassiz faced by coming to Munich to study medicine, yet working and thinking as a student of natural history. One result of the Brazil enterprise was that he earned the title of doctor, not in medicine, but under the faculty of philosophy in subjects of natural history. The doctorate in this field, Von Martius had urged, would give the title page of *Brazilian Fishes* distinction. In May of 1829, therefore, eighteen months after his arrival in Munich, Agassiz prepared written examinations on the comparative anatomy of vertebrates and invertebrates. His performance was superior, accomplished without much formal preparation except for private instruction from Döllinger, lectures from Johann Wagler on herpetology, and a course from Oken on *Naturphilosophie*. He took his oral examinations at the University of Erlangen, because the schedule for these at Munich had been filled before he decided to take the degree. So impressed was the Munich faculty by Agassiz's written effort, however, that they awarded him the diploma without waiting to hear the results of his oral performance. Auguste Agassiz was the first to learn of this new tri-

umph; furthermore, his brother told him that the dean "hoped
before long to see me professor, and no less the ornament of my
university in that position than I had hitherto been as student."[21]

Achievement on the road to becoming the "first naturalist" of
his time did not fulfil his obligations to his patient relatives. He
had promised faithfully that he would spend two years at Munich
studying medicine and then take the Swiss examinations for medical practice. Contemplation of the physician's profession, however,
became a dull prospect as Munich's intellectual potentialities unfolded. Besides, as Agassiz frankly confessed to Dinkel, medical
studies not only failed to interest him, except for work in anatomy
and physiology, but he hated to see people in pain.

Shortly after coming to Munich, therefore, Agassiz faced the
problem of convincing his parents that he could actually make a
living as a scholar and professional student of nature. This was a
formidable undertaking, as his family had been understanding and
liberal in granting his wishes for more education, agreeing to the
abandonment of the business apprenticeship plan, and in providing financial support.

He approached his mother first, confessing to her that he was
bored with medical studies and asking that he be allowed to devote his life to natural history. Rose Agassiz supported her son's
inclination, only asking that he finish his medical education and
then, if still so motivated, take up the naturalist's profession. He
should not forget, however, that a good son had an obligation to
fill an honorable niche in society, to marry, work hard, and plan
an orderly life. "To do all the good you can to your fellow-beings,
to have a pure conscience, to gain an honorable livelihood, to procure for yourself by work a little ease, to make those around you
happy—that is true happiness. . . ."[22] Louis was quick to assent to
his mother's request. He would of course get his medical degree;
it was the thought of a career in this field that he detested. He respected the values his mother emphasized. But his future life
would allow few comforts and less security. He had to travel, to
explore Siberia and the tropics, to gather large collections of specimens so that he could rank as one of those rare individuals who

"have enlarged the boundaries of science." As for tranquillity and domesticity, the youth seeking his freedom observed that his role in life would necessarily be unique, almost misanthropic, because "the man of letters should seek repose only when he has deserved it by his toil, for if once he anchor himself, farewell to energy, to liberty, by which alone great minds are fostered. . . . A young man has too much vigor to bear confinement so soon. . . ."[23]

When Rodolphe Agassiz read this exchange between son and mother, he was unimpressed and refused to allow any alteration in expected performance. "Begin by reaching your first aim, a physician's and surgeon's diploma. I will not for the present hear of anything else, and that is more than enough." Louis was to curb his "mania for rushing full gallop into the future"; if this was impossible, then he should not ask his parents to share such wild dreams. There would be no more talk about seeking mammoths in Siberia or explorations to the tropics but, instead, application to the work at hand for the purpose of making his way in life as a doctor.[24]

Agassiz was distressed that he had given his father cause for anxiety, concerned at the reproof his exuberance had earned. He proposed a compromise. "If, during the course of my studies I succeed in making myself known by a work of distinction," would his father then agree to approve a year's study in the natural sciences and to employment as a professor of natural history, providing that the medical degree was also earned?[25] Agassiz could make such a proposal with assurance, because by this time, in the spring of 1828, he had already undertaken work on *Brazilian Fishes*. He did not tell his father of this enterprise, wanting to surprise him when the book was done. Pastor Agassiz hastened to consult François Mayor before deciding on his son's proposition. Both men were by now convinced that Louis would not be confined by tradition or the wishes of his family. They understood, Rodolphe wrote, not only how much he disliked the idea of a medical career but also how he would equally dislike "any other profession by which money is to be made." They settled for a promise that Agassiz would earn the medical degree, thus insuring a means of livelihood should "some revolution fatal to your philosophy" destroy his hopes, and then he could go on to study and ultimately teach natural history.[26]

Agassiz had at last gained family approval for activity that would emancipate him from the bonds of tradition. That he reversed the promised educational effort and earned the doctorate in natural history first was no token, in his eyes, of failure to act as a good son. Moreover, parents and uncle were soon caught up in the excitement of Louis's success when they learned of the Brazilian project through Professor Schinz. The family was more than pleased; their son was being talked about as a promising scholar. Auguste reported:

I can assure you that the stoutest antagonists of your natural history schemes begin to come over to your side. Among them is my uncle, who never speaks of you now but with enthusiasm. . . . He has asked me a dozen times at least if I had not forgotten to forward the remittance you asked for. . . .[27]

Agassiz could now feel confident in rushing full gallop into the future, even confiding his deepest ambitions to Rodolphe early in 1829. Conscious of unfulfilled promises, he still wanted to bring honor to his family and homeland. But he also aspired to the rank of "first naturalist" of his age, and conscience seemed to demand the reconciliation of duty and scholarship. He thus reaffirmed his intention to earn the physician's diploma, reminding his father that when he had promised to do this upon coming to Munich, Rodolphe had also agreed to permit advanced medical studies in Paris once the degree was earned. As for the promise to take up medical practice in Switzerland once his education had been completed, he asked reassurance of release from this obligation. His ultimate goal was a professional naturalist's career, and medicine simply did not interest him. He knew he could advance toward the twin ambitions of teaching and research in natural history if he could make his talents known to the world, and *Brazilian Fishes* would gain him just such renown. He would therefore soon earn the doctorate in natural history.

But a naturalist needed more than a degree and a book to gain distinction. Once more, the compelling image of a grand exploration rose before him, and he asked permission to journey for two years to such far reaches of the world as Siberia in the quest for experience. Duty could then be combined with scholarship, because he would return to Switzerland and dedicate his life to instructing "my young countrymen" in the mysteries of nature, awakening in them "the taste for science and observation so neglected among us." It followed that service of this sort would make him "more useful to my canton than I could be as a practitioner." Family obligations, the cause of Swiss cultural nationalism, and the advance of science would all be served.

Agassiz's conception of a "good son" had changed since the memorandum of 1822. Now he felt he would be acting the part of a good son if he were given complete freedom to do the things that his parents could agree to only with reluctance. A good son was, by Agassiz's latest definition, a good citizen; and a good citizen had to go to Siberia before he could teach zoology to the youth of Switzerland. While the great journey did not materialize, and while Rodolphe might understandably view his son's future plans as irrational, for Agassiz such a conception was as logical as the imagination and self-confidence that inspired it.[28]

The Agassiz family understood, from this proposal, that they had lost the son who might have been a prosperous physician but had gained a distinguished savant who would glorify his family and country. "Your father and mother," wrote Rodolphe, "while they grieve for the day that will separate them from their oldest son, will offer no obstacles to his projects, but pray God to bless them."[29] It was a short step from this attitude to one of overflowing pride when Louis sent home his diploma in natural history and a copy of *Brazilian Fishes*. Rose Agassiz read words of high praise for her son's accomplishment in the Lausanne newspaper. Friends and well-wishers congratulated the Agassizes for having such a promising scholar in the family. Rodolphe could not find words appropriate to convey his pride and pleasure. He only wanted Louis to come home, sometime, to "the old father who waits for you with open heart and arms."

But the twenty-two-year-old author had important things to do. The success of the Brazil volume encouraged him. He was convinced that only Cuvier, of all the masters of natural history, was worthy of reverence and emulation. The Frenchman's pioneering investigations of fossil bones, his system of classification whereby the animal kingdom was ordered into exclusive areas of identity (mollusks, radiates, articulates, and vertebrates), and his new work on fishes all commanded Agassiz's admiration. Carl Linnaeus, the Swedish naturalist responsible for significant reforms in taxonomy, seemed to Agassiz to have made many errors in his efforts at classification. The work of earlier ichthyologists left much to be desired; the descriptions of Guillaume Rondelet were often naïve and incomplete, while the work of B. G. E. de Lacépède, one of the few men before Cuvier to attempt a natural history of fishes, was often erroneous, with too little attention to the relation of animals to their environment.

It required little effort to imagine that the prospective study of central European fresh-water fishes would be superior to anything done before. In September, 1829, only three years after leaving Heidelberg, Agassiz returned to the scene of early student days to attend the annual meeting of German naturalists and physicians. He announced to these assembled scholars that he planned a monumental treatise in ichthyology. The enthusiastic praise and many compliments received from older men on this occasion were so inspiring they could not even be reported in a letter; they were "better told than written." In high spirits because of his triumph at Heidelberg, Agassiz, in the manner of a professional naturalist, printed a prospectus describing his planned publication and asking for the help of his fellows, and distributed it to those who could aid him.[30] Two years before, he had consulted Cuvier; now he asked others to send him data and specimens. In a short time, the knowledge thus gained would make him an authority in this branch of natural history.

The Heidelberg experience had to be communicated to friends dear to him, so, before returning to Munich, Agassiz visited Carlsruhe. Cécile Braun had missed the handsome, vivacious youth. It was exciting to hear Louis talk about Munich professors, great explorations, and how he was making a name for himself in the world of science. Pleasure could be combined with scholarship, and Cécile sketched many fishes under Agassiz's direction during the early fall of 1829. She drew with skill and precision. She was attractive and intelligent, her family eminently respectable and congenial. Yet a youth had to be the vagabond a bit before settling down; he had "too much vigor to bear confinement so soon," else "he gives up many pleasures which he might have had and does not appreciate at their just value those he has."[31] Apparently, Louis and Cécile reached an understanding at this time, to marry when future conditions seemed most propitious. Alex Braun was in a good position to judge the alliance and to evaluate a friend who he felt showed proclivities toward a frightening egoism, and at the same time was a warmhearted, deeply emotional person with a fine concern for friends when they needed spiritual or material support.

"You ask me what I think of the bond connecting the friend and the sister," Alex confided to his mother a year later.

Formerly I hesitated to talk about this; I could not suppress a certain fear because I knew Agassiz from all sides and always saw two natures in him between which there was as yet no decision. Therefore I am most happy that now I see only goodness in him. . . . You can see how serious is his love from the copy of a letter I send you and which you will please not show to anybody.[32]

In January, 1830, Agassiz returned to Bavaria to earn his degree in medicine, with pleasant recollections of a short visit home, his first in two years. Having promised to become a doctor, Agassiz did not find this a difficult task. He paid for the services of a senior medical student who helped him review subjects he should have paid more attention to while working in ichthyology. He attended Döllinger's lectures faithfully, visited the Munich hospital occasionally, and found his prior training in comparative anatomy, physiology, and embryology excellent preparation for his oral and written examinations in medicine. These were taken and passed in April, 1830, with typical competence and distinction.

The Dean said to me, "The Faculty have been *very much* pleased with your answers; they congratulate themselves on being able to give the diploma to a young man who has already acquired so honorable a reputation." . . . The Rector then added that he should look upon it as the brightest moment of his Rectorship when he conferred upon me the title I had so well merited.[33]

This report elicited deep gratitude and pride from Rose Agassiz. "I cannot thank you enough, my dear Louis, for the happiness you have given me in completing your medical examinations, and thus securing to yourself a career as safe as it is honorable."[34] Mrs. Agassiz was apparently just as prone as her son to wishful thinking.

Even though he was simply fulfilling his parents' wishes and gaining a diploma he did not mean to use, Agassiz was nevertheless exceptional. One of the theses he defended in his examination excited great interest from both his professors and his audience, an argument that the physiological organization of woman was more complicated than and superior to that of man.

Having earned two degrees in such a short space of time, Agassiz needed the stimulus of still another goal. He could not return to Switzerland until he had explored the full potentialities of German scholarship. If he was to be a professor, he needed more to distinguish his talent than just one publication in ichthyology. Thus still another project was born. At Heidelberg the previous fall the paleontologist Bronn had been enthusiastic in his praise of his former student's accomplishments. The directors of the Strasbourg and Munich museums had been equally complimentary. Each of these gentlemen superintended a rich collection of fossils. Museums at Carlsruhe, Vienna, and Heidelberg were also notable as depositories of ancient organic remains. Since Cuvier's major work on vertebrate paleontology in 1812, this science had been placed on a new and modern foundation. Knowledge of comparative anatomy could reveal much, it seemed, about the character of extinct vertebrates. The natural history of the earth in ancient times, the succession of animals in geological time, the identity and duration of different periods in the earth's history, and comparisons of extinct and living animals were all fascinating aspects of this view of nature.[35] While not yet prepared to generalize about the cosmic significance of paleontology for an all-embracing insight into the history of creation, Agassiz was sufficiently impressed with the contribution of Cuvier to want to distinguish himself in this realm. Just as he had outlined a natural history of living European fishes in emulation of Cuvier, he now planned to write a history of the fossil fishes of Europe, unaware of the French master's current interest in this field.

Youth and the relative paucity of his geological knowledge did not deter Agassiz. Since there was so little known about fossil fishes and since museum directors from all over Germany had promised him the full use of their collections, he would once again be able to identify himself with research in the forefront of knowledge. With his ability to conceive vitally important enterprises, the future seemed bright indeed. "I feel sure of success," he wrote confidently to Auguste,

the more so because Cuvier, who alone could do it (for the simple reason that every one else has till now neglected the fishes) is not engaged upon it. Add to this that just now there is real need of this work for the determination of the different geological formations. . . . Now that I have it in my power to carry out the project, I should be a fool to let the chance escape me.[36]

In May, 1830, the ambitious scheme was launched with a trip to Vienna to examine the private collections and materials in the museums of that city. "Everything was open to me as a foreigner, and to my great surprise I was received as an associate already known . . . welcomed and sought by all the scientific men, and afterwards presented and introduced everywhere. . . ." Impressed by the depositories of natural history he saw, he was only troubled by the realization that "the thing I most desire seems . . . for the present, farthest from my reach,—namely the direction of a great Museum."[37] Thus Agassiz demonstrated his striking ability, one that never left him, to soar from one peak of imagination to another, and steadfastly, with all seriousness, to believe that what he wished for would come true.

If his publication dreams about living and fossil fishes were to be realized, Agassiz knew he needed money. A loan from Swiss relatives and an advance from a Stuttgart publisher who agreed tentatively to publish both studies, solved this problem for the time being. Agassiz could continue paying Dinkel to sketch the fossils he examined and described and the fish he found in market places. Back in Munich, the museum director allowed him to examine rare fossils at home, a great convenience in research. Mornings were spent preparing skeletons of living fish, afternoons were allocated to fossil fish, and the days and months passed pleasantly. Braun, Schimper, and other friends were amazed at such industry. Agassiz was quick to share his knowledge with his fellows. He was aware of the special competence possessed by Braun, Schimper, himself, and others, and his talents for leadership and organization were put to good use in the establishment of a select scientific society. This was the "little academy," where Agassiz and his friends delivered lectures to one another on their particular interests, sometimes joined by an invited faculty member. Knowledge, to be of value, must be shared, and in this instance it was disseminated by the infectious enthusiasm of a young and remarkably talented guiding genius.

By December of 1830, Agassiz knew the time had come to leave the university that had done so much for the development of his talent in the past three years. He knew now that nature, if it meant anything at all, was to be understood as a whole, a historical and contemporary unit of experience. "I cannot review my Munich life without deep gratitude," he recalled later in life, his appreciation just as great for the opportunity to understand the meaning of nature as for knowledge of techniques. The philosopher-naturalist Lorenz Oken was especially influential in stimulating Agassiz's philosophical perception. According to his student, Oken's instruction left a profound impression:

Constructing the universe out of his own brain, deducing from *a priori* conceptions all the relations of . . . living beings. . . . It seemed to us . . . that the slow, laborious process of accumulating pieces of detailed knowledge could only be the work of drones, while a . . . commanding spirit might build the world of its own powerful imagination. . . . The young naturalist of that day who did not share, in some degree, the intellectual stimulus given to scientific pursuits by physio-philosophy would have missed a part of his training.[38]

Agassiz, however, unlike Braun and Schimper, did not become a disciple of "physio-philosophy" or *Naturphilosophie*. The romantic, speculative approach to nature typical of German thought early in the nineteenth century was nevertheless instructive to him because through men like Oken he learned that the deductions of philosophy could exist side by side with the techniques of empirical investigation. The lectures of Oken, and the teaching of Döllinger, a man also concerned with the metaphysical implications of natural science, made Agassiz understand that facts, by themselves, were only partial insights. The facts of nature were indications of profound cosmic significance.

"We go once a week to hear Oken on "Natur-philosophie," Braun wrote his father, "but by that means we secure a good seat for Schelling's lecture immediately after. A man can hardly hear twice in his life a course of lectures so powerful as those Schelling is now giving on the philosophy of revelation. . . ."[39] Schelling served as the reflection in history, philosophy, and logic of what Cuvier meant for Agassiz in zoology. Change and development were everywhere in nature—they could be observed in the life-history of the embryo—but this process was fixed and determined by the cosmic purpose that underlay all creation. While the eye, and not the imagination, was the best observer of the embryo, the data of experience were perforce reflective of ideal representations of final purpose and fixed cause. Since the Absolute Being was present in all nature, Agassiz saw his task as the discovery of the manner in which the individual life form attempted to approximate that history and development stamped by the Deity upon the universe from the initial creative act.

Agassiz, therefore, felt no contradiction between the sciences of embryology or anatomy and the assumptions of *Naturphilosophie* and the philosophy of revelation. He learned that it was perfectly possible to observe a tissue under a microscope, minutely and precisely, leave the instrument, and listen to lectures affirming that the entire world was ordered according to a divine plan of which living creatures were the earthly duplications of a transcendental ideal. For the present, Agassiz was primarily concerned with the life of exact research. The study of nature could proceed, internally, from precise attitudes of observation and induction. No one could deny the objective validity of a fish skeleton, carefully prepared and accurately described. "A physical fact is as sacred as a moral principle," Agassiz affirmed in later years. Yet in the future, too, he would not forget the compelling motivation to build, like Oken, a temple of nature from the deductions of mind and intellect. Gradually, in years to come, Agassiz learned to straddle the two worlds of empiricism and idealism, a talent which made him the spokesman of an older generation while a newer age moved on to a more objective view of nature.

For the present, however, Germany had given Agassiz all he had anticipated. It had taught him to observe through the microscope and at the same time to realize that the embryo being studied signified an unfolding natural drama in which it was but an individual representative of a divine scheme. Since he could not, like Döllinger or Braun, combine these influences into one systematic view of the world, he did the next best thing; he accepted each of them as separate compartments of his intellect, each to exert its sway as the occasion demanded. Since the contemporary world of science was grounded on both objective investigation and cosmic assumptions, his education had prepared him well to meet its demands.[40]

Agassiz returned to Switzerland a person much different from the young man who had left it. Not only did he wear the clothes of a German student, smoke a pipe, and carry volumes of Goethe and Schiller in his bags, but he bore the stamp of a philosophical as well as a scientific education. Rodolphe Agassiz may have been able to understand the general importance of his son's planned research on fossil and living fishes, but talk of the philosophy of nature made Louis appear almost strange, a doctor who talked of things properly the province of doctors of theology, not of medicine. Moreover, Louis came home determined to push forward in his self-imposed quest for authority in his chosen career. His instructions to his parents, now moved to the parsonage of Concise on the west shore of Lake Neuchâtel, were specific: they were to provide a room for the faithful Dinkel and to secure the services of a young boy as a laboratory assistant. With publication prospects that might yield substantial monetary rewards, Agassiz was determined to transform his parents' home into a research establishment. Any lingering hope the elder Agassizes entertained that

their son might settle down and take up the physician's craft was
shattered by these plans. Rodolphe, understandably, was somewhat bewildered by his son's demands. Cécile Agassiz was soon to
be married; the house would be in an uproar, hardly able to contain Louis, his assistant, his artist, and the quantities of specimens
he planned to bring home. Hoping to deter his son from such determined activity, Rodolphe wrote him before he left Munich:

Where in Heaven's name, will you stow away a painter and an assistant
in the midst of half a brigade of dress makers, seamstresses, lace-makers,
and milliners . . . ? Where would you . . . put under shelter your posses-
sions (I dare not undertake to enumerate them) among all the taffetas
and brocades, linens, muslin . . . ? Give all possible care to your affairs
in Munich. . . . Leave nothing to be done, and leave nothing behind
except the *painter.* . . .[41]

Louis, however, had been responsive to family wishes when he
had had no choice and had to be certain of financial support. Now,
an author with two doctorates, he gave his parents to understand
that "the painter" was a necessity for his scholarly progress. He
must have no interruptions from family or friends, and he promised to make some arrangements for his "museum" that would not
interfere with household activity. Somewhat awestruck by such
seriousness of purpose, the Agassiz family welcomed their German
scholar to the beautiful parsonage of Concise. Artist Dinkel was
housed at the Agassiz home, and a young boy hired to keep the
specimens in order. Friends were sent out on collecting expeditions, fossils sent from Zurich were studied, and even a few pa-
tients, timidly inquiring if Dr. Agassiz would treat them, received
care.

But despite all his activity, Agassiz found Switzerland uninspir-
ing after Munich and Vienna. He was anxious to visit the great
museums of the Continent where he might increase his knowledge
of paleontology. Of these museums, the most alluring was the National Museum of Natural History in Paris, then the great center
of European science. Established in the early seventeenth century,
the museum was first known as the Jardin du Roi and then as the
Jardin des Plantes until 1794, when its name was changed to Mu-
séum National d'Histoire Naturelle. By the 1820's, the museum

had become in actuality a natural history center. It occupied a tract of seventy acres, an expanse that contained botanical gardens, laboratories, galleries of zoology, geology, and mineralogy, and, most important for Agassiz, a building that housed a museum of anatomy and paleontology. The National Museum had professors, appointed by the government, who worked in such special fields as botany, zoology, paleontology, and anatomy. George Louis Leclerc de Buffon had been perhaps the most notable of these scholars, serving as "keeper" of the Jardin du Roi. Lamarck, who had died in 1829, had also been a professor at the museum. In 1830 naturalists of such world renown as Antoine Laurent de Jussieu, Étienne Geoffroy Saint-Hilaire, and Cuvier all held professorships at the museum. Humboldt visited the museum often, as did naturalists from all over the world. Agassiz's anticipations of Paris, however, centered on the opportunity to meet Cuvier and to work in the museum building housing the fascinating fossil materials that Cuvier and others had gathered and studied with such important results. The Jardin des Plantes, as Agassiz and many of his contemporaries continued to call it, would provide vital experience and would always symbolize for Agassiz the supreme idea of a national research center in natural history.

The Paris of Louis Philippe was suffering from a serious cholera epidemic. Agassiz, as if to appease familial desires that he still practice medicine, argued that it was important for a good doctor to study this dread malady. Besides, Rodolphe had previously given permission for a sojourn in Paris at some future date. Money for the journey and a brief stay in the city was generously provided by François Mayor and a family friend, the Reverend Charles ("Papa") Christinat. Accordingly, in October, 1831, after only ten months at home, Agassiz set out for that center of learning he had long ago determined to know and master.

It was December before he arrived. Knowing that familiarity with the contents of German museums would provide the key to unlock the cabinets of the Paris museum and win the support of Cuvier, Agassiz, somewhat to the consternation of his mother, decided upon a roundabout route. The Strasbourg museum with its interesting fossil fishes was his first stop; Stuttgart, also with significant ichthyological remains, his second. Next he visited Heidelberg to consult with Bronn and examine more treasures of natural history. Then he doubled back to Carlsruhe to see its museum, visit the elder Brauns, and consult with Alex about the relationships of geology to paleontology. Here he spent some delightful weeks with Cécile before finally taking off for Paris, where he was to receive a note from Cécile written on Christmas day of 1831:

A quite small greeting to the far-away friend, and my heartfelt wish that in the big *Lutetia Parisiorum*—in the last days of the year now ending . . . he may have many happy and beautiful hours—but—besides—think often . . . of small quiet Carlsruhe, that is, only of the corner house on the left side of the street, and once in a while of Cécile. . . .⁴²

With over two hundred pages of his manuscript completed, and with many fine sketches by the faithful Dinkel, Agassiz was certain that the paleontological knowledge he had just acquired was bound to win Cuvier's admiration, since it was unlikely that the Frenchman had studied these materials. Contrary to Agassiz's earlier impression, Cuvier—now Baron Cuvier—had recently begun to study fossil fishes; at his personal disposal he had, of course, the great collection of vertebrate fossils housed at the gallery of paleontology of the Museum of Natural History. It was vital that Agassiz see these specimens and finish his research; if Cuvier were to publish first, "the completion of his work would destroy all chance for the sale of mine."

As he had hoped, Agassiz's knowledge of the fossils of Germany proved to be an entering wedge in Paris. After establishing himself in the Hôtel du Jardin du Roi, the residence of many visiting naturalists because of its proximity to the museum, he immediately sent a polite note to Cuvier requesting an audience. He was received the following day. The author of *Recherches sur les ossemens fossiles* greeted the twenty-four-year-old author of *Brazilian Fishes* with kindness but with reserve. When Agassiz showed the Frenchman his drawings and his manuscript and asked permission to inspect the fossils of the museum, Cuvier was impressed. He knew Agassiz to be a promising young man, but until then he had no idea that this energetic Swiss was interested in paleontology. Cuvier gave Agassiz full use of the specimens in his charge.

Agassiz set to work with Dinkel, somewhat bothered that Cuvier had not greeted him as more of an equal. He worried that Cuvier might press forward with his plan to publish a monograph on fossil fishes. In this case the best he could hope for was that his long-admired master might propose a joint publication venture, as he had recently done with Achille Valenciennes in his work on the natural history of living fishes. As the days passed, Cuvier left his private workrooms now and then to inspect the work of the determined young man. Pleased with what he saw, Cuvier gave Agassiz the use of part of one of the museum laboratories. This mark of esteem was soon followed by an invitation to Cuvier's home, a salon that welcomed statesmen, artists, and scientists, "the gathering place of all the most original thinkers in Paris," in Agassiz's estimate. In the presence of men like Valenciennes, Élie de Beaumont, Alexandre Brongniart, and others of similar stature, Agassiz was not at all uneasy. He could not dress as these gentlemen did. "I have no presentable coat," he wrote Auguste almost proudly, "but at M. Cuvier's only am I sufficiently at ease to go in a frock coat."[43]

Agassiz was soon a regular guest on Saturday evenings at the Cuvier home. This was the only diversion in long, intense weeks of work for fifteen hours each day, and then only the inadequate lighting at the museum made him stop. In February, 1832, any doubts Agassiz entertained regarding Cuvier's estimate of him were set at rest. A typical evening at Cuvier's found the sixty-three-year-old master alone with the young disciple. Cuvier asked his secretary for a certain portfolio. He spread drawings of fossil fishes on the table, affirming that these were the materials gathered at the British Museum and other centers of learning for his planned work on paleoichthyology.

He said that he had seen with satisfaction the manner in which I had treated this subject; that I had indeed anticipated him, since he had intended at some future time to do the same thing; but that as I had given it so much attention, and had done my work so well, he had decided to renounce his project, and to place at my disposition all the materials he had collected and all the preliminary notes he had taken.[44]

This legacy from Cuvier had been earned by hard work both in France and Germany. Agassiz was frank with his family in telling of the background of this triumph. He wrote Uncle Mathias Mayor:

It was . . . with the view of increasing my materials and having thereby a better chance of success with M. Cuvier, that I desired so earnestly to stop at Strasbourg and Carlsruhe, where I knew specimens were to be seen which would have a great bearing on my aim.[45]

Rose Agassiz quite probably forgave her son the long trip through Germany when he wrote her proudly:

M. Cuvier . . . has been led to make a surrender of all his materials in my favor. I foresaw clearly that this was my only chance of competing with him, and it was not without reason that I insisted on having Dinkel with me in passing through Strasbourg and . . . Carlsruhe. Had I not done so, M. Cuvier might still be in advance of me. Now my mind is at rest on this score.[46]

Thus Georges Cuvier, like others before and after him, surrendered to the determination of Louis Agassiz and was pleased to be of service to him. The more Cuvier came to know Agassiz, the more he liked him. He found traits in the young man that struck a responsive chord. Like Cuvier, Agassiz had many projects before him; he was always planning something greater than the task of the moment; he enjoyed polite society and the company of cultured laymen. Finally Agassiz, like Cuvier, thought of himself as a unique person.

The winter and spring of 1832 were times of significant intellectual development for Agassiz. Cuvier's friendship prompted Élie de Beaumont and Brongniart to offer him the use of their private collections in paleontology; all public resources for the study of natural history in Paris were his to use. Cuvier proposed that he submit his monograph when completed to the Academy of Sciences for evaluation, since a favorable report would greatly enhance his chances of publication. He was even offered the editorship of the zoology section of a scientific journal but declined because he knew his time in Paris had to be spent in productive research. He saw little of Paris aside from its scientific centers, but the solitary life did not make him unhappy. Rather, he reveled in the opportunity to work at the museum with Cuvier available to guide him. Here were fossil fishes to be found nowhere else in the world, and he was determined to master all the data about them. He patiently explained the process of his research to his father:

The aim . . . is to ascertain what beings have lived at each one of these . . . epochs . . . and to trace their characters and their relations with those now living. It is especially the fishes that I try to restore for the eyes of the curious, by showing them which ones have lived in each epoch, what were their forms, and, if possible, drawing some conclusions as to their probable modes of life. . . . In many species I have only a single tooth, a scale, a spine as my guide in the reconstruction of all these characters, although sometimes we are fortunate enough to find species with the fins and skeletons complete. . . .[47]

"We" in this context meant Dinkel, always on hand to sketch the fishes his master reconstructed by the talented use of Cuvier's principles.

The intensity of Agassiz's studies was such that fossil fish even became the subject of his dreams. The complete details of a partially preserved fish imbedded in stone had puzzled him for weeks. For two nights in dreams he saw the fish, fully restored, but could not recapture its form the next day. On the third evening, he went to bed with paper and pencil near him, and, when the image returned again, half-awake he sketched it. Returning to the museum the next day, he soon managed to separate the fossil from the stone, discovering as he did so that here was the original, exactly as he had sketched it.[48] This experience was to provide countless admirers in later years with the kind of material that they wove into the Agassiz legend.

Just as Germany had been valuable both for experience in science and exposure to philosophy, so, in a way, was Paris. For two years Cuvier had been engaged in a dispute with Geoffroy Saint-Hilaire about the nature of species and the plan of creation in the animal kingdom. The debates between these two naturalists were still going on when Agassiz arrived in Paris. On occasion, he attended Cuvier's lectures on the history of the natural sciences at the Collège de France and heard his mentor contradict the interpretations of Geoffroy. To Agassiz, it seemed that Geoffroy's ideas of the unity of all animal creation were much closer to the spirit of Goethe and *Naturphilosophie* than to the rigid insistence of Cuvier on the permanence of species and the primary importance of precise knowledge.

These debates were a significant intellectual experience for Agassiz. Geoffroy's ideas represented the idealized view of natural processes analogous to the Munich teachings of Oken and Schelling. The affirmation of a common unity of plan that bound together the different substructures of the animal kingdom was conceptually parallel to Goethe's ideas regarding the metamorphosis of plants and Oken's speculations classifying all creation according to an ideal form based on the structure of man. In the later stages of the debate—those arguments that Agassiz heard personally—the basic discussion shifted to a consideration of whether or not only one branch of the animal kingdom—vertebrates—could be identified as exhibiting homologous anatomical relationships. In taking the affirmative position, Geoffroy unfortunately supported his position with weak and inconclusive data of a highly speculative character. In his rebuttals Cuvier time and again pointed to the insufficiency of the evidence for Geoffroy's position.

Agassiz never reported his immediate reactions to these argumentations. Nevertheless, his exposure to such contrasting views was important for his future philosophic outlook. Experience of this sort tended to make Agassiz impatient with the metaphysics of his German education and more receptive to that aspect of his intellectual past which, like the attitude of Cuvier, emphasized exactitude and precision in the study of nature. As he analyzed Oken's attitude of mind in later years, such views encouraged the imposition "of one's own ideas upon nature, to explain her mysteries by brilliant theories rather than by patient study of the facts as we find them. . . ."[49]

Germany had opened grand vistas of philosophical speculation, but Paris, by contrast, meant intense, exact, and methodical labor. *Naturphilosophie* slowly faded from Agassiz's vision as he busied himself in the workrooms of Cuvier. The intellectual and personal assurance derived from Cuvier's friendship seemed to confirm the verities of precise analysis learned under Döllinger's tutelage. No longer need Agassiz feel awed by the generalizations so confidently advanced by a Schimper or a Braun. If reading Lamarck and listening to the sweeping statements of Oken and Geoffroy seemed to imply that nature was but one unbroken series of relationships, all its forms linked together by a common unity, Agassiz could take certain strength from his growing conviction that "he is lost, as an observer, who believes that he can, with impunity, affirm that for which he can adduce no evidence."[50]

Cuvier taught Agassiz the essential skills of a scientific craft that were to provide a firm foundation for the youth's future original investigations. He showed his Swiss disciple how to reconstruct fragmentary fossil remains according to the principle of the correlation of parts, wherein the proficient scientist could infer from one part of the body or segment of bone the entire structure of an animal and its place in the natural order. Agassiz learned, also, to make specific identifications of fossils from an anatomical knowledge of other living forms and fossil remains, and thus to reconstruct the probable life form of an ancient creature. As with Döllinger and the world that unfolded for him under the microscope, paleontology, as Cuvier taught it, would always excite and inspire Agassiz's searching intellect and would seem to him the noblest kind of objective study.

Empiricism was not the sole lesson Agassiz derived from Cuvier's instruction. "There is a great distance, between the man who, like Oken, attempts to construct the whole system of nature from general premises and the one who, while subordinating his conceptions to the facts, is yet capable of generalizing the facts, of recognizing their most comprehensive relations."[51] That singular man, for Agassiz, was Cuvier. The French naturalist's assertion that there was no evidence of animals having developed from other animals, or having produced animals higher than themselves, was to Agassiz a viewpoint grounded on evidence and observation. He displayed the same admiration for Cuvier's theory of types, which was based on the vitally significant concept that the animal kingdom could be divided into four great branches: vertebrates, articulates, mollusks, and radiates. A comparison of animals within one branch or type would yield knowledge of their rank and specific character. No comparison, or genetic relationship, was possible between individuals of different branches, each type being distinguished by an independent and mutually exclusive "ground plan" or order of identity. Classification, under Cuvier's system, proceeded from the categorization of individuals according to the type they represented, not from an assumption that all life shared a common unity in a progressively ascending scale of nature. For Agassiz, this kind of systematic analysis provided both a valuable tool for a specific understanding of nature and an intellectual antidote to the historic and contemporary unity of relationship suggested by *Naturphilosophie*.

Cuvier was certain that the history of creation had been marked by great catastrophes wiping out large portions of the animate world which had then been replaced by other forms characteristic of different periods of geological time. Agassiz came to share this view. Fossil animals were not related to animals living in the present; there was no genetic connection between species of different geological epochs. Species, under this view, were fixed and immutable, since it was impossible to discover transitional forms between ancient and living individuals. While Cuvier's system of classification represented a distinct advance over previous systems, notably in the idea of a "ground plan" or type scheme for each of the branches of the animal kingdom, such a conception did not allow for the kind of intellectual curiosity that would encourage the discovery of relationships of plan and structure between representatives of the major types. Cuvier thus represented a particular kind of typological thinking, an attitude of mind that also reflected an a priori conceptualization about the natural world.[52]

For Agassiz, however, Cuvier seemed a model of precision and formalism compared to the speculation he had known in Germany. He thus found himself in complete sympathy with Cuvier's insistence on the necessity for exact knowledge before generalization. At the same time, he also accepted Cuvier's views regarding the lack of relationship between fossils of different periods, for such an interpretation was in keeping with the evidence Cuvier had derived from exact, empirical observation. As for the meaning of nature, Agassiz adopted Cuvier's affirmation of the scientific reality of catastrophes, even improving on his master by asserting that all life had been wiped out at such times and created anew, successively, in each instance. He also accepted Cuvier's taxonomic system without reservation. Classification of animals had to be grounded on knowledge of embryology, paleontology, and comparative anatomy, but this approach was not simply a human construct to expedite investigation. Classification was an inductive process whereby the order of nature might be empirically understood. There were, as Cuvier had asserted, four grand structural plans characterizing the branches of the animal kingdom, but these plans were the discovery, not the invention, of the natural scientist. They existed in nature, and the investigator who discerned their character was in fact discovering the meaning of God's plan for the natural world, since nature was the work of thought, the production of intelligence, a premeditated scheme carried out according to divine intention.

In Agassiz's mind, then, it was possible to formulate a synthesis of Cuvier's empiricism and metaphysics, supported by the intellectual impulses that had once made Oken and Schelling seem such attractive philosophers. Agassiz was able to embrace an attitude toward nature that recognized a unity of plan in the animal kingdom, a unity stemming from supernatural design and intention, not from the characteristics of animals and plants themselves. Before generalization, however, the study of evidence and facts was primary, and here his study proceeded according to the patterns marked out by Döllinger and Cuvier. The interpretation of nature should not be "speculative" in the presumptuous manner of Oken, however. Rather, nature illustrated a profound arena for the assertions of theology. Exposure to the devout Protestantism of Cuvier helped make Agassiz receptive to a synthesis of science and theology. It would be his task, patiently and with exactitude, to study the data of nature. As he did so, he would discover the essential meaning and order that the Deity had impressed on organic creation. This intellectual unity would underscore the ideal, spiritual quality of all natural phenomena, but the idealism would be based upon empirical knowledge. Agassiz, then, did not discard a single item of his education: no matter how contradictory Oken and Geoffroy might seem in contrast to Cuvier, in Agassiz's mind the motivation of *Naturphilosophie* to view nature as a whole was substantiated by the techniques and presuppositions learned from the French naturalist, the only man whom he ever acknowledged as his intellectual master.

Philosophies of nature were inspiring, as was research on fossil fishes and the thrill of attending gatherings at Cuvier's home. But an allowance of two hundred francs per month made Agassiz fearful that his small store of capital would soon disappear. His parents became concerned that he was virtually living the life of a pauper in Paris, especially that he was burdened with having to pay Dinkel's salary. They urged he dispense with the artist's services, return to Switzerland, and sell the large zoological collections he had acquired in order to gain at least temporary means for support. Writing Louis in March of 1832, Rose Agassiz was on the whole unimpressed by the enthusiastic reports her son had made about Cuvier and his Paris reception. She wondered whether all this had left much time for studying the epidemic that had brought him to France. "With much knowledge, you are still at twenty-five years of age living on brilliant hopes, in relation . . . with great people. . . . Now all this would seem to me delightful if you had an income of fifty thousand francs." He should learn to accept the advantages of his home and family and secure a teaching position in a city such as Geneva, Lausanne, or even Neuchâtel, where the Mayor family was not without influence.[53]

But Agassiz had undertaken an ambitious research and publication venture in paleontology, one, to be sure, requiring considerable investments of energy and money before it would be finished. He thought of himself as working under Cuvier's beneficent influence. But, in May of 1832, the man who had taught him so much in such a short time died from the very disease Agassiz had come to Paris ostensibly to study. This was a vital event in Agassiz's life. With Cuvier's death, whatever sense of intellectual dependence Agassiz had known disappeared. The fact that the great naturalist had turned over important fossils to him for description and publication made Agassiz think of himself as Cuvier's disciple. He determined to model his intellectual efforts after the pattern set for him by Cuvier. The boy who had been remarkable for flights of fancy, who had rolled in the snow at the thought of going on a great expedition, was capable of dramatizing and romanticizing this event. He had a vast responsibility before him now, because his future life would be a testimony to Cuvier's faith. Just as Cuvier had been supreme in his self-confident authority, so too would Agassiz henceforth recognize no master, for now he was a master himself, with only the remembrance of Cuvier's practical and spiritual legacy to inspire him.

The practical problems attendant upon carrying on with his research were soon settled in a manner almost predictable throughout Agassiz's charmed life. There was one person in all Paris who equaled Cuvier in power, political influence, and commanding rank in natural history—Alexander von Humboldt. Councilor of state and court chamberlain to the Prussian monarchy, this geographer, geologist, world traveler, and philosopher of nature had just returned from the expedition to Russia and central Asia that Agassiz had hoped to join. At the height of his scientific fame and political influence, Humboldt was visiting Paris in an official capacity to report on the government of Louis Philippe to his Prussian sovereign, Frederick William III. Like Cuvier, Humboldt had been one of Agassiz's heroes. Humboldt knew of the talented Swiss through testimonials from both Oken and Cuvier. In the early spring of 1832, before Cuvier's death, Agassiz had received an invitation to visit the Prussian statesman-naturalist in his working quarters in the Rue de la Harpe.

Humboldt, like other men of position before and after him, was immediately attracted by Agassiz's charm, sincere manner, and intense ambition. Soon after this first meeting he wrote his impression to Mrs. Agassiz:

You are happy in the possession of a son as distinguished by his talents, by the variety and the solidity of his knowledge, as modest as if he knew nothing in these days when youth is characterized by a cold and scornful *amour propre*. One might well despair of the world if a person like your son, with information so substantial and manners so sweet and prepossessing, should fail to make his way. . . .[54]

Agassiz had much to learn from Humboldt. In many respects, the months of 1832 spent under the tutelage of this influential man were times of great importance in shaping Agassiz's outlook toward the world.

Humboldt had visited Agassiz's room at the beginning of their friendship and inspected his library. He approved of Aristotle, Linnaeus, and Cuvier, and of course his own *Views of Nature,* prominently displayed. He was touched to see so many books copied by hand. A publisher had sent Agassiz a twelve-volume encyclopedia. Humboldt did not approve: *"Was machen Sie denn mit dieser Eselsbrücke?"* Nonplused, Agassiz explained he needed this "ass's bridge" because "I have not had time to study the original sources of learning, and I need a prompt and easy answer to a thousand questions I have as yet no other means of solving."[55] The modest young man could not fail to make his way indeed, with answers such as these.

It was Humboldt who took Agassiz to the Collège de France to hear Cuvier's lectures attacking Geoffroy's interpretation of nature. Agassiz wanted to listen carefully so as not to miss a word, but Humboldt persisted in whispering sharp criticisms of Cuvier's analysis. Of course there was something to what M. Humboldt said, especially concerning the clear relationship of Geoffroy's ideas to those of Goethe and the spirit of *Naturphilosophie.* One should certainly respect the breadth of view and powerful conceptions of this attitude toward the world and listen with care to Humboldt's strictures on the formal, unimaginative approach to science typified by Cuvier's critique. One could not very well disagree with the opinions of the eminent man even though one's intellectual predilections were leading in another direction.

But the older man and his young friend did not talk often of science or of metaphysics. Humboldt's teaching was more personal, its implications as challenging to Agassiz's imagination as thoughts of great explorations. With Humboldt he went to dinner at fine restaurants, where the bill of fare was far beyond the reach of an impecunious student. He had permission to visit his new friend as often as he pleased. "How much I learned in that short time!" Agassiz fondly recalled. "How to work, what to do, and what to avoid; how to live; how to distribute my time; what methods of study to pursue—these were the things of which he talked to me. . . ."[56] These were lessons not to be learned in any museums. They were modes of behavior, and Agassiz absorbed with uncommon interest the manner of a man of science and the world, a man who was an adviser to kings, a friend of artists and poets, and a respected figure in polite society. "He was as familiar with the gossip of the fashionable and dramatic world as with the higher walks of life and the abstruse researches of science," Agassiz noted with pride.[57] By personal example and intimate advice, Humboldt taught Agassiz what he could not have learned from former teachers and only dimly understood from his short acquaintance with Cuvier. Humboldt taught him the forms and attitudes he thought essential for a young man who wanted to succeed in the world. He became Agassiz's model of the scientist who knew fame and distinction because of his understanding of the larger world, of power, prestige, personal influence, and the social and institutional support requisite for intellectual activity.

Humboldt had been a student of Abraham Gottlob Werner, and numbered among his close friends such masters of science as Goethe and Leopold von Buch. He had traveled to Mexico, South America, and the United States. To Agassiz he seemed the personification of the cosmopolitan man of science. His brother Wilhelm was Prussian minister of education, and Humboldt had an influential role in such national academic institutions as the University of Berlin. Humboldt enjoyed discovering talent. Justus Liebig, chemist, Achille Valenciennes, zoologist, and Friedrich Gauss, mathematician, had all benefited from the friendship of this singular individual, as a word here, a letter there, and occasional financial aid helped smooth their way. The Prussian scientist performed the same service for Agassiz. He wrote Agassiz's Stuttgart publisher, who was also his own publisher, pointing out that Agassiz's poverty made an advance of money necessary if the young man was to continue his research in Paris. When he received no immediate reply, Humboldt simply wrote Agassiz a note inclosing a check for one thousand francs because, "A man so laborious, so gifted, and so deserving of affection as you are should not be left in a position where lack of serenity disturbs his power of work. . . ."[58]

Agassiz's joy and optimism were unrestrained. He could remain in Paris and continue to enjoy Dinkel's valuable services. "Imagine what must have been my feeling," he wrote Humboldt in gratitude, "after having resolved on renouncing what till now had seemed to me noblest and most desirable in life, to find myself unexpectedly rescued by a kind, helpful hand, and now I have again the hope of devoting my whole powers to science. . . ."[59] This was only the first of many similar letters to powerful men to come from Agassiz's pen. The pattern was to become familiar: despondence, help from an unexpected source, and, then, high enthusiasm over prospects for the future.

But one could not stay in Paris forever without prospect of permanent means of support. Cuvier's death had deprived Agassiz of a powerful friend, but it had also opened a number of opportunities. There was, for example, the matter of Cuvier's planned volumes describing the natural history of fishes. Cuvier's publisher approached Agassiz with the proposal that he join Valenciennes in that naturalist's effort to bring the work to completion. This would have meant joint authorship, and Agassiz was reluctant to seek fame in this fashion. There was, moreover, the obligation to publish the research materials gathered on fossil fishes, an effort that would not prosper under such conditions. Research in paleontology was proceeding very satisfactorily. With the money from Humboldt, Agassiz knew that he would soon be able to complete his work in Paris. This fact, added to the growing awareness that he would have to seek some permanent way of earning money, led Agassiz to consider what the future might promise.

Agassiz's primary requirement was a secure existence in an environment where he could bring together his research materials, study Dinkel's drawings, and begin writing and publishing his major work. Typically, this goal was to be achieved through concentrated effort, a firm belief in personal capability, and the help of influential people. Sensitive to her son's needs, Rose Agassiz had suggested a teaching position in some Swiss academy. Now she pointed out to Louis that the town of Neuchâtel would soon establish a new *collège* and public museum of natural history. Seeking a place to settle down and begin writing, Agassiz saw a fine opportunity before him. The new institution would require a professor of natural history who could also organize the museum. The museum needed specimens, and these he had in ample number from the days of youthful collecting and the activity of recent years. Uncle François Mayor enjoyed the friendship of Louis Coulon, a wealthy Neuchâtel aristocrat who took a deep interest in natural history.

The affairs of canton and town were under the control of two influential sources of power: the local aristocracy and the Prussian monarchy. The aristocracy would provide the primary financial support for the new school and museum. The canton also had a dual political status, under federal Swiss authority and also under the constitution granted in 1814 by Frederick William III. This sovereign appointed the Prussian governor and nominated members of the ruling council of state and a portion of the membership in the legislative body. From considerations of diplomacy and ties of sentiment, the king took a benevolent interest in Neuchâtel affairs, occasionally supporting public projects with grants of money.

Any public appointment would need the approval of both monarchy and aristocracy. In aspiring to a position in the new school and museum Agassiz did not think of himself as conforming to familial wishes for a career in Switzerland. Rather, he saw himself as a cosmopolitan intellectual, strengthened and emancipated by the experiences of Germany and France, coming to bring enlightenment to his countrymen. His entrance into Neuchâtel society, moreover, would be under the beneficent influence of Prussia and the local aristocracy, important recommendations for a rural pastor's son. Above all, he would enjoy an excellent opportunity for reflection and writing.

Once imagined, the actual achievement of this end proceeded with remarkable speed and dispatch. Although the idea had been inspired by Mrs. Agassiz, Louis soon thought of it as his alone, a project to challenge his ability and new sense of power. There was first a letter to Coulon, whom Agassiz had met on his previous visit to Switzerland. A simple inquiry, it told of his interest in a professorship of natural history, should one be established at the new institution. It mentioned his friendship with Humboldt and the wish of the Prussian savant that he receive the appointment. There was another matter:

I have made a very fair collection of natural history. . . . Should an increase of your zoological collection make part of your plans for the Lyceum, I venture to believe that mine would fully answer your purpose. . . . I have spoken of this also to M. de Humboldt, who is good enough to show an interest in the matter, and will even take all necessary steps with the government to facilitate this purchase.[60]

It was hardly possible to ignore a request and proposal that spoke so confidently of the sympathy of the king's councilor of state and promised a collection along with a professor. Coulon replied as he had to: Agassiz had been recommended for the professorship of natural history, a position that had actually been established as a result of his inquiry. Unfortunately, there were no funds as yet for a salary, but perhaps a public subscription of money from influential townsfolk might provide a subsidy for this purpose.[61]

Agassiz informed Humboldt of Coulon's letter. His Prussian mentor was delighted with the opportunity now apparent to employ his talents of diplomacy and his influence in Berlin. He wrote his brother Wilhelm, recommending Agassiz and urging prompt action by the king. He wrote General von Pfüel, Prussian governor of Neuchâtel, indicating his and the monarchy's interest in the appointment. He wrote Leopold von Buch at the University of Berlin to exercise influence from that quarter. Agassiz did not sit idly by waiting for important men to make his position secure. Early in June of 1832, with something less than complete frankness, he wrote Coulon that he was flattered that the Neuchâtel citizenry wished him for their professor, but he had serious offers in Paris to consider, notably the proposed joint venture with Valenciennes on the Cuvier ichthyology, and he could not keep the publisher of this series waiting for his decision. However, he was a Swiss, not a Frenchman, and would really like to return to his native land. He hoped he would be free to come to Neuchâtel, if "the urgency of the Parisians does not carry the day." He would try his best to hold out against their entreaties, but since the new term would start soon and lectures would have to be prepared, he urged Coulon to act promptly because, "if your people are favorably disposed toward the creation of a new professorship we must not let them grow cold." He did not care for a very lucrative position, only the opportunity to instruct the youth of Switzerland.[62]

Coulon and other wealthy townsfolk acted quickly. They raised a sum of money which would guarantee a modest salary for Agassiz. Coulon pleaded the case of Neuchâtel in writing the young naturalist. "I can easily understand that the brilliant offers made you in Paris strongly counterbalance a poor little professorship of natural history at Neuchâtel." But the influential men of the town had rushed to fill the subscription list for his salary; they were all anxious that he come to Switzerland and dignify the intellectual life of Neuchâtel by teaching natural history.[63] One month later, in July, Coulon received another letter from Humboldt, with words that buttressed his recent action. Agassiz was a young man "distinguished by his talents, by the variety and substantial character of his attainments, and by . . . his natural sweetness of disposition." He was engaged in a study of fossil fishes, "the most important ever undertaken." More than this, "the illustrious Cuvier, also, whose loss we must ever deplore, would have recommended him with the same heartiness, for his faith, like mine, was based on those admirable works of Agassiz which are now nearly completed."[64]

Agassiz never doubted that he would accept the Neuchâtel offer if it were made. Thus, as early as May, 1832, he turned down the proposed joint publication with Valenciennes but wrote Coulon of its attractiveness in June. He could be frank with Humboldt. He wanted to go to Neuchâtel and was anxious to succeed there, although he felt somewhat concerned that his supporters had pictured him as a kind of "wonder from the deep," and he hoped he would not disappoint them. Neuchâtel would provide the quiet atmosphere he needed for writing. "Whatever may be growing up within me will have a more independent and individual development than in this restless Paris." The indirect association with

Prussia suggested an advantage. "From Paris, also, it would not be so easy to transfer myself to Germany, whereas I could consider Neuchâtel as a provisional position from which I might be called to a German university."[65] Agassiz therefore accepted the Neuchâtel position and immediately wrote out a course of lectures appropriate for an institution that would be both a secondary school to prepare students for the university and one that offered public instruction to the adult population.

After less than a year in Paris, a period of primary importance in his personal and intellectual development, Agassiz returned to Switzerland in September, 1832. He came home as an independent scholar, supported by the Neuchâtel aristocracy and the Prussian monarchy. Agassiz had told Coulon he was not a Frenchman; neither was he a Prussian; he was a Swiss with excellent French and German training and recommendations. Munich, Cuvier, the Museum of Natural History, and Humboldt had all provided spiritual and intellectual resources for the future. The Agassizes were pleased with their son's new position. He was now a professor, with a small salary but excellent prospects, admired by the great of Neuchâtel. His formal education was at an end. He had two degrees, a book to his name, and projects of scientific importance before him. At twenty-five, the young man who aspired to be the first naturalist of his time could settle down to writing and research.

3

FROM SWITZERLAND
TO BOSTON

1832-1846

Great was the emotion at Neuchâtel when the report was spread abroad
that Agassiz was about to leave for a long journey. It is true he promised
to come back, but the New World might shower upon him such marvels
that his return could hardly be counted upon. The . . . students regretted their beloved professor not only for his scientific attainments,
but for his kindly disposition, the charm of his eloquence, the inspiration of his teaching; they regretted also the gay, animated, untiring
companion of their excursions, who made them acquainted with nature, and knew so well how to encourage and interest them in their
studies.[1]

THIS WAS THE MOOD, as recaptured later by a Neuchâtel citizen,
on a March morning in 1846 when Agassiz left Switzerland. Students and well-wishers gathered at his carriage to bid him farewell,
and some thought the town was now seeing the last of the man
who had brought it such fame. As he prepared to face a new life
in a new world, this was one scene that Agassiz carried away with
him.

The Neuchâtel years were good ones for Agassiz; they were in fact the most productive of his intellectual career. He was in the full bloom of life, driven by a consuming desire to reach the top in his field, and his accomplishments during these years brought him full rewards. His self-conception, molded long before 1846, was built not only upon the triumphs of the present but, as ever, upon the promise and the challenge of the future. This looking toward the future was not without its penalties in the present, but in the picture of the world Agassiz fashioned for himself and his purposes, these were no hindrance to him. Beyond the horizon, there always beckoned a dream such as the one Humboldt had once inspired, the heroism of a great exploration. And so this romantic hero, living in an age whose romanticism still had tangible rewards, believed that the dreams of the present could well be the realities of the future, if one wished hard enough for them and worked for them with enough faith and intensity.

Agassiz began his academic duties in Neuchâtel in November of 1832. The town of six thousand people, prosperous, complacent in its traditions, proud of its new school then being constructed and even more so of its new professor of natural history, turned out its most prominent citizens to hear Agassiz's first public lecture. Agassiz made a striking appearance. His father, seated in the company of wealthy Neuchâtel officials, the Prussian governor, Louis Coulon, and members of the council of state, was undeniably elated by his scientist son. The handsome young naturalist who spoke so feelingly and with a winning smile about the "Relations between the Different Branches of Natural History and the Then Prevailing Tendencies of the Sciences" certainly seemed to justify all the years of education he had received. The professor rose to the occasion with artistry and dignity; his expressive hands, facile mind, and powerful voice gave meaning and significance to the most complex facts of nature for the citizenry of his new home. A vision of Humboldt—the interpreter of the secrets of the universe to an earlier generation—flashed across his mind. The occasion over, he was quick to tell his Prussian mentor of his feelings:

I have been received in a way I could never have anticipated, and which can only be due to your good-will on my behalf and your friendly recommendation. You have my warmest thanks for the trouble you have taken about me, and for your continued sympathy. Let me show you by my work in the years to come . . . that I am in earnest about science, and that my spirit is not irresponsible to a noble encouragement such as you have given me.[2]

In the month following his first public lecture, Agassiz organized a Neuchâtel Society of Natural Sciences. It had only six members at first, but under Agassiz's direction proceedings were published, discussions of recent discoveries held, and membership soon increased. Soon thereafter a select group of prominent Neuchâtelois found themselves engaged in studying the secrets of natural history, informally and pleasantly, under the guidance of the energetic professor. Botany, zoology, geology, the philosophy of nature, all such subjects were topics for discussion. These people were grateful for the opportunity thus provided to learn what college students studied so intensively.

No age group was denied the pleasures of Agassiz's instruction. The shores of Lake Neuchâtel, the surrounding hills, fields, and mountains were soon invaded by scores of children led by a gifted and enthusiastic teacher. The formation of lakes, rivers, valleys, and streams was patiently and carefully explained to these youngsters; books were unnecessary; nature had to be experienced to be appreciated, and long walks in the country during spring and summer became the favorite occupation of the young people of Neuchâtel. Lessons were conducted indoors during inclement weather with plants, coconuts, and various exotic flora and fauna spread out on a large table to be studied. When appropriate, the end of the lesson meant that one could eat the material just examined. Since all children were asked to begin collections of minerals, fishes, fossils, birds, small mammals and similar objects as part of their instruction, the balance of nature in the Neuchâtel region had a new factor added to its composition.

The Neuchâtel museum, already established according to plan, found in Agassiz an able director. With young and old alike serving as both collectors and patrons, the establishment grew and prospered. Lectures to students followed the pattern established for their fathers and brothers and sisters. Botany, zoology, and

geology were studied both formally in the classroom and informally in the surrounding countryside. Agassiz found that his lectures did not take much time or preparation and that independent
investigation and description of fossil fishes could proceed as
planned, especially when Uncle Mathias Mayor made a contribution of $250 to pay the expenses of Dinkel, who was still in Paris
faithfully sketching relevant material.

The energetic Agassiz had descended suddenly upon Neuchâtel,
turning the town into a veritable beehive of scientific pursuits—
college instruction, informal discussions, field trips for children,
as well as advanced investigation under his direction. Republican
and aristocrat, merchant and noble, young and old, student and
teacher, all were now united in the quest for scientific knowledge.
The wealthy Louis Coulon became Agassiz's champion and spokesman in the councils of Neuchâtel government and society. He was
president of the scientific society and sponsor of the museum and
enjoyed the prospect of a new stimulus to intellectual activity in
the town.

In the midst of all this inspired effort a letter arrived from Professor Tiedemann in Heidelberg bringing the news that Agassiz's
former zoology teacher, F. S. Leuckart, had accepted a post at the
University of Freiburg. Only six years after entering Heidelberg
as a student, Agassiz was asked to return as a professor. He would
be the only instructor in zoology, have full use of the collections,
receive a higher salary than Neuchâtel could pay him, and his research would benefit. Tiedemann advised his former student that
he should not restrict himself by a career in an isolated, provincial
community but should come to Heidelberg, where his full talents
could be appreciated.[3]

This proposition, testimony to the reputation Agassiz had already made in the scientific profession, seemed to promise all he
had expected from Neuchâtel. He had hoped to go from there to
a German university, and this was one of the reasons for not staying in Paris. Now, his aspiration seemed to be realized almost immediately. Agassiz, however, found it impossible to accept the offer from Heidelberg. Writing to Humboldt, he affirmed that his
lectures did not take much time, that the publication of the fossil

fishes could be directed from Neuchâtel as easily as from any other place, and that the quiet life of the town was beneficial to his physical condition, which, it seems, had been somewhat taxed by the strain of close inspection of Paris fossils.

There were other factors originating in the character of the local cultural environment which influenced his decision. "I am satisfied that the people here would assist me with the greatest readiness should my publications not succeed otherwise," he confided to Humboldt.[4] Since the conditions of labor were equal, since he could write as easily in Neuchâtel as in Germany, there was no reason to leave a place where he was the sole representative of his profession to compete in a wider academic setting with such figures as Bronn and Tiedemann at his side. While Heidelberg respected his talents, a young professor of twenty-five would hardly receive the special acclaim Agassiz enjoyed at Neuchâtel. The citizens of the small community were his ardent supporters. With his scientific society and the backing of Louis Coulon, he had become a pied piper of natural history, leading an admiring public away from the humdrum concerns of politics and business to an appreciation of the grandeur of the universe. Furthermore, Neuchâtel seemed to promise tangible support as well as emotional security.

In writing Humboldt of his decision to remain at Neuchâtel, Agassiz sought not only the older man's approval and advice but something more. There was the matter of Agassiz's personal collection of natural history materials.[5] Humboldt, hovering like a guardian angel over the young naturalist's career, appreciated and seconded the reasons for his decision to remain in Neuchâtel. He praised Agassiz's energy in gaining the friendship and support of the populace. In Neuchâtel he would have less money for the present, but affluent friends held promise of future material assistance. Once he had published his paleontological researches, there would be many German universities seeking his services. The important thing was to publish them as soon as possible, and for this he needed a calm atmosphere and some financial assistance. Humboldt acted quickly on this last matter. He persuaded the monarchy to grant funds to Neuchâtel for the purchase of Agassiz's collections. This would provide money to keep Dinkel at work in

Paris, and for the support of another artist working in Neuchâtel.[6] Then he spoke frankly to Louis Coulon, as Agassiz could not.

The Prussian diplomat of science told the Neuchâtel business-man in direct terms of the attractiveness of the proposition made to his young friend. "I hope that he will refuse it. He should re-main for some years in your country, where a generous encourage-ment facilitates the publication of his work."[7] Since the royal treasury would pay most of the cost of Agassiz's collections, the Neuchâtel citizenry need do no more than defray the remainder, and by doing so they would be guaranteeing the security of their initial investment in Agassiz. By the spring of 1833, Agassiz, his collections, the Neuchâtel museum, and local enthusiasts of na-ture were all united in a common cause. Germany could wait. Three thousand dollars was made available for research from the sale of the collections, ties to Neuchâtel had been strengthened materially and spiritually, the aid of the Prussian monarchy had been secured in behalf of Agassiz's services, and Professors Tiede-mann and Bronn could only regret the failure of their effort.[8]

Before coming to Neuchâtel in the fall of 1832, Agassiz had vis-ited Carlsruhe once more, but this time he was not at all con-cerned with the fossils of the museum. It had been six years since Cécile and Louis had met each other, and intervals of separation, while reinforcing their affection, were not to be endured. Louis would see what kind of future Neuchâtel had in store for him. Since Cécile's parents and the ever faithful Alexander had given their approval, it only remained for Agassiz to be sure of his pros-pects. By the spring of 1833 these seemed secure indeed; a visit to Carlsruhe at this time confirmed prior promises, and, in October of 1833, Louis and Cécile were married in her home. Agassiz could indeed marvel at his good fortune. In Cécile he had not only merged his fortunes with that of a devoted and uncommonly at-tractive young woman of twenty-four, but had acquired member-ship in a distinguished family. His friend Alexander would now be even closer; Cécile's father and mother, prosperous, cultured, representing the best in German upper-middle-class society, were in-laws to be honored and respected. On their part, while the

Brauns might have wished their son-in-law to have greater wealth, they well understood the life of science through the occupations of both Alex and his younger brother, Max, who had embarked upon a career as a mining engineer. When Agassiz returned to Neuchâtel with his new bride and settled in a small apartment in the town, he could feel that his German ties were far from broken, and, besides, Cécile was a capable artist who would be of great assistance in her husband's research.[9]

The naturalist and his pretty young wife were happy and content in their surroundings. The constant support and good wishes of townsfolk, the warm friendship of Rose and Rodolphe Agassiz, the congenial presence of François Mayor, and quiet dinners in the home of Louis Coulon and other Neuchâtel notables all combined to make Cécile's life with her brilliant husband full and interesting at first. In addition to social activities, there were many plates of fossil fishes to be drawn by Cécile's fine artist's hand. The small Neuchâtel apartment soon became a scientific workshop, with the requirements of publishers' deadlines making work and more work a daily occupation for the young wife. Soon, however, intellectual labors had to be replaced by domestic ones, for in December of 1835 a son was born to the young couple; he was named, appropriately, Alexander, in honor of a brother and a fellow student.

Humboldt watched from afar the progress of domestic tranquillity and intellectual maturity in his naturalist-ward:

> It is a pleasure to watch the growing renown of those who are dear to us; and who should merit success more than you, whose elevation of character is proof against the temptations of literary self-love? I thank you for the little you have told me of your home life. It is not enough to be praised and recognized as a great and profound naturalist; to this one must add domestic happiness as well.[10]

Humboldt's words of praise were well deserved. Beginning in the fall of 1833, Agassiz commenced a decade of publication and research that amply repaid the faith and support of such older men as Schinz, Tiedemann, and Cuvier. By 1843, *Recherches sur les poissons fossiles* had been published in complete form, the work being issued in separate parts over the course of the decade. These volumes were the first modern effort to depict this significant branch of paleontology. No other research or publication effort by Agassiz ever equaled the talent he displayed in this description and analysis of over 1,700 species of ancient fishes. This remarkable contribution to natural history was the primary inspiration for a new area of inquiry in the history of organic creation. Agassiz's work was so singular and definitive that its scope and originality were not matched until late in the nineteenth century.[11]

The character and subject of this publication had profound implications for the development of Agassiz's private and public personality. Between the ages of twenty-six and thirty-six he became "a master," and in his own view achieved that which would dignify him as "the first naturalist of his time." The fact that it was never Agassiz's lot to surpass or even duplicate this performance is incidental to the primary personal self-evaluation that resulted. Agassiz would always think of himself as a master naturalist because of this effort.

Colleagues capable of judging Agassiz's work agreed with this personal estimate. The publication of *Poissons fossiles* transformed Agassiz from a young man of great promise to a professional of established competence. "I think it is by far the most important work now on hand in the geological field," the Reverend Adam Sedgwick wrote to Charles Lyell. Lyell replied that Agassiz's "knowledge of natural history surprises me the more I know of him. . . ."[12] That two geologists of sharply divergent theoretical persuasions, the first an avowed catastrophist and the second just then beginning to transform geological conceptualization with his *Principles of Geology*, could both marvel at Agassiz's effort reflects his stature in these years. In France, Germany, and England, natural scientists of such rank and differing views as Élie de Beaumont, Alexandre Brongniart, Achille Valenciennes, Leopold von Buch, Humboldt, Bronn, William Buckland, and Sir Roderick Murchison joined in the swelling chorus of praise for Agassiz's achievement. The Geological Society of London, the British Association for the Advancement of Science, the Royal Society of London, the Academy of Sciences of the Institute of France, the Prussian monarchy, and an impressive number of local scientific societies all honored Agassiz with grants, prizes, medals, and similar marks of distinction in recognition of this contribution.

It is instructive to note the full title of these five volumes: *Researches on the Fossil Fishes, comprising an introduction to the study of these animals; the comparative anatomy of organic systems which may contribute to facilitate the determination of fossil species; a new classification of fishes, expressing their relations to the series of formations: the explanation of the laws of their succession and development during all the metamorphoses of the terrestrial globe, accompanied by general geological considerations; finally, the description of about a thousand species which no longer exist and whose characters have been restored from remains contained in the strata of the earth.*[13] Agassiz had taken upon himself a most ambitious task. Paleontology, geology, ichthyology, comparative anatomy, and taxonomy all furnished special tools by which he studied his beloved fossil fishes. The fact that of necessity Agassiz brought different specialties to bear on the problem of defining the character and rank in time of fossil fishes was significant in that it showed a conscious effort to assume the authority of a wide-ranging intellect. The methods and contributions of *Poissons fossiles,* emphasizing the importance of comparative anatomy for an understanding of geology and paleontology, marked Agassiz as the true heir of Cuvier. But the volumes were dedicated not to the French master but to the man who most personified that universal approach to natural history to which Agassiz aspired, "A Son Excellence M. Alexandre de Humboldt."

From 1833, when the first part of *Poissons fossiles* was published, each year of this decade saw Agassiz grow more convinced that it was his mission to command the entire world of natural history. The extinct animals of the past, the living beings of the present, the history of the earth, and even the role in nature of man himself were all caught up in the same compulsive intellectual quest. The multitude of new orders, genera, and species of fossil fishes carefully described by Agassiz and realistically delineated by his artists thus came to mean more than a noteworthy examination of a special branch of natural history; the volumes of *Poissons fossiles* represented a first step toward their author's intellectual dominance of natural science.

The substantial basis for this synthesis was the describing of a vast number of fossil fishes, here, through Agassiz's careful analysis, depicted for the first time. These portrayals, models of the kind of precise investigation Agassiz had learned from Döllinger and Cuvier, were in themselves primary contributions to knowledge. They showed the results of painstaking investigation and patient detection during the thirteen years since the idea for this research had first taken hold of Agassiz's imagination. In these volumes, fellow naturalists could see spread before them the rich holdings of the great and small museums and private collections of Europe, specimens Agassiz had studied and analyzed with understanding and comprehensiveness. The exactitude of individual descriptions and the characterization of orders and genera were one side of this intellectual monument. The fidelity and beauty of the illustrated plates accompanying each description added much to the distinction of the volumes. Most of these were in color, some drawn by Dinkel, others by Cécile, others by artists who worked for Agassiz from time to time. The color illustrations represented a new departure in the reproduction of natural history data, the process of lithochromatic duplication, an engraving art first developed and refined by men who worked under Agassiz's direction at Neuchâtel.[14]

Agassiz knew that facts by themselves were meaningless. He would not be a mere antiquarian, content simply to describe every species he found and to spend his time amassing more and more data. In order to be a master, one had perforce to interpret the materials of nature. Thus, he devised a new system of classification for certain vertebrate fossils. His taxonomy, a division of ancient fishes into orders he called Cycloids, Ctenoids, Ganoids, and Placoids, was based upon an effort to understand the distinctions among fossil forms by inferences drawn from his knowledge of living fishes. This procedure was entirely appropriate for the man who thought of himself as Cuvier's successor and represented a major contribution to natural history. But Agassiz's knowledge of comparative anatomy was hardly as encyclopedic as the demands of his system of classification required. His taxonomic innovation, consequently, did not entirely stand the test of time, based as it was on the character of the scales of fossil fishes.[15] This deficiency, almost inevitable because of Agassiz's determination not to admit genetic relationships between certain fossil forms or between these and living animals, was hardly sufficient to detract from the excellence of his generalizations. He demonstrated, moreover, a constant awareness of the need to identify the particular stratigraphic association of various species insofar as this was possible. Furthermore, his characterization of the Ganoidei, today designated as a superorder, was an enduring contribution both to ichthyology and paleontology.

Classification represented the naturalist's understanding of the data of his experience. The basis for an objective taxonomy was thus of vital importance. As revealed in his insistence upon the significance of comparative anatomy as the guide to identifying ancient fishes systematically, Agassiz was concerned with understanding nature from the perceptions provided by direct experience. The facts of nature could be the only guide to true knowledge. Agassiz would not be guilty of "speculation," of "constructing the universe out of his own brain." The writing of *Poissons fossiles* carried him farther and farther away from the spirit of *Naturphilosophie*. The naturalist could not invent a scheme of nature, or impose his intellectual predilections upon the creative process.

Agassiz's interpretation of the meaning of natural facts in his major work represented, in his view, a mirror-image of nature itself. With patience and precision, he let the facts speak their essential import through the agency of his intellect. His discoveries, however, yielded generalizations crucial to the understanding of the entire creative process, and thus Agassiz began to assume the role he would delight in more and more as the years passed: analyst of nature's cosmic significance as well as model scientist working to uncover new truths in a special realm of natural history.

In assuming this dual function, Agassiz was inspired by important intellectual resources. If Oken could embrace a synoptic view of the entire animal kingdom, and Schelling extol the ideal form that all species strove to achieve, Agassiz could also set forth an all-embracing view of creation. In *Poissons fossiles* Agassiz gave expression to an idealism that sought to embrace the entire natural world in its assumptions. This idealism stretched beyond the natural philosophy of his German education to encompass a fundamental and deep-seated internal piety that sought to synthesize natural theology with natural history. The student who had found Aristotle "charming" and had read and reread his works "every two or three years" was compelled to see nature as the product of thought, planning, and intelligence. With this attitude of mind, Agassiz was not, of course, unique in history or in his own time. The generalizations he derived from research in paleontology would have found Cuvier in strong agreement with their major suppositions. But Agassiz went one step further, and here too he had Cuvier for his model rather than Oken. The natural theology that emphasized the history of creation was real and vital because it was a simple reflection of the data of experience. The facts demonstrated that species were immutable, that ancient forms bore no relationship to living animals, "that species do not pass insensibly one into another, but that they appear and disappear unexpectedly, without direct relations with their precursors. . . ."[16] This was satisfactory verification of the permanence and immutability ordained by the timeless wisdom of the Creative Power.

Agassiz emancipated himself from *Naturphilosophie,* not because he distrusted its assumptions of spirituality and the cosmic implications of natural history, but because such a doctrine seemed to him suspiciously close to a denial of the ultimate and transcendent power of the Creator. *Naturphilosophie* seemed to suggest that the types of animate creation made forever permanent by the Deity could in fact blend insensibly into one another and that they bore relationship and resemblance to one another in a scale of development and derivation. This attitude of mind, exemplified for Agassiz by the assertions of Geoffroy Saint-Hilaire, was thus to be condemned on two counts: it was "speculative" and imaginative, and it posed a threat to the concept of permanence in the universe.

Agassiz, as Cuvier's self-appointed successor, would use the evidence of nature to counter such fanciful assertions and, through intense examination of fossil fishes, would be able to proclaim aloud the marvelous operations of the Creative Power.

A few years after *Poissons fossiles* appeared, Agassiz gave succinct expression to this view:

The first notion of progressive development of the animal kingdom, of an agreement between the order of succession of types and their structural gradation, has been brought forward by that school of philosophers who in Germany take the name of nature-philosophers. . . . But with them the idea of a gradual development of the animal kingdom was by no means the result of investigation—was not the expression of facts, but was an *a priori* conception, in which they made their view of the animal kingdom the foundation for a particular classification.[17]

He would investigate nature first and then come to conclusions about its meaning. "Nature-philosophers" might assert that lower animals were first introduced in the scale of creation and then gradually replaced by higher animals. But Agassiz, who knew the facts, could show from his discoveries in paleoichthyology that vertebrate animals, in this case fishes, had existed in the "oldest epochs." Moreover, he could affirm that representatives of all the four great types or divisions of the animal kingdom existed simultaneously in all the most ancient geological formations. There could be no development from "lower" to "higher" types, since the latter existed at the same time as the former.

With these intellectual justifications for his operational scheme that denied the vague and inconclusive assertions of physical unity and genetic relationship, Agassiz could proceed with authority to generalize on the meaning of facts. There was, for example, unquestionable "development" in the animal kingdom. This occurred within any of the classes comprising the larger branches of animate nature, past and present. But development was rigidly categorized; there was no utility or meaning in comparing or relating types of one branch to those of another. Agassiz was too familiar with embryology and the observed distinctions in vertebrate paleontology, however, not to recognize change and development as primary facts of natural existence. "One single idea has pre-

Birthplace of Louis Agassiz, Motier, Switzerland

Agassiz at the age of nineteen.
Engraving from a pastel
drawing by Cécile Braun
Agassiz.

Cécile Braun Agassiz in 1829; a self-portrait.

Alexander Agassiz at the age of twelve; from a drawing by his mother.

Ignatius Döllinger in 1837.
(Courtesy of the National Library of
Medicine, Washington, D.C.)

Lorenz Oken. (The Bettmann
Archive.)

Baron Georges Cuvier about 1832.
(The Bettmann Archive.)

Alexander von Humboldt. (The
Bettmann Archive.)

Agassiz in 1844

Hôtel des Neuchâtelois

The fossil fish of Agassiz's dream (*Cyclopoma spinosum*, Agassiz)

Agassiz in 1847

Elizabeth Cary Agassiz in 1852.
(Courtesy of Radcliffe College.)

"Zoological Hall." The original building in a modern setting

John Amory Lowell. (From A. Hunter Dupree, *Asa Gray, 1810–1888,* The Belknap Press of Harvard University Press, 1959. Photograph courtesy of the Gray Herbarium.)

Benjamin Silliman. (From George P. Merrill, *The First One Hundred Years of American Geology,* 1924. Courtesy of the Yale University Press.)

Asa Gray in 1867

James Dwight Dana

Cornelius Conway Felton about
1855. (Courtesy of Edward W. Forbes
and the Harvard University [Felton]
Archives.)

Spencer Fullerton Baird in 1873

The Agassiz home, Quincy Street, Cambridge

Agassiz in his study

Museum of Comparative Zoology (*on the left*) on Divinity Avenue in the 1860's

REGULATIONS

MUSEUM OF COMPARATIVE ZOÖLOGY.

At a meeting of the Faculty, held Nov. 5, 1863, it was voted to adopt, until otherwise ordered, the following Rules and Regulations for the administration of the Museum, viz. : —

1. That all assistants in the Museum be annually appointed by the Faculty, upon nomination of the Curator, and that their compensation be fixed by the Faculty upon recommendation of the Curator.

2. To obtain an appointment in the Museum a candidate is expected to furnish satisfactory evidence of his ability to co-operate in the general work of the institution, and of his desire faithfully to devote himself to this task. An original investigation, or a series of preparations requiring exceptional skill; may be taken as such evidence, if coupled with a detailed account of the candidate's scientific pursuits up to the time of application.

3. The Curator is authorized to employ temporarily, in the Museum, individuals who may not be entitled to a regular appointment as assistants, or may be unwilling to accept such a position.

4. No one connected with the Museum shall be allowed to own a private collection, or to traffic in specimens of Natural History, except for the benefit of the Museum. If an officer of the Museum, or a student working for the Museum, possesses a private collection with which he is unwilling to part, he must deposit the same in the Museum during his connection with the institution.

5. No one connected with the Museum is authorized to work for himself in the Museum during the working hours fixed for Museum work. Whatever is done by any one connected with the Museum, during that time, is to be considered as the property of the Museum, but due credit is to be given him by the Curator in his Annual Reports. Any claim or grievance concerning this kind of work may be submitted to the Curator, or to the Faculty through the President of the University, at any time within three months after the publication of the Annual Report.

6. Every one admitted to work regularly in the Museum is expected to be at his work within the walls of the Museum at least seven hours every day, unless duly authorized to be absent. Vacations from Museum work are to be considered as a reward for special application and effective work ; but not as a right.

7. No one is authorized to publish, or present to learned societies, anything concerning his work at the Museum, without the previous consent of the Curator. All such contributions are to be submitted to the Curator for examination.

8. While intrusted with a special department, an assistant or worker in the Museum shall have the privilege of freely examining every specimen belonging to that department (but not those of other departments), and of taking specimens out of the cases for special investigations; but he shall not dissect or alter the condition of any specimen without special leave of the Curator. He shall further enjoy the privilege of using freely the Museum Library, and taking down to his desk, under the prescribed regulations, all the books needed for his work. No books or specimens are to be taken out of the Museum building without special leave of the Curator.

9. All the specimens and books temporarily removed from their proper place for use shall be returned to their shelves at the close of every month, unless special leave for an extension of time has been obtained from the Curator.

10. Every one is expected to keep his working place in the Museum clean, and himself to do what is necessary to this end. He may, however, call upon the Janitor to remove the offals of his work.

11. The rights and duties of the assistants, not specified above, shall be determined by the Curator, his determination being subject to the revision and final decision of the Faculty.

The museum regulations of 1863

Francis Calley Gray

Jeffries Wyman. (Courtesy of Edward W. Forbes and the Harvard University [Wyman] Archives.)

Henry Wadsworth Longfellow. (Courtesy of the Harvard University Archives.)

Ralph Waldo Emerson. (The Bettmann Archive.)

Agassiz in 1855

CONTRIBUTIONS

TO

THE NATURAL HISTORY

OF THE

UNITED STATES OF AMERICA.

BY

LOUIS AGASSIZ.

FIRST MONOGRAPH.

IN THREE PARTS.—I. ESSAY ON CLASSIFICATION.—II. NORTH AMERICAN TESTUDINATA.—
III. EMBRYOLOGY OF THE TURTLE; WITH THIRTY-FOUR PLATES.

VOL. I.

BOSTON:
LITTLE, BROWN AND COMPANY.
LONDON: TRÜBNER & CO.
1857.

Title page of the *Contributions*

Turtles drawn by Jacques Burkhardt for Volume II of the *Contributions*

Drawing by Joseph Dinkel for Agassiz

Caricature by Agassiz of Thomas
Henry Huxley, drawn about 1861.

Agassiz's penciled rhyme reads:
    Idée solitaire!!
    Idée unique!!
    Ecole Polytechnique
    FRS; ASS!

Sir Charles Lyell

Charles Darwin about 1854

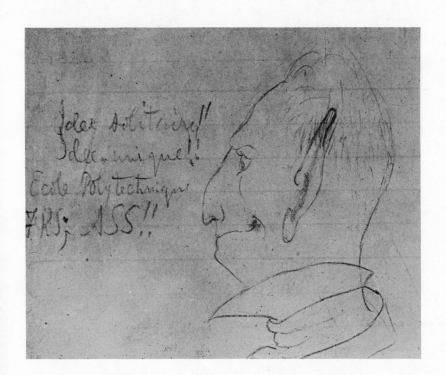

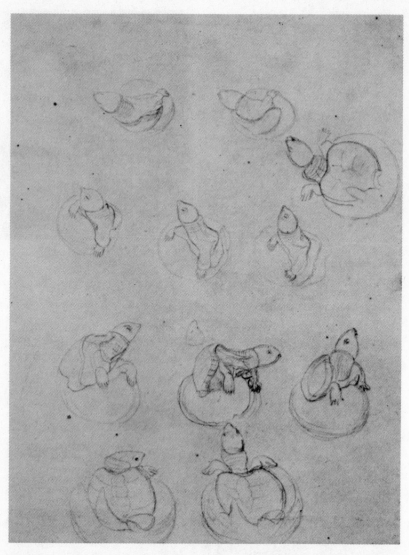

Sketches by Agassiz of the development of the turtle

Agassiz in 1863

Painting depicting the founding of the National Academy of Sciences. *Left to right:* Benjamin Peirce, Alexander Dallas Bache, Joseph Henry, Agassiz, President Lincoln, Senator Henry Wilson, Admiral Charles Henry Davis, and Benjamin A. Gould.

Agassiz and Benjamin Peirce in 1869

The Agassiz cottage at Nahant

Sketch by William James of the Thayer Expedition to Brazil. The caravan depicts the adventures of James's friend Newton Dexter and other aspects of the journey. The stick figure is Agassiz carrying a sign reading "4,00000000000 new species of fish." The figure labeled No. 2 is described by James as "Large Diamond from the 'Emp.'" The policeman and other irrelevant sketches are James's pencil doodlings. (Courtesy of the Houghton Library, Harvard University.)

Agassiz in 1861

The Anderson School of Natural History on Penikese Island, 1872

Sketch of Agassiz in the lecture room of the Anderson School, Penikese

Sketch of the dining hall at the Anderson School

Agassiz in 1872

Elizabeth Agassiz in 1872.
(Courtesy of Radcliffe College.)

sided over the development of the whole class [fishes] and . . . all
the deviations lead back to a primary plan, so that even if the
thread seemed broken in the present creation, one can reunite it
on reaching the domain of fossil ichthyology."[18]

While physical unity between different type-schemes was impossible and while there was no evidence linking ancient and modern
fishes, there was a supreme unity in nature that far transcended
the inventions of philosophers like Oken. Out of all the deviations,
from all the diversity, a single idea or thread led the empirical investigator back to the perception of a primary plan, the substance
of an intellectual connective tissue that bound together all dispar-
ity, all life, past and present. Most important, this overwhelming
power was discovered, not imagined, from close, analytical examination. Agassiz could discover a unity imposed from without, not
the result of gradations and relationships. The modern naturalist
was thus the true interpreter of God's plan for the universe.

Such facts proclaim aloud principles not yet discussed in science but
which paleontological researches place before the eyes of the observer
with an ever-increasing persistency. I speak of the relations of the creation with the creator. Phenomena closely allied in the order of their
succession, and yet without sufficient cause in themselves for their appearance; an infinite diversity of species without any common material
bond, so grouping themselves as to present the most admirable progressive development to which our own species was linked, are those not
incontestable proofs of the existence of a superior intelligence whose
power could have established such an order of things?[19]

"Development," then, proceeded from its causal relationship to
a superior intellectual power, a thoughtful planner, operating
from without upon the data of nature. This superior intellectual
power united all creation in a bond that "unwinds itself throughout all time, across this immense diversity and presents . . . a definite result, a continual progress in the development of which man
is the term, of which the four classes of vertebrates are intermediate forms, and the totality of invertebrate animals the constant
accessory accompaniment."[20] A student who had observed the
growth of the individual from the egg to the adult form could not
fail to be impressed with "progressive development." This un-

folding, whether discovered in the embryology of living things or in the parallel series of "embryological" succession revealed by the life-history of extinct forms, was always apparent to Agassiz's trained eye. It was discernible in adults, embryos, and ancestors. Change had occurred over long periods of time and had often been sudden, violent, and catastrophic. Successive faunas had been created, destroyed, and replaced in full or in part by other faunas.

The all-embracing unity which made this development significant operated "according to plan" and was present from the beginning. Individuals changed and varied, of course, but Cuvier's student was quick to affirm that change in individuals in no wise implied change in species. Development in the various classes of the animal kingdom was wisely confined to each class itself, an internal process of change repeated both in the individual and in the class to which it belonged. For one who aspired to be a master, what was more masterful than the discovery of a plan of creation that had been consciously ordained from the beginning, with each great type in the animal kingdom developing according to a predetermined plan? Change was everywhere, but this was the sort of change Döllinger had taught Agassiz to look for through the medium of microscopic analysis. It was change, moreover, that saw each individual conform to its specific, planned identity once its adult stage had been achieved. Consequently, any concept suggestive of the transmutation of species was both irresponsible speculation and contrary to observed fact. Agassiz's close examination of a great number of fossil fishes gave him the assurance to proclaim:

No one will seriously pretend that the numerous types of Cycloids and Ctenoids, almost all of which are contemporaneous with one another, have descended from the Placoids and Ganoids. As well might one affirm that the Mammalia, and man with them, have descended directly from fishes. All these species have a fixed epoch of appearance and disappearance; their existence is even limited to an appointed time. And yet they present . . . numerous affinities more or less close, a definite co-ordination of a given system of organization which has intimate relations with the mode of existence of each type, and even of each species.[21]

It was patently self-evident that physical relationship and material unity were contradicted by the very data of Agassiz's experience of

nature. That there could be genetic affinity even among representatives of one class—fishes—was a false assertion because it suggested descent. At the same time, each class of animals was bound
together by physical resemblances, but these signs of alliance were
imposed from without, by the superior intellectual unity that gave
inspiration and meaning to all existence, past and present.

If such a view was not optimistic or compelling enough for a
world where *Naturphilosophie* had painted an inspiring canvas of
the grand scheme of creation, Agassiz could offer equally romantic, positivistic visions. There was the ennobling contemplation of
embryonic development that demonstrated the progressive unfolding from egg to adult, according to plan. Progressive development,
moreover, was illustrated in the life-history of classes within the
branches of the animal kingdom, activity demonstrative of the supreme idea exemplified in the type-plan. Finally, there was the
solace and confidence to be derived from contemplating the intentions of the Creator with regard to the vertebrate division: "There
is a manifest progress in the succession of beings on the surface of
the earth," Agassiz affirmed in 1848.

This progress consists in an increasing similarity to the living fauna,
and among the vertebrata, especially, in their increasing resemblance to
Man. But this connection is not the consequence of a direct lineage
between the faunas of different ages. There is nothing like parental
descent connecting them. . . . The link by which they are connected is
of a higher and immaterial nature; and their connection is to be sought
in the view of the Creator himself, whose aim, in forming the earth,
in allowing it to undergo the successive changes which geology has
pointed out, and in creating successively all the different types of animals which have passed away, was to introduce Man upon its surface.
. . . In the beginning the Creator's plan was formed, and from it He
has never swerved in any particular. . . . To study . . . the succession of
animals in time, and their distribution in space, is therefore to become
acquainted with the ideas of God himself.[22]

Progressive development, as preached by advocates of *Naturphilosophie,* was dangerous because it tended to speculation, was
grounded on a priori assertions, and suggested physical unity. The
march of progress as defined and heralded by Agassiz reflected a
simple correspondence between the perceptions of the naturalist

and the fundamental plan of the universe and denied physical relationship through the substitution of a higher intellectual unity.

In *Poissons fossiles* Agassiz showed marked powers of intellectual adaptation. He synthesized a world view from those aspects of his religious beliefs and secular education that could be integrated into a cosmic philosophy of nature. Once the basic assumptions were accepted, generalizations followed with a logic as unassailable as the affirmations of permanence and divinity derived from empirical study. Agassiz thus came to think of himself as a passive reflector of the essential reality of nature. To be able to explain this reality by reference to transcendental purpose was the inevitable result and meaning of knowledge. Idealism and empiricism were thus comfortably joined in Agassiz's mind, and he felt no contradiction from the merger.

The intellectual result of this marriage was that Agassiz was able, in a work like *Poissons fossiles,* to produce the most careful, exact, and precise descriptions and then, without apparent effort, to indulge in flights of idealistic fancy, often within the compass of the same analysis or chapter. As fame, success, and distinction came to him because of his superior analytical and descriptive powers, he felt more and more able to formulate and believe in grandiose generalizations. These assertions, especially when made before public audiences or announced in a popular vein, were often entirely unsupported by evidence. This tendency in Agassiz surprised even colleagues of the same temperament and intellectual persuasion. Thus Adam Sedgwick could write in 1835:

Agassiz . . . read a long stupid hypothetical dissertation on geology, drawn from the depths of his ignorance. . . . He told us that each formation . . . was formed at one moment by a catastrophe, and that the fossils were by such catastrophes brought from some unknown region, and deposited where we find them. . . . I hope before long we shall be able to get this moonshine out of his head. . . . [23]

But such criticism was rare in these years of personal and intellectual triumph. The primary ingredient of Agassiz's authority was his superior ability to describe and analyze the materials of vertebrate paleontology. The distinction that came from this achievement gave him the personal security to assess the entire scheme of creation. The consequence was that he became identified with a view of the world equally idealistic, equally a priori in its assumptions, as were the generalizations of Oken and Schelling. It was as if, by 1843, Agassiz had assumed the wisdom and authority of a Moses of science, leading his people out of the wilderness of false doctrine into a new land whose ecology was based on observed fact. If examined too closely, however, the landscape looked strangely like the sinful world so miraculously left behind.

In undertaking the writing of *Poissons fossiles,* Agassiz was from the first conscious of the enormity of the task before him. "I have certainly committed an imprudence in throwing myself into an enterprise so vast in proportion to my means as my Fossil Fishes," he confided to Humboldt in a moment of despair over lack of funds. "But having begun it, I have no alternative; my only safety is in success. I have a firm conviction that I shall bring my work to a happy issue."[24] He pursued his self-imposed task with compulsive determination. In the preface to his first volume he described some of the difficulties he encountered:

Possessing no fossil fishes myself . . . I have been forced to see the materials for my work in all the collections of Europe. . . . A serious inconvenience has resulted from this mode of working, namely, that I have rarely been able to compare directly the various specimens of the same species from different collections, and . . . I have often been obliged to make my identification from memory, or from simple notes, or, in the more fortunate cases, from my drawings only. It is impossible to imagine the fatigue, the exhaustion of all the faculties, involved in such a method. . . . [25]

When he began to publish in 1833, Agassiz was immediately discontented with the character of his work. Although, by his own estimate, his knowledge of the museums of France and Germany was, in 1833, already comprehensive, he felt he needed to know much more about fossil fishes, and therefore he traveled all over Europe in the next decade, seeking an encyclopedic knowledge of his subject. The result of his travels was that Agassiz became known everywhere as the young scholar who was doing important work on fossil fishes. England supplied him with particularly valuable materials. Some of these were incorporated into his major study, others described in separate papers, and others—the fossils of the "Old Red Sandstone"—formed the basis for a separate study of major importance published in 1844–45.[26]

England also provided Agassiz with lasting personal friendships. In visits there in 1835 and 1840 he met such men as Charles Lyell, Sir Roderick Murchison, Richard Owen, Adam Sedgwick, and William Buckland, all of whom became lifelong friends and admirers. English interest in Agassiz's research was very strong. Men like Sedgwick were impressed by Agassiz's seeming empirical justification for the concept of successive, special creations superintended by divine intelligence. Men like Lyell were struck by the insights Agassiz provided into the complexities of the ancient history of the earth and its inhabitants. Such men, and the scientific institutions they represented, made Agassiz's work easier by material support. Sir Philip Egerton and Lord Cole, Earl of Enniskillen, both contributed the means for Joseph Dinkel to live in England for various periods and to draw the specimens of their valuable private collections. In the years 1834–36, Agassiz received grants totaling £240 from the Geological Society of London and the British Association for the Advancement of Science. In 1836, the Wollaston Medal of the Geological Society was awarded him for his publications thus far in paleoichthyology. In 1835 he had been elected an honorary member of the British Association and a foreign member of the Royal Society of Edinburgh. In 1838, he was elected a foreign member of the Royal Society of London.

On the Continent, Frederick William III was a constant source of support, through grants for research from the Prussian monarchy. Achille Valenciennes, Élie de Beaumont, and Paul Deshayes were but a few of the naturalists who helped spread Agassiz's fame in France. Indeed, Élie de Beaumont wrote him from Paris after receiving the first part of the first volume:

I have read your first number with great pleasure. It promises us a work as important for science as it is remarkable in execution. Do not let yourself be discouraged by obstacles of any kind; they will give way before the concert of approbation which so excellent a work will awaken.[27]

Poissons fossiles was only the most ambitious and best publicized of the works that brought Agassiz fame as a paleontologist. Other forms of marine life attracted his interest, notably sea urchins, starfishes, sea lilies, and forms of shellfish. These years thus witnessed publications comparing living and fossil forms of echinoderms, as well as a volume on fossil mollusks.[28] It seemed that no form of ancient marine life failed to fascinate Agassiz.

This renown and increasing diversity of interests were essentially symptomatic of Agassiz's growing self-confidence. But so intense was his desire to master all known materials relating to paleoichthyology that he drew back from the task of bringing his research on fossil fishes to a close. By 1837, Agassiz had gained a commanding knowledge of over one thousand species of ancient fishes. He had published a portion of this material in the first parts of *Poissons fossiles,* issued between 1833 and 1837. A prudent understanding of physical and monetary limitations might have dictated a steady effort toward publishing descriptions of materials already at hand. Agassiz, however, was determined to write a comprehensive study of all known fossil fishes. The result was that his labors were not completed until 1843.

The lasting value of the work made his decision valuable for posterity, but it did not entirely please his colleagues in science. *Poissons fossiles* appeared during the decade 1833–43 in different sections, some describing the species and families of one order, others dealing with another order, still others comprising generalizations about the entire subject. Moreover, the illustrations, which made an understanding of individual descriptions meaningful, appeared at different times from the text. By 1843, Agassiz had to synthesize his entire effort into five volumes of co-ordinated text, rewrite his introductory remarks to the first volume, recast his generalizations about the implications of his research, and issue a separate volume of additions and corrections for the whole series. For ten years the scientific world had before it the name of Agassiz and the character of the work he was doing; but it was not until the end of this period that his work had any real utility for fellow investigators. Actually, the final result, depicting over 1,700 species of ancient fishes, contributed little more than additional documentation to the fundamental picture of vertebrate history that Agassiz had known in 1837.

Rodolphe Agassiz had once chided his son for what he termed "a mania for rushing full gallop into the future." He had perceptively identified Agassiz's primary personality trait—envisioning grand projects with little reference to practical realities. Rodolphe died in 1837, an event that brought great sorrow to his son, who through the years had come to value, if not to follow, the cautious and level-headed advice of his father. At the time of Rodolphe's death, Louis was reaching toward the heights of fame, but the earlier assessment was as accurate as ever. Agassiz was plainly unhappy doing one thing at a time. He was most excited when he had three, four, or more projects and grand schemes under way at once. He had come to Neuchâtel with two goals: a history of central European fresh-water fishes and a study of fossil fishes. By 1837, he was not only engaged in these two monumental activities but was also directing the translation of scientific monographs into German and French, compiling a volume on zoological nomenclature, and beginning to consider a comprehensive treatise on the action of glaciers.[29]

This diversity of activity called forth the first known rebuke, mild though it was, from Humboldt, who in this year of 1837 was concerned that Agassiz's health had begun to suffer under the strain of his varied occupations. As always, Humboldt understood Agassiz with a perception only equaled by Rodolphe. "For mercy's sake," Humboldt pleaded,

take care of your health. . . . I am afraid you work too much, and (shall I say it frankly?) that you spread your intellect over too many subjects at once. I think you should concentrate your . . . strength upon this beautiful work on fossil fishes. . . . In accepting considerable sums from England, you have . . . contracted obligations to be met only by completing a work which will be at once a monument to your own glory and a landmark in the history of science. . . . You will say this is making you a slave of others; perfectly true. . . . We men of letters are the servants of an arbitrary master, whom we have imprudently chosen, who flatters and pets us first, and then tyrannizes over us if we do not work to his liking. You see . . . I play the grumbling old man, and, at the risk of deeply displeasing you, place myself on the side of the despotic public.

Humboldt urged Agassiz to stop work, recall his artists from the museums of England and the Continent, and realize that the plan of work he envisaged would not be completed for a long while and at tremendous sacrifice. He concluded with a prediction and a further plea:

If you continue . . . you will struggle with domestic difficulties, and at sixty years of age . . . you will be as uncertain as you are today, whether you possess . . . all that is to be found. . . . Finish, first, what you have this December, 1837, and then . . . publish the supplements in 1847. . . . Little will it avail you should I vanish from the scene of this world with your fourteenth number! When I am a fossil in my turn I shall appear to you as a ghost, having under my arm the pages you have failed to interpolate. . . . [30]

But this was not to be Agassiz's way, in 1837, 1847, or 1857. That he finally did conclude the *Poissons fossiles* was due only to constant prodding by friends such as Humboldt.

The sources of Agassiz's constant striving for large projects to pursue, and his parallel inability to conclude many of them, are rooted in the character of his early social and intellectual life. From the time he penned the memorandum setting forth his hopes and dreams at fifteen, Agassiz never knew failure. He had an uncommon talent for discerning those areas of zoology that were most fruitful for pioneering research efforts. Competent analysis and description in such a field as paleoichthyology was bound to bring rewards even in excess of the objective value of the investigation itself. As accomplishment piled on accomplishment, Agassiz's imagination of future achievement soared higher and higher. The youth of fifteen who wanted to begin to write at twenty-five and the adult of thirty who wanted to publish all there was to know about fossil fishes was the same free-spirited, intellectually restless, ambitious person. This romantic soul could thus never be satisfied with the activities of the present. And, whenever the present seemed dull or predictable, he had one unfailing resource that rescued him from the real world of former obligations: there was always a new project born in his mind, and, following this, there was always the moral and material support of powerful men and institutions to help launch the project in a grand manner. Such habits may have displeased Rodolphe Agassiz and Humboldt, but they supplied the personal and public qualities of success.

In the summer of 1836, in the midst of writing *Poissons fossiles*, Agassiz received an invitation to vacation at the home of Jean de Charpentier, director of the Salt Works at Bex in the Rhône Valley. Agassiz spent a delightful summer here in the canton of Vaud, made especially interesting because his friend took him on walking tours to inspect the glaciers of the valleys of Chamounix and Diablerets, and the moraines of the Rhône Valley. The outcome was that one year later, in an address before the Swiss Society of Natural History in July, 1837, Agassiz announced his discovery that in a prehistoric Ice Age in the Pleistocene epoch ice had covered the world from the North Pole to the borders of the Mediterranean and Caspian seas.[31]

The "Eiszeit" theory was associated with Agassiz's name from that time on. Countless professional and popular audiences were to hear Agassiz describe how the earth, in the most recent period of its history, had been covered with a gigantic ice sheet, of which the glaciers of the present creation were remnants of the events of the past. As these ice masses receded, they left traces of their existence by many signs. These evidences of the past consisted of giant boulders, scratched and polished rocks, masses of rocky gravel forming moraines, loose beds of sand and gravel or "drift," erratic blocks, and numerous other geological features of unusual appearance in their contemporary setting. The glacial theory rested on two primary sources of data: First, observations derived from contemporary investigations of the movement of glaciers in Switzerland, providing the modern student with an insight into the process that had taken place in former times. Second, evidence derived from the land forms of such areas of Europe as Switzerland, Scotland, England, and Germany, configurations that could be explained by the action of a universal mass of ice.[32]

Research on fossil fishes distinguished Agassiz as a careful investigator. Advocacy of the glacial theory furthered his renown, increased his public stature as a universal naturalist, embroiled him in scientific controversy, and cost him valued personal friendships. Agassiz visited the Rhône Valley in 1836 conscious of the fact that his friend Charpentier had advocated the glacial concept as an explanation for the presence of such unexplained phenomena as moraines and erratic blocks in the Alpine regions of Switzerland. He came to Bex a skeptic; he left an avowed convert, eager to gain acceptance for Charpentier's interpretation by conducting his own field research all over Europe and by stating his views publicly. Thus, in the midst of publishing the *Poissons fossiles*, Agassiz spent every summer but one during the years from 1837 to 1845 in the Swiss Alps, with visits to England, Scotland, and Germany, observing the movement of glaciers, inspecting their internal composition, and discovering new evidence of former glacial action. In 1840 he published an epoch-making work, a book-length summary of his findings to date, *Études sur les glaciers*. Here, he expanded the assertions made in his address of 1837 in a work notable for its clarity of presentation, the excellence of its illustrations, and the evidence presented from three years of personal investigation. In 1847, he published still another primary contribution, *Système glaciaire*, which reported the results of personal investigations in many parts of Europe, contained precise descriptions of physical measurements, and comprised the end product of a decade of labor on a question of fundamental importance.[33]

The work of these years identified Agassiz with the Ice Age concept in both the public and the professional mind. High adventure and romance were personified in the popular image of a young, energetic, naturalist-explorer who led a band of devoted followers in the scaling of Alpine peaks and firsthand observations of glacial movements, often at great personal danger to himself. Agassiz, his assistants, and fellow scientists who came to observe and be instructed, spent nine summers in the Bernese Alps, the Valaisian Alps, the Rhône Valley, visiting nearly every significant mountain chain and glacial site in Switzerland. The great glaciers of the Rhône, Monte Rosa, Aletsch, and the Grindelwald were all explored and mastered.

By far the most important of these was the Unteraar glacier, formed by the junction of the Lauteraar and Finsteraar glaciers. This great glacier site contained an excellent example of a large medial moraine, a landmark signifying the height, extent, advance, and retreat of the glacier in former times. In 1840 Agassiz and his helpers established an observation station here so that exact studies of glacial movement and internal structure could be made. Called "Hôtel des Neuchâtelois" by its explorer-inhabitants, this edifice became the gathering place for many geologists from England and the Continent who came to observe Agassiz's work and make their own independent studies. The Agassiz station became well known also to travelers not engaged in research. A Scotch lady touring in the Alps mistook the Hôtel for a public establishment, stopped at the rustic camp, and politely inquired about accommodations for the evening. The amused but gallant scientists of course provided lodging for her. A Swiss gentleman of substance stopped at the Hôtel and asked to meet the great Professor Agassiz of Neuchâtel. When Dinkel pointed out a very youthful, handsome young man in his early thirties laughing uproariously at the banter of his friends, the traveler was furious at what he believed to be a joke at his expense. He demanded of Agassiz that he be introduced to his father immediately. When the man learned the truth, his astonishment provided excellent data for the Agassiz legend.

Agassiz looked upon this most recent occupation as the pleasurable, leisure-time activity of a scientific amateur. He never pursued his glacial researches with the fanatic intensity lavished upon his work with fossil fishes. It was relaxing, healthful, and personally invigorating to breathe the clear air of the Alpine heights, to marvel at a snowstorm seen from high above the ground, to relax, drink, and laugh around the nightly fire at the Hôtel after a long winter writing descriptions of fossil fishes. It was not that Agassiz spurned the professional renown that came to him from his advocacy of the glacial concept, or that he did not enjoy defending his thesis in intellectual interchange or relish the opportunity to find new evidence. It was as if he had carelessly turned to another occupation to relieve the tedium of paleontological research, only to find that this almost casual interest yielded bright and unexpected public rewards.

Agassiz's geological researches made significant contributions to man's knowledge of the past and the present. The Ice Age concept provided a key to understanding the process by which animal and plant life, in certain regions of the world, exemplified particular patterns of geographical distribution. The southern movement of glacial ice transformed tropical areas into temperate zones, pine forests, frozen wastes, and large ice sheets. This process was reversed as the ice retreated. With the ice and cold, arctic forms advanced southward, retreating north again as the warmth returned, thus explaining the close relationship of species existing in different geographic regions of the world in the recent period. The glaciers of the modern period served as contemporary evidence for the Ice Age of the recent past. They also explained phenomena that had long puzzled observers. Until the glacial theory was advanced, the erratic boulders strewn over the surface of the land in clearly unnatural positions, gravel ridges, kettles in sand plains, polished and striated rock, and surface deposits of drift, were variously explained as due to the action of weathering, ocean currents, waves, wind, icebergs, and floods. The most favored explanation referred such phenomena to the effects of catastrophes, notably the Noachian deluge and floods of water that were suddenly and violently released from the interior of the earth or caused by the upheaval of mountains. The Ice Age concept and the glacial theory together provided a rational explanation to account for the appearance and distribution of such formerly mysterious land configurations: ice masses had advanced and subsequently retreated, grinding the earth into different contours, depositing drift, boulders, rounded masses of pebbles and primitive rocks, and thus leaving permanent signs of their former existence, duration, and distribution.

These were the permanent contributions of glacial geology to the understanding of natural history. But for Agassiz, the primary significance of the Ice Age idea and glacial theory resided in their cosmic implications. Although he had advanced them himself, he refused, consistently, to interpret these concepts as secondary mechanisms or agencies which could explain in a rational manner the geographic distribution of animals and plants and the genetic relationship of past and present species. As with his singular con-

tributions in paleontological science. Agassiz was forced to contradict the really meaningful insights that other men, with different philosophical assumptions, derived from his work. More than
any other intellectual trait, it was this characteristic that identified
Agassiz both with the past and with the present. In an era of transition in the interpretation of nature, Agassiz lived as a man who
provided basic insights and discoveries that helped effect such
change and at the same time as one who fought against the implications of these insights for the new framework of natural history.
It was for these reasons that Agassiz was such an admired yet perplexing intellect to men like Charles Darwin and Charles Lyell.

After presenting and publishing his views in Switzerland in
1837, Agassiz went on to publicize them in France, Germany, and
England. Having effectively removed the agency of the flood from
geological speculation with the glacial concept, Agassiz replaced
this catastrophic explanation with another. Glaciers were to him
"God's great plough," destructive and chaotic natural forces which
signified a beneficent supernatural plan for the universe. The
glacial period, to Agassiz, was thus a magnificent, albeit chilly,
demonstration of the power of the Deity in causing great catastrophes, events which had wiped out life in previous epochs and
replaced it with new flora and fauna. As Agassiz dramatized the
natural history of the Pleistocene epoch:

The long summer was over. . . . For ages a tropical climate had prevailed
. . . and . . . gigantic quadrupeds, the Mastodon, Elephants, Tigers,
Lions, Hyenas, Bears . . . possessed the earth. . . . But their reign was
over. A sudden intense winter, that was to last for ages, fell upon our
globe; it spread over the very countries where these tropical animals
had their homes, and so suddenly did it come upon them that they were
embalmed beneath masses of snow and ice, without time even for the
decay which follows death.[34]

For Agassiz the glacial theory provided further substantiation for
a belief in the fixity of species. He could thus write in a light but
revealing mood to his sympathetic colleague, William Buckland
of Oxford: "Since I saw the glaciers I am quite of a snowy humor,

and will have the whole surface of the earth covered with ice, and the whole prior creation dead by cold."[35]

In public expression, this intention took the following form:

There is then a complete break between the present creation and those which precede it; if the living species of our times resemble those buried in the levels of the earth, so as to be mistaken for them, it cannot be said that they have descended in direct line of progeniture, or what is the same thing that they are identical species.[36]

Cuvier would have been pleased with this new way of justifying catastrophism. Agassiz had substituted glaciers and ice for the flood, but the meaning was the same. The action of water was too irrational an explanation for many scientists. Glaciers, whose action could be precisely observed and demonstrated, provided an empirical justification for the idealistic view of creation. Just as fossil fishes served Agassiz as evidence against genetic affinity in species, the glacial theory provided him with an effective intellectual explanation of disturbing resemblances between past and present animal forms. He was able to brush all thoughts of affinity from his mind with the reassuring conviction that ice had intervened to prevent extinct Ganoid forms from sharing genetic relationship with contemporary fishes. Paleontology and glacial geology, studied carefully, furnished the data for a world view that denied change and relationship in natural history. It would have taken a hardy fish to survive Agassiz's ice.

The glacial theory and the Ice Age idea, however, were both concepts that established Agassiz as a pioneering intellect; he was breaking new ground in geology, just as he had in paleontology. His ideas constituted an obvious contribution to geomorphology, to the understanding of the earth's history through knowledge of its present character, and to the crucial subject of geographical distribution. "You have made all the geologists glacier-mad here," wrote Edward Forbes of the English scientific climate in 1841,

and they are turning Great Britain into an ice-house. Some amusing and very absurd attempts at opposition to your views have been made by . . . among others, poor Sir George MacKenzie who has read a paper at the Royal Society here maintaining that all the appearances you refer to glaciers were caused by blocks of ice which floated this way in the deluge.[37]

"Poor Sir George" represented the antiquated notions that attempted a synthesis between biblical history and the observed data of natural history. For Agassiz and his followers the scientific study of nature now supplied the materials for an idealistic world view that transcended mere recorded assumptions about the powers of the Deity. The essence of the Creative Power was to be discovered in the book of nature itself, not in the Bible.

The glacial theory disturbed the convictions of believers in the deluge but found strong supporters in men of catastrophist convictions like the Reverend Adam Sedgwick and William Buckland. Sedgwick delighted in the opportunity to hold dinner parties for his Swiss friend whenever Agassiz came to England. Richard Owen, Sir Roderick Murchison, and Sir Philip Egerton found Agassiz's visits occasions for excellent conversation on fossils and ice. Other men, of other views, were also impressed. "Lyell has adopted your theory *in toto*!!!" Buckland reported to Agassiz in 1840. "On my showing him a beautiful cluster of moraines, within two miles of his father's house, he instantly accepted it, as solving a host of difficulties that have all his life embarrassed him."[38] To a man of Lyell's convictions, the actions of glaciers that could be observed in the present were a reasonable explanation for phenomena previously understood through the medium of speculations that did not stand the test of investigation. Like Lyell, Charles Darwin was similarly impressed with Agassiz's glacial theory. He wrote him a diffident note after they had met at the annual meetings of the British Association for the Advancement of Science in 1840. "I have enjoyed reading your work on glaciers, which has filled me with admiration," the young Englishman informed his Swiss colleague.[39]

Darwin's evaluation reflected a personal appreciation of Agassiz's arguments. On his third visit to Great Britain, in 1840, Agassiz spent much time in the Scotch Highlands, a region yielding as great a preponderance of evidence in support of the glacial theory as the valleys of Switzerland. In the Grampian Hills, the valley of Glen Roy offered a particularly fertile and unusual field for one concerned with discovering traces of former glacial action. As Agassiz described the scene, it was a winding valley trending in a northeasterly direction, and some ten miles in length. Across the mouth of this valley, at right angles with it, runs the valley of Glen Spean, trending from east to west, Glen Roy thus opening directly at its southern extremity into Glen Spean. Around the walls of the Glen Roy valley run three terraces, one above the other, at different heights, like so many roads artificially cut in the sides of the valley, and indeed they go by the name of the "parallel roads." These three terraces . . . are repeated for a short distance at exactly the same levels on the southern wall of the valley of Glen Spean, just opposite the opening of the Glen Roy valley; that is, they make the whole circuit of Glen Roy, stop abruptly, on both sides, at its southern extremity, and reappear again on the opposite wall of Glen Spean.[40]

The "parallel roads" had previously been explained as the shorelines of ancient lakes or as the result of the action of the sea, rising and subsiding at a period when the ocean had penetrated far into the interior. Darwin was one of those who had advocated the action of the sea as responsible for them. He soon abandoned this explanation in the face of Agassiz's informed analysis. The phenomena of Glen Roy, said Agassiz, were the results of glacial action. The parallel terraces had been the shores of a glacial lake, restricted in its bed by surrounding glaciers that descended from the other neighboring valleys. The terraces represented the different levels of the water, which was released from its confinement as the glaciers receded.

Investigation of this sort made Agassiz a respected figure in the councils of English geology. Similar triumphs were recorded in the scientific centers of Germany, France, and Agassiz's native Switzerland. By 1843, he was as notable a figure in geology as in paleontology. But Agassiz paid a price for his fame; it exacted both professional and personal penalties.

Some friends and colleagues, like Humboldt and Leopold von Buch, were both opposed to the glacial theory and unhappy that Agassiz had delayed his work on fossil fishes by involving himself in glacial research. Von Buch thus wrote him in 1837, from Berlin:

M. de Humboldt tells me that you are seeking a better climate here in the month of February. You may find it . . . thanks to our stoves. But as we shall still have plenty of ice in the streets, your glacial opinions will not find a market at that season. . . . I am expecting the numbers of your Fossil Fishes, which have not yet come. . . . Ah! how much I prefer you in a field which is wholly your own. . . .[41]

It was understandable that Von Buch, admiring as he did Agassiz's work in paleontology, thought he was wasting time and energy in geological speculation, especially since this master naturalist was convinced that heat and fire had been the primary agents shaping the rock formations of the past. Plutonists like Von Buch, and Neptunists who believed, conversely, in the action of water in determining land forms were, however, to find that, in the decades following the 1830's, Agassiz's glacial ice gained wide acceptance as a partial explanation for the appearance of certain regions of the earth.

In his address and published paper of 1837, Agassiz had advanced four basic concepts. There was, first, the significance of glacial action in Switzerland that accounted for the boulders and other erratic materials of the contemporary scene; such action could be understood by observing the movement and character of present glaciers. Following from this was a second assertion, that the Alps had been raised to their present heights by an upheaval of the earth, a great convulsion that took place under the ice, with boulders rolling down the slopes from the upheaval. The ice itself had been the result of a sudden and mysterious drop in temperature, part of a cyclic climatic pattern that had characterized the entire history of the earth. Finally, ice had covered not only Switzerland but great parts of Europe in one vast Ice Age.

Of these concepts, only the Ice Age idea was unique with Agassiz, and here he was partially indebted to another scientist. Jean de Charpentier, a geologist of considerable reputation, had listened to what he thought was the fantastic assertion of a Swiss mountaineer who told him in 1815 that the boulders of Alpine valleys had been deposited by glaciers. The reasonableness of this explanation was evidenced by the fact that in 1826 Peter Dobson, a Connecticut cotton manufacturer, advanced the same concept as an explanation for the scarred and scratched boulders in the ground moraine of his region. Three years later, the Swiss engineer E. T. Venetz presented a paper to a gathering of Swiss naturalists wherein he affirmed that the entire Valais region and the canton of Vaud had once been covered by an immense glacier which had deposited the boulders and erratic materials of the

Valais valley in their particular locale. Charpentier, by this time impressed with the validity of the glacial concept, arranged for the publication of an 1821 paper by Venetz that advanced the same proposition as the 1829 address. In 1835 Charpentier established the glacial doctrine on a firm basis by publishing a paper that substantiated the theoretical conceptions of Venetz with evidence derived from his own field investigations in the Rhône Valley.[42] Then Agassiz came to Bex in 1836, listened to Charpentier's ideas, examined his evidence, and came away a convert to the glacial theory. In his paper of 1837, Agassiz acknowledged the priority of Venetz and Charpentier in regard to the concept of glacial action.

When *Études sur les glaciers* appeared in print late in 1840, Charpentier's volume describing his glacial researches was almost ready for publication. Agassiz had freely acknowledged Charpentier's priority in *Études,* even dedicating the work to Charpentier and Venetz. Fully aware of Charpentier's forthcoming publication, Agassiz might have waited to publish his own work until his friend's book had appeared. Such propriety would not have detracted from Agassiz's volume, since it alone contained the concept of the Ice Age. When Charpentier's work did appear early in 1841, it contained an almost emotional reference to Agassiz's volume, giving the clear understanding that the geologist's glacial researches had been superseded by Agassiz's publication. The result of Agassiz's urge to put his ideas before a larger public in book form was that he lost a friend.[43] The *Études* appeared in a German translation at the same time as the French edition, so that Agassiz's ideas were more widely circulated than were Charpentier's. The fact remained, however, that Agassiz's conception of the Ice Age was unique and novel at the time, and his book therefore had a lasting value.

But the Ice Age concept itself resulted in another personal involvement for Agassiz. In February of 1837, Karl Schimper, the old friend of Munich days, arrived in Neuchâtel to work with Agassiz. The brilliant German delighted the intellectual community of the town with mystical discourses on phyllotaxy and the morphology of plants. Understandably, the two friends talked much about natural history, especially Agassiz's recent excursion in the Alps. Once, in jesting fashion, Schimper suggested the term "Eiszeit" for Agassiz's developing awareness that the history of the earth had been marked by a long age of ice. Schimper even wrote a short poem on the subject which he read to the Neuchâtel Society of Natural Sciences. In his 1837 paper, Agassiz acknowledged Schimper's contributions. He gave Schimper credit for the concept of the upheaval of the Alps and the consequent transporting of boulders, an idea which proved entirely incorrect and which Agassiz did not refer to again after his book of 1840. The 1840 volume, however, did not mention Schimper's "Eiszeit" suggestion. There is no doubt that the Ice Age concept was Agassiz's intellectual creation, although his vague awareness was clarified by the stimulus of Schimper's poem.[44] Once again, however, a friendship was destroyed. Schimper, who now despised Agassiz, published a poem in 1840 that contained a bitter accusation of plagiarism against his former friend. Greatly distressed, Agassiz was forced to respond in 1842 with a pamphlet reciting his intellectual relationships with both Charpentier and Schimper and detailing the genesis of the Ice Age idea.[45] Agassiz had never claimed originality in regard to the concept of glacial action, but Charpentier thought he deserved better treatment from the man he had initiated into glacial geology. Schimper felt equally mistreated, and a simple acknowledgment in the *Études* would probably have prevented the personal animosity that ensued.

The Charpentier and Schimper affairs were not the last of the personal results of Agassiz's glacial research. In 1841, James D. Forbes, a Scottish physicist of some distinction, visited the Hôtel des Neuchâtelois and was guided through the Alpine regions by Agassiz and his companions. Forbes's visit resulted in still another controversy, one carried on in the scientific press of England and Scotland during the years 1842–43. The dispute between Forbes and Agassiz concerned, once more, the question of priority of discovery, in this case discovery of the laws regulating the motion of glaciers. Forbes claimed that he had observed the peculiar ribboned structure, like veins or bands, discernible in glacial ice and that, from measuring these bands at different times, he had determined a method that revealed the uniform rate of glacial movement. Agassiz asserted that he and his assistants had done the same thing as early as 1838.[46]

Agassiz was furious at Forbes's claims. He pursued the man relentlessly, writing letters to powerful English friends condemning the Scotsman and attacking Forbes publicly in a pamphlet which was distributed to leading naturalists of the Continent and England. Forbes had inspired retaliation by public claims of his priority,[47] but the bitterness of Agassiz's attacks was probably conditioned by his unhappy experiences with Charpentier and Schimper. He was becoming intolerant of colleagues who worked in areas related to his interests and incapable of admitting others to his world on terms of equality. He defended his actions to Sir Roderick Murchison, saying:

I wanted to make known to my friends that Mr. Forbes allowed himself to commit a gross indiscretion against me. . . . My publications show that I give justice to anybody who has communicated to me the smallest observations, and I defy anyone to impugn the slightest anticipation of my work to Mr. Forbes.[48]

As for the merits of each man's position, the most reasonable evaluation was that of Lyell, that each investigator had discovered the same phenomena independently, although Agassiz's was probably the first effort.[49]

Forbes's friends never forgot Agassiz's behavior. In six years, Agassiz's work in glacial geology had cost him the friendship of three scientists, a loss of esteem that slowly began to influence some professional opinion concerning his personal relationships. The Ice Age concept nevertheless remained the singular contribution of Agassiz's imagination. The concept itself was both refined and advanced in future decades. Glaciation was shown to be confined almost entirely to the northern Arctic and Temperate zones, with great regions in these areas unaffected by it. Agassiz believed there had been one single advance and retreat of the ice mass. Subsequent study indicated the existence of successive glacial periods, perhaps as many as four or five, with long interglacial periods of warm climates after glacial retreats.[50] Had such facts been known in Agassiz's lifetime, he would quite probably have interpreted the phenomena of successive glacial epochs as additional demonstrations of magnificent, divinely inspired catastrophes.

The dominant view of Agassiz was now that of an almost universal naturalist, capable of intellectual efforts in marine biology, vertebrate and invertebrate paleontology, ichthyology, and geology. The Munich dream of a history of central European freshwater fishes was not entirely fulfilled, but one monograph detailing the embryology and anatomy of a particularly important family, the Salmones, was ultimately published in these years of accomplishment.[51] Another notable effort comprised a series of works dealing with the tools of the naturalist's profession. In 1842 Agassiz published *Nomenclator zoologicus,* a guide, index, and dictionary of zoological terminology written in an effort to bring order and unity to an essential research technique. Even more useful was a series of four volumes, completed in 1846, that testified to Agassiz's encyclopedic knowledge in all realms of natural history and the range of his professional competence and associations. This was the *Bibliographia zoologiae et geologiae,* a comprehensive listing of books and articles relating to all phases of natural history published up to 1846, together with a guide to scientific periodicals and to the publications of societies and institutions of natural science. A monumental effort, its publication was made possible by the Ray Society in England, which subsidized publication and provided the editorial services of H. E. Strickland, who was responsible for the ultimate form of the volumes that appeared during the years 1848–52. The sum total of this diverse intellectual involvement was the publication of eleven separate studies during the Neuchâtel years, ranging in scope from monographs in ichthyology to bibliographies, including the notable endeavors in paleontology and geology, all of which helped develop the attitude of mind that viewed all things as possible.

Inquiry into the personal resources, attitudes, and relationships that underlay Agassiz's accomplishments in these years of achievement is pertinent to an understanding of the man whose mind leaped from starfish to glacier to trout to fossil with apparent ease, confidence, and competence. Humboldt wrote Agassiz his evaluation of the sources of success:

You are happy . . . in the more simple and yet truly proud position which you have created for yourself. You ought to take satisfaction in it as the father of a family, as an illustrious savant, as the originator and source of so many new ideas, of so many great and noble conceptions. . . . You have shown not only what a talent like yours can accomplish, but also how a noble courage can triumph over seemingly insurmountable obstacles.[52]

Humboldt's image was accurate as a public view but overly sanguine as an assessment of Agassiz's personal relationships. In 1837, a daughter, Ida, was born to the Agassizes, and, four years later, Pauline arrived. With Rose Agassiz a frequent visitor after Rodolphe's death, the Neuchâtel apartment no longer served domestic requirements, and Louis rented a house to provide for his family. Meanwhile, Agassiz's role in the Neuchâtel community grew in power and stature. The international reputation Agassiz was gaining made his services more expensive for Neuchâtel and the Prussian monarchy. Beginning his career at a salary of four hundred dollars, Agassiz received one thousand dollars a year by 1838, the result of attractive offers from the academies of Lausanne and Geneva. Agassiz was interested in neither proposition because Neuchâtel supplied his wants. Frederick William III was as generous with funds to support glacial studies as he was with money for paleontology. In March of 1838 the aristocracy of the city dignified Agassiz with the title "Bourgeois de Neuchâtel," and as an honored citizen he had a part in civil administration and could command additional financial support. Soon thereafter, the Prussian king, partly in response to the intellectual aspirations of the community and partly in recognition of Agassiz's efforts, constituted the Neuchâtel school a formal academy, with a sizable grant of money and the authority to confer higher degrees. Appropriately, Professor Agassiz served as rector of the academy during the years 1838–42.

Agassiz felt compelled to create an environment that reflected his social, public, and professional distinction. Late in 1836 he decided that a private secretary and his own publishing house were primary requirements for intellectual achievement. These decisions had permanent consequences for his personal relationships. The early parts of *Poissons fossiles* had been published in

Munich, but Agassiz was dissatisfied with the time that elapsed between writing and publication. He was unhappy, too, at the quality of the reproductions of his illustrations. Hercule Nicolet, a Neuchâtel printer and engraver highly skilled in the process of lithochromatic printing, owned an establishment which, by 1837, had in effect been taken over by Agassiz. All the work done was for Agassiz; he guaranteed the salary of the required staff, supplied the manuscripts to be printed, and hired new assistants when necessary. From this time forward, all of Agassiz's books published before 1846 bore the imprint, "Aux frais de l'auteur." Agassiz was understandably disturbed by the problems incidental to scientific publication elsewhere. He needed the assurance and authority given him by controlling his own publishing enterprise and was willing to undertake the tremendous financial responsibility entailed. To meet this obligation, he determined that austerity would be the guiding principle of his domestic life; he would live by the most frugal means and spend all he could spare for the advance of science, since this was the only important goal in life.

The advance of science was a dynamic process. Artists were a primary requirement, and although Dinkel stood by, faithful as ever, more artists were needed to meet the demands placed on Agassiz the publisher by Agassiz the author. The publishing establishment had to operate continuously, and projects such as the bibliography and guide to zoological nomenclature were undertaken for this purpose. As these proceeded, new ones, such as the translations of English monographs, were begun. These efforts required financing, and money was supplied by the citizens of Neuchâtel in the form of loans and gifts, and also by grants from the Prussian government. As more money became available and new projects were born, Agassiz felt in desperate need of an assistant who could function as secretary and manager of the rapidly growing enterprise. For this post, Edward Desor, a young German law student with multilingual skills, was hired in 1837. Desor, under Agassiz's tutelage, learned the rudiments of natural history, worked hard in his employer's behalf, and soon assumed the role of superintendent of the artists, clerks, engravers, and printers who labored to further the purposes of their employer. By 1839 there was a total of twelve individuals working at scientific pursuits under Agassiz's direction at Neuchâtel. These included Arnold Guyot, now a geologist and professor at the academy, Jacques Burkhardt, a highly skilled artist, Karl Vogt, a young German naturalist who worked at embryology, anatomy, and paleontology, and Agassiz's students, Count François Pourtalès and Charles Girard.

Thus a veritable "scientific factory," as Vogt described Agassiz's Neuchâtel headquarters, was organized and operated. The Agassiz house had a room set aside as a laboratory for anatomy and embryology; another room served as a studio for a modeler who made casts of sea animals and shells and relief maps of mountain ranges. Vogt and Desor each had their own rooms, as did Rose Agassiz. Mealtimes were occasions when the entire band of Agassiz assistants gathered at the house to discuss their various enterprises, report on progress to their employer, drink much good wine, and enjoy the hearty good humor of men who scaled alpine peaks and felt they were contributing to the intellectual progress of their age. Conversations went on well into the evening hours, as more work was planned to keep the printing establishment busy. In the summer, Agassiz and his followers searched for evidence of former ice action in the Alpine regions. When the group returned, they would often find foreign naturalists come to Switzerland to meet Agassiz and live for a time in his house. James Forbes described the man who would later quarrel with him as a person who was "cheerful, kind, and frank . . . always ready to contribute to the cheerful companionship of a party, the chief of which he justly considers himself."[53]

The intellectual achievements of all these enterprises were magnificent; the personal toll they exacted was unfortunate. It was impossible for those close to Agassiz to accept his ambitions, drive, and energy passively; one either went along wholeheartedly in the spirit of great adventure or opposed the course of progress as Agassiz defined it. Fellow scientists were often amused at the excesses of Agassiz's enthusiasm, but they did not have to live day by day under the constant impact of his striving for dominance. At Neuchâtel, some who knew Agassiz in these years of triumph found they could not tolerate the conditions of life he imposed upon all around him.

Karl Schimper was the first to grow restive under the conditions of life at Neuchâtel, and in 1840 he left, convinced not only that Agassiz had appropriated the "Eiszeit" concept without due acknowledgment but that he had not given Schimper credit for work done in describing a number of fossil fishes. Then, the year 1841 witnessed the dispute with Charpentier, and the Forbes controversy occurred during 1842–43. As if these events were not intolerable enough, in 1844 Karl Vogt joined the ranks of those who felt maligned by Agassiz. Like Schimper, Vogt was convinced he had been wronged on intellectual grounds. He felt that the embryological research on the family of Salmones was his alone and that he should receive rights of authorship to the publication. Agassiz was equally certain that the work had been done as a result of his inspiration and direction and would only agree to the joint authorship of one part of the monograph. As in other instances, Agassiz was incensed at what he considered evident ingratitude for friendship and material support. One reason for Agassiz's anger with Vogt was that the German naturalist had already published part of the Salmone monograph under his own name. All Agassiz had been able to do was to insert in the text his explanation of the character of the collaboration. The Vogt affair tarnished another young dream, as Vogt went so far in his animosity toward Agassiz in later years as to claim authorship of substantial parts of the *Poissons fossiles*. A patent falsehood, the allegation is significant only as a reflection of the larger pattern of Agassiz's inability to collaborate with others.[54]

In the light of Vogt's feelings, his recollection of Agassiz is of more than ordinary interest:

Agassiz was the most amiable companion one could find; cheerful, usually in a pleasant mood, adjusting with ease to the moods of others, and a thoroughly congenial character. He grasped the greatest tasks with almost playful ease, overcame difficulties without any particular effort, and developed an incredible energy when the problem was to get a project underway. . . . He was very shrewd in procuring material for study. When, however, it had been assembled, fleetingly looked over, and put in order, and then methodical work on it was to begin, he collapsed and could only be kept to his task by great effort.[55]

Even if one felt wronged, it was impossible to ignore the almost irresistible force of Agassiz's personality.

Despite these unfortunate experiences with colleagues and assistants, Agassiz did not question his personal goals, his behavior, or his determination to live only for science. "Whatever befalls me, I feel that I shall never cease to consecrate my whole energy to the study of nature; its powerful charm has taken such possession of me that I shall always sacrifice everything to it; even the things which men usually value most." These sentiments, written to a colleague in 1845, were the deeply sincere expression of Agassiz's complete emotional and intellectual devotion to natural science.[56]

But this dedication, in addition to making life difficult for his co-workers, was the cause of grievous personal experiences during the years from 1842 to 1845. It was during these years that Cécile Agassiz began to feel discontented with her life at Neuchâtel. She had been happy in the first flush of a new life as a professor's wife in a small community. But as time passed, her German home, with its environment of art, music, and cosmopolitan culture, seemed much more attractive than the commonplace, narrow world of Neuchâtel, whose inhabitants impressed her as pretentious and provincial. Cécile thus began taking ever more frequent trips to Carlsruhe, and with each visit Neuchâtel seemed less bearable when she returned. She became ill frequently; her spirit and constitution could not tolerate the social atmosphere, the climate, the dusty roads of the Swiss town. Alexander was the one real joy in her life, and she lavished care and affection on her son, resolved to raise him as a German sensitive to the fine manners of worldly life. Cécile found it difficult to confide her unhappiness to her husband. Louis was always busy with some scientific project; he was often away in England, France, Germany, or on Alpine excursions, and it seemed that a solid family life did not appeal to him.

Cécile appreciated the talents of her husband. She knew him to be brilliant, impulsive, kind, generous, and good-humored. But his scientific ambitions seemed to dictate an utter disregard for domestic matters, for family finances, for concern about his wife's welfare. There were three growing children to be schooled and supported. Even the ordinary expenses of a household were not easily met by an academic salary. But Agassiz planned and worked as if he had unlimited wealth. Dinkel, Desor, Vogt, Nicolet, and, it seemed, countless others were all paid from his limited funds. Income grew larger with the years, and handsome grants came from outside sources, but all this was expended on new projects. Cécile was increasingly aware that her modest and normal wants were vain aspirations. Family enjoyments, a comfortable home, the pleasures of a full and varied life, all appeared incompatible with the pursuit of science.

Rose Agassiz, the one person who could understand Cécile's unhappiness because she knew the habits and mental attitudes of her son, and Auguste, now a prosperous businessman, pleaded with Louis to give up the publishing establishment that was driving him deeper and deeper into debt. Both mother and brother urged that Louis's family responsibilities should take precedence over the production of beautiful books of science. Agassiz would not heed such appeals, and Cécile grew more and more unhappy. Her discontent focused on her plain distaste for the men who worked for Agassiz and shared her home and dinner table. Particularly she despised Edward Desor because she believed that he had come to exert an unwholesome influence over Agassiz's affairs, urging expenditures that were unjustified in the light of family finances. What was worse, Cécile thought Desor to be a crude, vain, irresponsible person. She also disliked Karl Vogt, but not with the intensity of her feeling about Desor. At the dinner table, Vogt's atheism, his biting jibes at religion, and Desor's frequent off-color remarks shocked and offended the shy and pious Cécile.

By 1845 all these circumstances had served to alienate Cécile from her husband. Although Vogt had left Neuchâtel, Desor remained a constant source of annoyance to her. The expenses of Agassiz's publication enterprise mounted, and, as they did, Cécile's feeling of estrangement grew. It was not that Agassiz was entirely unresponsive to Cécile's feelings. But as his public reputation grew and recognition came his way, he lived more and more in the pleasant world of popular renown, neglecting or ignoring the pressing problems of domestic existence. He loved his wife and children deeply. But protestations of affection were small compensation for the frugal, almost pauper-like existence that intellectual ambitions forced on his family, and they provided little compensation for the unpleasant life Cécile knew in her Neuchâtel home. Agassiz could wear the Cross of the Red Eagle of Prussia with pride; he could delight in the company of General von Pfüel and wealthy Swiss businessmen at the Hôtel des Neuchâtelois; he could feel self-important enough to dispute publicly with local theologians over the proper interpretation of the Bible in the light of scientific knowledge; he could not, however, plan his intellectual undertakings realistically or give his wife the attention and security her nature demanded. Prepared to sacrifice "the things which men usually value most" for scholarly goals, Agassiz was faced with a choice between personal security and public ambition.

Desor's very presence in the Agassiz home had by now become intolerable to Cécile. Both she and Rose Agassiz had repeatedly warned Louis that his secretary was not to be trusted, that he had squandered money, indulged himself in material pleasures at Agassiz's expense, and had subtly become his master, dominating the Neuchâtel establishment by devious manipulations that had shaken the confidence of the men who worked there. Agassiz was unwilling to heed these warnings, convinced that he needed Desor's services for the success of his scholarly ventures. Feeling strongly that Desor was the essential cause of all her domestic unhappiness, Cécile reasoned that all would be well again if the German left her house. She equated Agassiz's domestic attitudes and expensive scientific undertakings with Desor's influence. Cécile evidently pleaded with Louis to get rid of Desor and, her pleas ignored, determined to take direct action. Accordingly, in the early spring of 1845, she left Neuchâtel for Alexander Braun's home in Carlsruhe, taking Pauline and Ida with her and leaving Alexander to finish his schooling in the academy.[57]

The departure of Cécile was in itself a calamitous event, but the world seemed even bleaker with the happenings that followed soon after. Early in 1845, increasing debts incurred in publishing ventures forced the closing of the printing house and the end of the scientific activity associated with it. Manuscripts were sold to a nearby publisher to meet some financial obligations. Other debts were guaranteed by Agassiz's family and Neuchâtel friends. The scientific factory, begun with such enthusiasm eight years before, was thus disbanded. Agassiz was in the depths of despair, contemplating with sadness a society that did not appreciate "my unhappy books, which never pay their way because they do not meet the wants of the world." He had resolved, even before the financial reversal of 1845, that if Neuchâtel could not supply the material means required to achieve his goals, he would "be forced to seek the means of existence elsewhere."[58] Now, he looked toward the larger world of science and its institutions to provide the basis for a future life. At thirty-eight, Agassiz faced a crucial juncture in his life.

The fact that immediately after Cécile's departure and the failure of the publishing enterprise an event occurred that restored Agassiz's optimism and self-confidence, could readily have convinced loved ones that it was futile to berate Agassiz for believing that society would never fail to fulfil his desires. This was a time when recent events might have dictated a realistic self-appraisal, but new plans made introspection impossible. In his hour of greatest insecurity, Agassiz received a letter of vital import from Humboldt. This intellectual godfather it seemed, could perform special and periodic feats, guaranteed to free Agassiz from the burdens of the present and launch him once again, happily, into the future. For Humboldt informed Agassiz in March of 1845 that Frederick William IV of Prussia, continuing the interest of the monarchy in Agassiz's career, would grant him the sum of three thousand dollars to undertake a journey to the United States, where he would study the natural history of the New World.[59] At last the opportunity for a great exploration was at hand, another early dream realized.

In the same process by which Cuvier had given Agassiz research materials on fossil fishes and Neuchâtel had made him a professor, the seemingly fortuitous American journey had been made possible by careful advance preparation. The first effort was a natural one, dictated by scholarly activity. In 1835, Benjamin Silliman, Yale University chemist, geologist, and editor of the *American Journal of Science,* received complimentary copies of the early parts of *Poissons fossiles,* Agassiz's object being to gain subscribers for his work in the United States and to borrow representative American fossil fishes from Silliman. Flattered by this sign of European regard, Silliman fulfilled both these requests and became Agassiz's publicist in the United States by virtue of his key position.[60] In 1837, Agassiz's growing fame resulted in his election as a corresponding member of the Academy of Natural Sciences of Philadelphia, a leading scientific institution. On the basis of these relationships, Agassiz began corresponding with such important American naturalists as John H. Redfield of New York, Augustus A. Gould of Boston, and Samuel H. Haldeman of Philadelphia. Writing to Silliman in 1838, Agassiz affirmed his high opinion of American research in paleontology and the keen interest in national science awakened in him by reading articles in the complete set of *American Journal of Science* volumes sent him by Silliman. He emphasized his enthusiasm by asserting that "for several years I have calculated the possibilities of making a tour in America, for I must say . . . that I have the greatest desire to see that country, and to make your personal acquaintance."[61]

The achievement of this ambition came about through the converging influence of Agassiz's powerful friends and supporters. By 1842 the idea of an American journey seemed close to realization when Charles Lucien Bonaparte, prince of Canino, Alpine companion of Agassiz, and a naturalist familiar with the United States, proposed to his Swiss friend that they both journey across the Atlantic and spend a summer studying the flora and fauna of the New World. Prince and professor would write a natural history of American fishes from the experience of this journey. In late 1844, as domestic problems and tensions increased at Neuchâtel, the prospect of the journey grew more attractive. The trip was planned for the summer of 1845, and Agassiz was naturally pleased that Bonaparte had offered to pay all his expenses.[62]

Agassiz determined to leave nothing to chance, and to enlarge the scope of the project even before it began. He wrote Humboldt in December of 1844. A summer in the United States would be wonderful, but it would be a shame that such a short visit would prevent study of American geological formations, a subject of deep interest to Agassiz and Humboldt. Would it be possible to interest the king of Prussia in supporting a sojourn of longer duration? Not content with just one source of possible aid, Agassiz wrote Charles Lyell in February of 1845, asking his assistance. Lyell, he knew, had undertaken a highly successful tour of the United States during 1841–42, when he had delivered a course of lectures in Boston to audiences of the Lowell Institute. Perhaps Lyell could intercede with Boston friends and arrange for Agassiz to lecture in their city, thereby making an American journey possible. He did not tell Lyell of his inquiry to Humboldt, nor did he inform Bonaparte of either of these appeals.[63]

Having planted these seeds, Agassiz waited for them to germinate. The appeal to Humboldt bore first fruit with the letter of March, 1845, about the king's intention to grant him funds that would allow a journey to the United States of two years' duration. The award was made public later this same month, and Neuchâtel citizens were anxious to hear the forthcoming public lectures of Agassiz, fearful that this would be their last opportunity.[64] On the first of May, 1845, Lyell answered Agassiz's inquiry by reporting that John Amory Lowell, trustee of the Lowell Institute, would be delighted with the opportunity of presenting such a distinguished foreign savant to Boston's literate society. Agassiz would be paid fifteen hundred dollars for a lecture series. Agassiz now learned that Bonaparte, who was ill, would not be able to accompany him. This was unfortunate, for it meant that Agassiz would have to prepare for the American visit by learning as much as possible about the country, instead of relying on Bonaparte, who had been there before.[65]

With the assurance of funds from Prussia and Boston, Agassiz could plan to have assistants like Desor and Dinkel join him in America to help in the effort to understand the natural history of this land. Dinkel, however, decided against joining Agassiz in this venture. The artist suspected that the latest Agassiz enterprise would be longer than two years, and he felt that nearly twenty years of his life was enough to devote to his Swiss friend. The time had come for Dinkel to establish himself as an independent creative person.

There was so much to do in this spring of 1845 that Agassiz hardly knew where to begin. The Prussian Academy of Sciences sent a long list of specimens its members wanted from America, including such items as dried plants, eyeless fish, and water from the Mammoth Cave in Kentucky. Agassiz would do better. He would make two collections, each the same, one for Prussia and one for the Neuchâtel museum. There was even some justification for hoping that Prussia might reward its traveling naturalist by an appointment at the University of Berlin if the American journey were the success it promised to be. In personal terms, a Berlin appointment might mean that Agassiz and Cécile could find new happiness in the land of her birth, and her present stay in Germany might improve her failing health. A professorship in Paris was another possibility.

Agassiz faced the future with characteristic exultation over its promise. "For pity's sake husband your strength," Humboldt warned, "you treat this journey as if it were for life."[66] Humboldt had detected a deeper ambition that transcended the Berlin or Paris opportunities. The compulsion to succeed in this new enterprise was so powerful that Agassiz repeatedly postponed his departure for America. There was, for example, the matter of public lectures in the United States. These were prepared carefully, delivered at Neuchâtel as a rehearsal for the Boston experience, and their salient ideas illustrated by a Paris artist in drawings that were forwarded to John Amory Lowell. Agassiz took Lyell's advice and began practicing his English pronunciation. Moreover, he determined to finish incomplete projects, so that he might go to America unburdened by former responsibilities. He took a last trip to the Unteraar glacier and made arrangements in Switzerland, France, and England for the publication of monographs in zoology and geology. All this made it impossible to leave Neuchâtel until March of 1846.

In New Haven, Silliman, anxiously awaiting news of Agassiz's plans, learned something of the reasons for his delayed departure:

My purpose is to become well up-to-date on everything new in the natural sciences here before leaving our old Europe for a long time. . . . There is something intoxicating in the prodigious activity of the Americans which makes me enthusiastic—I already feel young through the anticipated contact with the men of your young and glorious republic.[67]

The romance between Agassiz and America had begun, and this first overture was immediately reciprocated. The May, 1846, issue of the *American Journal of Science* carried intelligence of first importance to its readers. "This distinguished naturalist, who is known wherever science is cultivated, is about to visit the United States," Silliman announced. A list of Agassiz's distinctions followed this introduction: the legacy of Cuvier, the friendship and support of Humboldt, the patronage of the king of Prussia, the glacial researches, the studies of fossil fishes—all these recommendations were detailed with high admiration. As for the man himself,

His devotion, ability, and zeal . . . his amiable and conciliating character, will . . . secure for him the cordial cooperation of our naturalists, and the favor of the public. We recommend him and his cause to all lovers of science and mankind. He is strongly impressed by the activity and success of our own naturalists.[68]

With success seemingly assured beforehand, there was still much to do. The road from Neuchâtel led first to Carlsruhe, where Agassiz visited Cécile and his daughters. The museums of Carlsruhe and other German cities were visited once more before Agassiz arrived in Paris in April, 1846.

In the city that had given Agassiz such lasting intellectual resources in 1832, the award of the Monthyon Prize by the Academy of Sciences, the good wishes and fellowship of French colleagues, and the pleasure of studying the fossils of the Museum of Natural History once more were all experiences vital for present self-confidence and future purposes. Publication arrangements, research, and writing kept Agassiz busy in Paris until August of 1846. He then left for England, where close friends such as Owen, Murchison, Sedgwick, and Lyell welcomed him and were happy to have him inspect the data of paleontology in their collections. In September the members of the British Association for the Advancement of Science were pleased to see and listen to Agassiz at their Southampton meeting. Lyell, a perceptive observer of American manners and culture, took this opportunity to tell Agassiz about the land he was visiting. On September 19, he accompanied Agassiz to Liverpool and watched him board a steamer for the Atlantic voyage, certain that his friend would meet the challenge of the future.

For Agassiz, the sixteen-month interval between the time the American journey became a certainty and the time he left England was a period of significant accomplishment. The trips to Germany, France, and England were more than the farewells of a traveler. Agassiz was storing up, as it were, impressions of Europe. As he told Silliman, he was anxious to know all that was important about the latest advances in his science. He wanted to be sure, too, that European museums, scientific societies, and intellectual activity would remain fixed in his memory. Thus Agassiz felt compelled to prepare and fortify himself for whatever the future might bring. Success in America was vital to him.

Contemplating his journey, Agassiz had hardly been able to contain his excitement. He had written Augustus A. Gould that a primary purpose of his visit would be to compare the living and fossil animals of America with those of Europe. In August of 1846 he enlarged this conception by confessing to Silliman that he wanted to study the ecology of the Great Lakes, the Ohio River, and the Mississippi, to explore the Rocky Mountains, to learn the fishes of the New World, to study anew the zoological identifications of that former student of American natural history, Constantine Rafinesque, and to look for signs of former glacial action. "All these objects . . . are well worthy to fix the attention of a naturalist; and . . . there are few countries . . . where one can study phenomena so varied," he told Silliman, "so I make in advance a *fête* to myself in the prospect that I now have a glimpse of."[69] Such goals were the work of a lifetime, and Agassiz knew too much natural history not to have wondered if they might not prove so.

The ship that left Liverpool on September 19, 1846, carried with it a scientist at the crossroads of his career. Captain Ryrie was pleased with his robust, broad-chested passenger, who smoked cigars incessantly and insisted on practicing English conversation with him. The passengers were struck by this decidedly handsome man of thirty-nine who talked so volubly and seemed to enjoy every aspect of the voyage, which, though uneventful, all knew to be a dangerous undertaking in this season of Atlantic storms. When the "Hibernia" docked at Halifax before going on to Boston, an incident typical of Agassiz occurred:

... eager to set foot on the new continent so full of promise for me, I sprang on shore and started at a brisk pace for the heights above the landing. . . . I was met by the familiar signs, the polished surfaces, the furrows and scratches, the *line engraving* of the glacier . . . and I became convinced . . . that here also this great agent had been at work.[70]

The New World had met a new explorer.

The portrait Agassiz painted of his aims, aspirations, and character to American correspondents prior to his arrival was revealing. He wrote Gould that he planned to bring many European specimens with him, pleased at the chance to increase the knowledge of his new scientist-friends. He did not mind any discomfort in his labors, for his was the serious business of science. "I have been a naturalist since my childhood, accustomed to all sorts of fatigue and privations. . . . What I fear most is to lose precious time." Gould was properly impressed with such dedication and told his fellow scientists of Agassiz's high purposes. These men, in turn, were delighted by their coming good fortune, the chance to have such a person in their midst.[71]

Benjamin Silliman also learned from Agassiz that he would have to live very frugally in America in order to provide for the assistants who were to follow him. Everyone knew that important explorations needed the aid of devoted helpers. Science, Agassiz informed Silliman, was a hard taskmaster. All he wanted to do was to learn all there was to know about American natural history.

This is what I desire . . . rather than to sparkle in the world, being neither entitled to being received in the circles of public life, nor having any pretension to figure in society in any manner whatsoever. My sphere . . . is completely the world of science. . . . I am by no means a misanthrope, but I learned early that, when one has no fortune, one cannot serve science and live at the same time in the world. If I have been able to produce numerous expensive publications, it has been only by following this system of economy and voluntary seclusion; and the results which I have obtained thus far have rewarded me so well for the privations which I have suffered, that I have no temptation to adopt another style of life, even should I have hereafter, and especially in your country, more trouble than I have had to sustain it in my own.[72]

In the light of future events, these words must stand as the most grossly inaccurate phrases Agassiz ever penned. If the image he drew was only partly reflective of the past, it served well the purposes of the present. America knew it was welcoming a man deeply dedicated to a self-appointed mission to uncover nature's innermost secrets. In an age of growing materialism, a soul of such nobility could only be admired. If America was prepared to admire Agassiz, Agassiz already had provided himself with the emotional basis for admiring America. He had freed himself from the courtly muses of Europe by finishing tasks long incomplete. This intensive effort could not have been undertaken without an unconscious hope that one career was ending and another beginning. Humboldt understood some of this. Silliman, when he knew Agassiz better, understood also.

4

THE AMERICAN WELCOME

1846-1850

WHEN JOHN AMORY LOWELL welcomed Louis Agassiz to his comfortable home in Boston's Pemberton Square, he greeted a man of energy and purpose whose thoughts were tuned to the promises of the present and always responsive to the potentialities of the future. This distinguished scientist, young for his attainments, with a broad, pleasant face and ready smile seemed all that Charles Lyell had said of him. More than politely cordial, he radiated friendship and good fellowship, his flowing conversation punctuated by a delightful French accent.

Lowell, cotton manufacturer, financier, prominent member of the Corporation of Harvard University, and patron of science, was an excellent person to welcome Agassiz to New England and the United States. Trustee of the Lowell Institute and president of the Boston Athenaeum, he was a dominant figure in the cultural life of his community, equally at home in the commercial world of State Street or the intellectual milieu of Quincy Street across the Charles River in Cambridge. Representative of the congenial alliance between material wealth and the life of the mind that distinguished Boston society in these years, Lowell was familiar with the world Agassiz typified. He had welcomed an impressive number of foreign scholars to these shores, and, meeting Agassiz on this morning of October 3, 1846, Lowell could count himself fortunate in having acquired the services of a man as widely known as Agassiz.

To Agassiz, Boston seemed both familiar and strange. Its streets and buildings reminded him of London and Paris; its October skies were clear and pure like those of Italy. Its autumn climate was colder than the land he had left, and he had noticed five different species of oak trees in his walk to Lowell's home as he admired their multicolored foliage. Lowell himself, with his sincere interest in natural history in general and botany in particular, seemed at once a familiar person, not unlike Louis Coulon or the Swiss businessmen who visited the Hôtel des Neuchâtelois.

Intent upon learning as much as possible about America's natural history, Agassiz spent the first month after his arrival traveling. Accompanied on part of the tour by the Harvard botanist and professor Asa Gray, he visited Albany, New Haven, New York, Princeton, Philadelphia, and Washington, meeting the most distinguished of native naturalists and visiting important centers of scientific activity. He was accustomed to whirlwind tours through places of scientific interest and knew well how to discern what fossil specimens of Philadelphia's Academy of Natural Sciences were unique, what charts at the United States Coast Survey in Washington were of particular interest, and what wares of New York fishmongers were to be delivered to him for closer examination. All the naturalists he met were interesting to him, all specimens he examined were fascinating and needed to be studied more carefully. "How a nearer view changes the aspect of things," he confided to his mother. "I thought myself tolerably familiar with all that is doing in science in the United States, but I was far from anticipating so much that is interesting and important. What is wanting to all these men is neither zeal nor knowledge. In both, they seem to compete with us, and in ardor and activity they even surpass most of our savants."[1]

The talents of men such as the New Haven geologist and zoologist James Dwight Dana, the Philadelphia anatomist Samuel George Morton, the Dickinson College ichthyologist Spencer F. Baird, the Philadelphia zoologist Samuel H. Haldeman, and the New York paleontologist William C. Redfield all held promise of significant accomplishment. The obvious ability of a man such as Gray, the wide knowledge of James Hall in geology and paleontology, the specimens gathered by the scientists of the United States Exploring Expedition, all these, and more, were ample evidence to Agassiz that he was not entering an intellectual wilderness. Institutions for intellectual pursuits might be inferior to those of Europe, but the American scientists were worthy of respect and admiration.

Wherever he traveled, Agassiz epitomized Continental science and played the role expected of him with dignity and pleasure. He praised highly Morton's collection of fossils and human skulls, correcting an identification now and then and urging increased activity in paleontological research. He was properly impressed by the publications of Hall and others describing the geology of New York State and was struck by the beauty of the New Haven countryside. The spirit of an America caught up in a material and idealistic effort to expand her borders and match physical progress with cultural and intellectual advance struck a responsive chord. He too was eager to support the progress of the American mind.[2]

Agassiz was torn by conflicting emotions and interests in these early days of his visit. He had come for an intellectual purpose, but wherever he went to seek its fulfilment, he was overwhelmed by the vigor and enthusiasm of the people and by the excitement of a new land and a new culture. He was attracted and repelled by "the irresistible power of steam, carrying such heavy masses along with the swiftness of lightning," as he traveled by railroad. He marveled at a scene of three thousand workmen and artisans gathered together to form a public library association.

Naturalist as I am, I cannot but put the people first. . . . What a people! . . . I should in vain try to give . . . an idea of this great nation, passing from childhood to maturity with the faults of spoiled children, and yet with the nobility of character and the enthusiasm of youth. Their look is wholly turned toward the future . . . and thus nothing holds them back, unless, perhaps, a consideration for the opinion in which they may be held in Europe.[3]

If Agassiz was overwhelmed by the Americans, Americans were no less impressed by the enthusiasm and energy of their visitor. Jacob W. Bailey, West Point naturalist, wrote James Hall that he had caught the "big fish Agassiz" and would not let him get out of his net until he was sure he could guide "this scientific whale" up the river to Albany.[4] Asa Gray thought Agassiz a "capital fellow" and wrote to William C. Redfield that he was "as excellent and pleasant a man as he is a superb naturalist."[5] Morton confided to Haldeman that the hours he had spent with Agassiz were priceless. "I am delighted with his astonishing memory, quick perceptions, encyclopedical knowledge of Natural History and most pleasing manner. There is no affectation or distrust about him."[6] F. E. Melsheimer wrote his friend Haldeman: "Before now . . . you must have seen Professor Agassiz, and have the reports of this big geologico-everythingo-French-Swiss gun; you will of course tell us all about him. . . ."[7] Benjamin Silliman, delighted to make Agassiz's acquaintance at last, arranged for the Swiss visitor to give an impromptu lecture on glaciers before his geology class. Agassiz "won us all by his affability, good-humor . . . accommodating disposition . . . and cordial manners."[8] James Dwight Dana, Denison Olmsted, John P. Norton, and Silliman's scientist son, Benjamin, Jr., were all similarly pleased. This last naturalist summed up what was probably a common reaction to Agassiz:

He is full of knowledge on all subjects of science, imparts it in the most graceful and modest manner and has, if possible, more of *bonhomie* than of knowledge. He has a more minute knowledge of his subject and at the same time a more wonderful generalizing power and philosophical tone than any man I have ever met. . . . It is not yet agreed whether the Ladies more liked the *Man* or the Gentlemen more admired the *Philosopher.*[9]

When Agassiz returned to Boston late in November, he found his reputation had grown during his absence. He complained that "here everybody considers himself authorized to display the interest he has in the sciences by the time he takes away from those who occupy themselves with them."[10] The only solution was to do what

Humboldt did when in Paris. Agassiz kept two residences, a "public dwelling" on Tremont Street and a private room of unknown location where he could work undisturbed. But Agassiz could not ignore requests for his company. He dined with President Edward Everett of Harvard and met such scholars of the university as Benjamin Peirce, the mathematician, and Cornelius Conway Felton, the classicist. At Felton's home, Henry Wadsworth Longfellow found his dinner companion "Mr. Agassiz. . . . A pleasant, voluble man, with a bright, beaming face."[11] It seemed that wherever he turned, there was another dinner engagement, another group of admiring New Englanders to be charmed. At Nathan Appleton's Boston home he met such distinguished men of commerce and letters as Francis Cabot Lowell, Abbott Lawrence, and George Ticknor. Longfellow aptly recaptured the mood of such gatherings: "Dinner party . . . for Agassiz. The recollection . . . is charming. Agassiz lounging in his chair . . . eagerly listening to what was said. . . . Agassiz extolling my description of the Rhône Glacier in *Hyperion*, which is pleasant in the mouth of a Swiss who has a glacier theory of his own."[12]

Other admirers, waiting impatiently for the promised Lowell Institute lecture series on the "Plan of Creation in the Animal Kingdom," were not disappointed. During the winter of 1846–47 thousands of Bostonians crowded into the Tremont Temple, on some evenings as many as five thousand of them, to hear the foreign professor. So great was public interest that Agassiz had to repeat his lectures each day to a second audience. Bostonians had often heard scientists popularize nature study from the Lowell Institute platform, but Agassiz was unique, his lectures a new and compelling experience for those who heard them.[13]

These lectures, planned in Europe as merely a means to increase funds for scientific research, became ends in themselves. Silliman journeyed from New Haven to hear Agassiz and reported: "Professor Agassiz gives great satisfaction and wins universal favor. The vast extent of his knowledge, his . . . winning affability and modesty, make him a great favorite. . . . His manner is calm, dignified, and yet engaged."[14] Asa Gray wrote his fellow botanist John Torrey that Agassiz "charms all, both popular and scientific! I observed to him that there was much quite new to me in his last lecture. . . . He replied that it would be equally new in Paris, much of it."[15]

Gray put his finger on the secret of Agassiz's appeal. Lecturing easily, ignoring his notes, Agassiz made his words all the more attractive by a charming Continental accent. He treated his subject "so that ladies may attend." His audiences were delighted by references to "God's leetle joke" in creating an odd looking crab or turtle, or to the perfect symmetry of the "Aacheenodaarms." This popular tone did not hide the fact that Agassiz presented in brief synthesis his intricate research on fossil fishes, his knowledge of embryology, and his glacial theory. Agassiz was thus the dual personification of the dedicated scientist and the inspired, cosmopolitan European teacher.

Besides introducing Agassiz to a wider world which he would conquer with the same determination that had seen him scale Alpine peaks and gather fossil fishes from all over Europe, the Lowell lectures were singular landmarks in the history of nineteenth-century national culture. In an age alive and responsive to the idealism of Emerson, Agassiz gave a scientific demonstration of the spiritual quality underlying all material creation. Men might know all there was to know about the facts of nature, but if they did not appreciate the magnificence of the master plan fashioned by its Author, and the complexity of the relationship that bound all organic creation to the Higher Power, they could know the world only partially. Man's soul, his intelligence, his divine nature, provided the connective link by which he could appreciate the power of the Deity in relating him to the process of nature. "We may . . . come to a full understanding of Nature from the very reason that we have an immortal soul."[16] The religion of nature that Agassiz illustrated by precept and by line drawings of animals demonstrated that the entire history of creation, beginning with the smallest radiated animal and ending with man, had been wisely ordained. The men and women who listened to Agassiz heard that their species was not only the highest form of vertebrates but represented the direction and the purpose to which all creation had moved from the beginning.

Gray once more characterized the force of Agassiz's appeal. "They have been good lectures on natural theology. . . . His admirable lectures on embryology contain the most original and fundamental confutation of materialism I ever heard."[17] Agassiz's arguments were almost impossible to resist. The anonymous author of the widely read *Vestiges of the Natural History of Creation* (1844) had suggested that a process of development from lower to higher forms had marked natural history; Agassiz affirmed that "species . . . do not descend from each other . . . have never been derived from each other."[18] Change in nature was only change in individuals, occurring according to a creative plan which made every animal conform to a permanent and fixed type intended from the beginning. Views such as those of the *Vestiges* were in fact unscientific, old-fashioned, and "entirely unworthy of notice by any serious scientific man."[19] New theories, then, were really old ones, but the facts about nature Agassiz described were new. With the proper spirit of reverence, all could study and appreciate them. Scientists and laymen alike found such a message both instructive and reassuring.

His obligations to John Amory Lowell and the institute audiences discharged, Agassiz might have been expected to proceed on his appointed mission. But his success had been too great. Offers for lecture engagements poured in from all over the East; New York City wanted him for a series, the Salem Lyceum begged his instruction, a group of Bostonians asked for a private course of lectures, Daniel Webster's Society for the Diffusion of Knowledge pleaded for words from the famous professor. The faculty of the Harvard Medical School asked for the same privilege; Benjamin Silliman wanted a special course in New Haven for Yale students. The needs of Prussia and Neuchâtel for specimens of eyeless fish were secondary. Americans cried for instruction, and Agassiz could not resist their pleas. He was living in the fullest enjoyment of his talent for imparting a love of nature to all who would listen, and their numbers seemed to grow every day. "Never did the future look brighter to me than now," he confessed to a Neuchâtel friend. "If I could for a moment forget that I have a scientific mission to fulfill, . . . I could easily make more than enough by lectures which would be admirably paid and are urged upon me, to put me completely at my ease hereafter."[20]

Agassiz accepted these lecture engagements, feeling he was justified in so doing because the fees would relieve financial obligations incurred at Neuchâtel. Boston thus had the first opportunity to hear more of Agassiz, and its appreciative audiences listened to lectures on glaciers, comparative anatomy, embryology, and variations of the theme that described so eloquently the plan of creation in the animal kingdom. Louisa Minot reported in regard to the lectures on glaciers that "they are quite unique, graphic, philosophical and instructive and I expect to have made, when they come to a close, a scientific and picturesque tour through this most grand and interesting part of Europe and to retain nearly as much as if I had actually been there in body, as I shall have been in spirit."[21]

Agassiz may have missed the museums and universities of Europe, but the consuming desire of Americans to learn about nature rewarded him in material ways that made the Old World seem distant indeed. In less than six months he earned, by the most conservative of estimates, the sum of nearly six thousand dollars from lectures in New England, money that helped defray the accumulated debts of European years and that provided funds for the support of assistants and artists expected soon in America.[22] The "unhappy books" of Neuchâtel days would be paid for by the worshipful audiences of America, drawn by the knowledge their author imparted so enjoyably.

"Boston is by no means dull," a young naturalist visiting the city in these days reported.

Agassiz took tea with us and amused us greatly. . . . He has barrelled up thousands of fish skeletons, which he boiled down at Albany and New York. All sorts of fish are welcome to his kettle. —Some one asked very earnestly the other day . . . where A. *kept shop.* Thought perhaps he had opened a *restaurant!* He has sent out standing orders to all the Michigan hunters for all they can send in. . . . He has a jolly disposition. . . . Says the fish are his prime collectors, sometimes finds two or three shellfish in one vertebral fish. . . .[23]

Asa Gray was also appreciative: ". . . his work-shop (looking like a wholesale butchery) goes on flourishingly. He has been of great use to science here, and we shall long feel the benefit of the impression he has made upon the minds of men here."[24]

Agassiz worked long hours of the day and night, intent on mastering the world of American natural history. Boston had indeed proved a congenial environment for such purposes, and he decided to make the city his American headquarters. Beginning in February, 1847, the Agassiz legions arrived to work under the command of their general. First Desor and Count François Pourtalès, a former Alpine assistant, arrived and were set to work looking for signs of glacial action. This initial contingent was reinforced by zoologist Charles Girard, artist Jacques Burkhardt, and lithographer Auguste Sonrel, all helpers from Neuchâtel days. These scientist-adventurers were soon comfortably established near the seashore in an East Boston house provided through the generosity of John Amory Lowell. The scientific factory was operating again in a new and more favorable climate.[25]

The spring of 1847 thus presented a delightful prospect to Agassiz, as he worked hard describing the marine life of the region, sent his minions out for more material, and welcomed turkeys, turtles, fossils, and beloved echinoderms to his quarters. He told Spencer F. Baird of his life:

. . . conceive of the position of a naturalist entirely devoted to his studies without any other object as this is and you will easily imagine how I have been carried away by the objects immediately around me. . . . The examination, anatomical as well as zoological, of every species of animal I could obtain from the market and from excursions on the beaches . . . brought me into such a state of excitement that I at last was taken sick so severely that I have not moved from my bed for the last three weeks.[26]

A trip to New Haven in May helped his convalescence. Here, amidst the friendly hospitality of Silliman, his son, and his son-in-law Dana, Agassiz regained his health and with it his energy and ambition. His fortieth birthday, celebrated in Boston, left him depressed because there was so much yet to do in life, and so little seemed to have been accomplished. He could not help reflecting, also, on the kind treatment received from the Dana and Silliman families, attention that he had not enjoyed even in his own family environment in Neuchâtel.[27] Apparently Americans were happy only when they could please him and help forward his purposes. An expressed wish to see Niagara Falls and look for signs of glacial action in the White Mountains resulted in an excursion during June to these regions, paid for by John Amory Lowell, who brought with him some Boston gentlemen and Harvard College students. These Americans enjoyed lectures in the field by the energetic and brilliant professor, who showed them fascinating signs in the moraines and on the scratched and polished rocks showing the ancient action of "God's great plough." The summer in Boston was by no means dull, for the East Boston house and laboratory served as an excellent location for finding and studying all kinds of fossil shells, marine life, and signs of glacial activity. On Sunday afternoons groups of curious Bostonians often took excursions to Agassiz's headquarters. Invited inside by a smiling and gracious host, these ladies and gentlemen heard Agassiz describe the anatomy of the turtle or the radial symmetry of the starfish in terms they could readily understand.

The federal government aided the purposes of the visitor too. Alexander Dallas Bache, superintendent of the United States Coast Survey, placed the U.S.S. "Bibb" at Agassiz's disposal for a cruise off the waters of Cape Cod and Nantucket during August. Lieutenant Charles Henry Davis, in command of the ship and a student of nature himself, reported to his chief that "Mr. Agassiz is a delightful companion. . . . His mind is devoted, with a considerable singleness of purpose, to his . . . science. . . . He labors assiduously, and appears to be perfectly happy."[28] The discovery of a new system of muscles in fishes, one new genus, and several new species, did indeed make Agassiz happy that summer.

Although the grant from Prussia was expended by September of 1847, Agassiz was not in the least perturbed. A lecture series in New York would take care of financial needs. And there was still much to do. He was planning a textbook on zoology in collaboration with Augustus A. Gould and seeking new European recruits who would make American explorations more productive.

Members of the Association of American Geologists and Naturalists heard of such plans from Agassiz at their September Boston convention, where he addressed them, detailing phases of his recent research. The assembled scientists asked him to help them transform their organization into one representing all phases of scientific study, and Agassiz, wise in the ways of European scholarly organization, was happy to be of service in founding the American Association for the Advancement of Science.[29] His new interests in comparative embryology, fossil botany, and American ichthyology quickened Agassiz's ever fertile imagination, and once more he delighted in having a series of projects before him. Furthermore, he had the sympathy and good fellowship of such admired Americans as Baird, Dana, Morton, Redfield, and Jeffries Wyman. Everything that had happened in this year had been good and productive and joyful.

The environment somehow always adjusted the forces of fate and progress to include Agassiz's purposes. Europe represented the past, America the future. Although he had written to a Paris colleague that "notwithstanding the gratifying welcome I have received here, I feel, after all, that nowhere can one work better than in our old Europe,"[30] he had tasted the fruits of a new and challenging world and could not bring himself to plan his return. Neuchâtel, with unpleasant recent memories, would be a poor replacement for worshipful Lowell Institute audiences. Paris or Berlin were potential environments for the future, but even these cities seemed less attractive by comparison with a land whose scientists looked on Agassiz with awe and adoration. Clearly, Humboldt was needed in this time of decision, but Humboldt was in Berlin, and Berlin seemed far away indeed. Agassiz was thrown on his own resources.

John Amory Lowell, Edward Everett, and a host of powerful Bostonians were determined not to lose the illustrious Agassiz, and Agassiz had decided not to leave his pleasant new home. The result of this mutual ambition was revealed in the November issue of the *American Journal of Science:* "Prof. Agassiz . . . has consented to remain in this country in connection with the scientific

corps of Harvard. . . . Every scientific man in America will be rejoiced to hear so unexpected a piece of good news."[31]

The forces that combined to unite Agassiz's future with Harvard, New England, and the United States stemmed in part from
the character of the national culture. In the 1840's, America was
more receptive to a proper appreciation of science than ever before.
The most obvious value of science, its utilitarian function in
improving the economic capacity of the nation, was a significant
factor in stimulating this awareness. Americans, moreover, were
impatient that their nation's contributions to theoretical knowledge rested on the reputation of only a few exceptional figures,
such as Benjamin Franklin. Eager to emulate the scientific achievements of Europe, they were delighted with the opportunity to
learn from such representatives of Old World culture as Agassiz
and Lyell. Consequently, Americans were beginning to realize the
need for increased social support for scientists and their profession,
a realization encouraged by the efforts of men such as Silliman
and Bache who recognized the necessity for high standards of scholarship, organization of knowledge, and institutional and private
financing of science. Agencies like the Smithsonian Institution,
recently established in Washington under the able direction of
Princeton physicist Joseph Henry, and the United Sates Coast
Survey under Bache's leadership were significant signs of a developing public awareness of the importance of scientific knowledge
for American culture.

Americans had been traditionally interested in the character
of their natural environment. During the early decades of the
nineteenth century, the necessity of exploring the continent and
discovering its special geographic, mineralogical, and zoological
features had prompted the great expeditions of Meriwether Lewis
and William Clark, Major Stephen H. Long, Lewis Cass, Henry
Rowe Schoolcraft, and John C. Frémont. These journeys had
yielded a tremendous amount of information detailing the natural
history of the new nation. Reports by foreign travelers, while designed for a curious Europe, also helped stimulate the interest of
Americans in their native landscape. The discovery of strange

fossil bones, the uses of herbs and plants as medicinal agents, or the existence of copper-bearing rock beds were all matters for public comment, speculation, and interest. On a smaller but more intensive scale, the numerous state geological surveys had yielded important data and encouraged a growing national interest in science. These investigations, moreover, had been responsible for the practical education of many first-rate geologists, among them James Hall, William Barton Rogers, and Lardner Vanuxem. On a world scale, the United States Exploring Expedition under the command of Lieutenant Charles Wilkes had surveyed the natural history of the entire Pacific Ocean area during the years 1838–42. The many valuable specimens collected by such expedition members as James Dwight Dana prompted public officials to promote the preservation, study, and description of important data, and professional naturalists like Asa Gray, John Torrey, John Cassin, Charles Pickering, and Dana were among the men who aided the government in this undertaking.

Concomitant with this interest in natural history and still more dramatic was the social impact of such inventions as the steam engine, the galvanic battery, the electric telegraph, and the rotary printing press. Efforts to harness the forces of nature gave rise to a popular conception of the "application of science to the useful arts" as one of the highest values of civilization. In the restless and changing America of the early nineteenth century, science and technology would, in the common view, soon effect a "progress in the ways of civilization . . . rapid and glorious beyond example in the history of the world."[32] Emerson might complain that Americans had faith in chemistry, steam engines, and turbines rather than a belief in divine causes, but this faith was a sign of the age in which they lived.

Giving shape and direction to the rapidly developing social prestige of science was a growing group of devoted professional men who earned their livelihood through work in the physical and natural sciences. These men immediately impressed Agassiz in his early tours of the country as far superior to their less professional colleagues of an earlier day. But Agassiz also saw significant deficiencies in scientific organization and procedure. Research laboratories were poorly equipped and few in number. There were scarcely any natural history collections of importance aside from those in Philadelphia, Boston, and Washington, and even these were deficient in comparative materials. Educational facilities to provide young men with professional knowledge in the sciences were also inadequate. Conservative university administrations, sensing a threat to the time-honored classical curriculum, had been reluctant to lend support to scientific instruction. Most American naturalists had received their educations in Europe, had been virtually self-educated, or had been formally trained as medical men.

Agassiz soon discovered that no region of the country promised as much for the improvement of science as New England. At Yale, Silliman and his colleagues had recently persuaded the administration to establish a separate school of science. In Boston and Cambridge, the location in one region of Harvard University, the American Academy of Arts and Sciences, and the Boston Society of Natural History served to provide important institutional encouragement for scientists and their craft. The pursuit of science had long been a distinguished and productive occupation at Harvard, as exemplified by the work of John Winthrop, Benjamin Waterhouse, Samuel Williams, and Aaron Dexter. In the 1840's, the university could boast of the presence on its faculty of such men as the physicist Joseph Lovering, the astronomer William Cranch Bond, the chemist John White Webster, the mathematician Benjamin Peirce, and the botanist Asa Gray. This science faculty would soon be distinguished by the appointment of Jeffries Wyman as Hersey professor of anatomy, the chemist Eben Norton Horsford as Rumford professor, and Oliver Wendell Holmes as Parkman professor of anatomy in the Medical School. The American Academy served as a common meeting place for most scientists in the area; Asa Gray and Jacob Bigelow were prominent in its activities. D. Humphreys Storer, Amos Binney, and Augustus A. Gould, all accomplished zoologists, were active in the affairs of the Boston Society of Natural History. The presence of such men and institutions gave this region a status in science approached only by Philadelphia.

Boston and Cambridge were also distinguished by an unusual body of laymen dedicated to the moral and material encouragement of scientific pursuits. At Harvard, President Everett, Treasurer Samuel Atkins Eliot, and Corporation members Charles Greeley Loring and John Amory Lowell were all active in the affairs of the American Academy, all firm believers in the importance of science in an advanced society. Boston friends of these men, such as Nathan Appleton, Francis Calley Gray, and Francis Cabot Lowell, shared their convictions and buttressed them by substantial support for scientific undertakings. Many local scientists also played a prominent role in the social and cultural life of Boston, thus strengthening a rapport between science and society that made this subject at once a fascinating and eminently respectable occupation.

One of the most important factors in the social encouragement of science was the economic transformation taking place in New England in these years. Finance capital was helping to change the Boston region from a mercantile and trading community to a powerful economic center thus supporting a nascent industrial society. The cotton textile industry and the railroads of New England and the Northeast were typical of the new enterprises bringing an unprecedented prosperity to the region.

Men of industry and finance were well aware that science could not prosper without institutional encouragement in terms of teaching and research. The impetus to establish a separate school of science at Harvard was a direct outgrowth of such an understanding. The Harvard Corporation had begun to plan for a school of this type in December of 1845. Early in 1846, Edward Everett mentioned the need for such an establishment in his inaugural address as president of the university, a need recognized and independently set forth by Benjamin Peirce in an organizational outline for a school of science drafted in February of that year and enlarged in June. By November of 1846, these proposals prompted the Harvard Corporation to appoint its members John Amory Lowell, James Walker, and President Everett as a special committee to formulate a definitive plan for this new branch of instruction. The result of their deliberations was announced on

February 13, 1847, when Harvard established a "School of Instruction in Theoretical and Practical Science."[33] However, only a hopeful framework was thus provided because no funds were available for buildings or for professorships beyond the existing science faculty of the college.[34]

President Edward Everett worked diligently to gain funds for the new school from the friends of Harvard. In June, 1847, his efforts were rewarded when Abbott Lawrence, cotton manufacturer and business associate of John Amory Lowell, offered the university a donation of fifty thousand dollars for the school of science. Like Lowell, Lawrence appreciated the value of scientific activity. His role in the new industrial economy of New England made him particularly sensitive to the importance of scientific education to material progress. Moreover, a Paris-educated engineer, Charles Storrow, was a close friend of Lawrence and helped convince him of the need for an American institution comparable to European schools of applied science.[35]

Lawrence's practical orientation was in keeping with the spirit of the age. In offering his bequest he wrote Harvard's treasurer, Samuel Atkins Eliot, that New England had "stronger motives . . . than . . . any other part of the country to encourage scientific pursuits, from the fact that we must hereafter look for our main support to the pursuit of commerce, manufactures, and the mechanic arts." "Hard hands" were ready to work upon "hard materials"; the need was for "sagacious heads to direct those hands."[36] Harvard now named the new branch of the university the Lawrence Scientific School. It also accepted Lawrence's proposed organizational scheme for the new establishment as laid down in his letter of donation. Lawrence emphasized the need for a school where young men might learn the sciences of engineering, mining, mechanical drawing, and the methods of constructing and working with machinery. He proposed that students learn the essentials of pure science through courses in chemistry, mathematics, and physics and that they study geology, natural history, and mineralogy. Only two new appointments would be required, with the distinguished faculty already at hand, and these were to be in the fields of engineering and geology. These new professors would be paid at least fifteen hundred dollars annually, and with student fees for instruction it was quite likely that their salaries would be as great as three thousand dollars. Lawrence, however, would guarantee only the smaller figure. More than half the donation was to go to construct buildings and equip laboratories. As for students, it was believed that they would come in sufficient numbers. Seniors, students over eighteen years of age, students attending other professional schools in the university, and college graduates could all be admitted to the school.[37]

By early June of 1847, therefore, a professorship of geology had been created at Harvard through the generosity of a Boston manufacturer, acting from his understanding of the economic tendencies of the age. This benefactor also made possible the realization of an equally popular public sentiment when he acted to insure Agassiz's continued presence in the United States by working to bring about the appointment of the Swiss naturalist to the faculty of the new school of science. It did not matter that Lawrence had stressed the educational importance of geology in its practical relations to mining, surveying, and mineralogy. Nor was it worthy of note that men who had had long experience with state geological surveys, such as the brothers William B. and Henry D. Rogers or James Hall, were all logical candidates for the professorship. Similarly, it was soon forgotten that the original plan for the scientific school had noted that instruction in natural history could be offered by present members of the Harvard College faculty. Agassiz had made too powerful an impression upon the minds of Lawrence, Lowell, and Everett for the opportunity presented by the available professorship to be ignored.

In sponsoring Agassiz for the position, Lawrence effectively transformed the character of the original scheme he had proposed for the scientific school. This was for the moment of no importance, since it was possible to envision a school of "practical" science and at the same time provide an American career for Agassiz. Agassiz was interested in applied science only to the degree that it might have a practical effect upon his future. His experience as a geologist had been limited to glacial studies, and these hardly filled the utilitarian requirements set forth in the original bequest.

Nevertheless, it is likely that Lawrence suggested to Agassiz the possibility of the professorship even before he made his bequest in June.

Many years later, Agassiz's version of these events was that he assented to the proposal because "I felt my affirmative answer would be an additional inducement for him to make his endowment."[38] Lawrence had inducement enough, but in the specific matter of Agassiz he had acted not only from his own conviction but under the stimulus of Lowell, who had urged the importance of such an appointment.[39] Lowell quite probably used the opportunity of his June excursion to the White Mountains with Agassiz to emphasize the benefits to be derived from an association with Harvard. Agassiz was concerned only that the title of the professorship did not reflect the actual range and extent of his scientific interests. Men of Harvard and Boston sought to remedy this situation. On July 19, Horsford reported to his friend Hall that "Agassiz says nothing, but it is generally expected he will take the professorship . . . [and] if he does not I am inclined to think Mr. Lawrence will so modify his plans to include him."[40] On the same day, Lawrence wrote another formal expression of his plans for the donation, a document he later affirmed to be the binding agreement on his part to establish the school of science. The only change from the earlier bequest of June was a more specific statement regarding the source of professorial salaries and a paragraph that could only have been written with Agassiz in mind; here Lawrence stated that the only rights and privileges he wished to reserve for the future related to "extending the School, by combining some other branches of science whenever I may deem it expedient."[41]

The way was now clear for Agassiz and Harvard to come to an agreement. President Everett wrote him late in July expressing the official interest of the university in his appointment, and Agassiz was thus prompted to write the Prussian envoy to Washington for permission to remain in the United States.[42] Returning to Cambridge early in September after spending August cruising on the "Bibb," Agassiz found that Lawrence and Everett were in agreement that he should have the professorship and that its scope should be his to define. Everett wrote him on September 13:

We shall in a few days have the satisfaction of definitively electing you to a professorship in our Scientific School. I will thank you to let me have your wishes as to the precise title. We have supposed that you might prefer simply "Geology," but that it would be better to make the designation a little more extensive. "Professor of Geology and Zoology" has occurred to us. . . . We might say "Professor of Natural History," but that happens to be precisely the title of Mr. Gray's department. . . .[43]

On September 25, the title Agassiz actually preferred was recorded in the proceedings of the Harvard Corporation, records that revealed only one name ever put forward in regard to the "geology" professorship. "Voted: That this Board do now proceed to the Election of a Professor of Zoology and Geology in the Lawrence Scientific School. . . . Whereupon ballots being given in it appeared that Professor Louis Agassiz . . . was chosen."[44] Horsford, while still not entirely informed, understood enough to bring Hall up to date: "Agassiz has formally accepted the appointment as Professor of Geology and Zoology (sic) in the . . . School. . . . It is as I said to you—obvious that the Professorship was created for him."[45]

The Lawrence gift had thus served a really practical goal by tightening the bond already existing between Agassiz and scholars and businessmen of Boston and Cambridge. Upon receiving official confirmation from Everett of his appointment—which contained no reference to its length of time but which Agassiz understood to be for a three-year period—he waited a short interval and, still having had no word from the Prussian official, accepted the Harvard position on October 3, the first anniversary of his arrival in the United States.[46]

Agassiz was merely acting on decorum in asking the Prussian government for permission to stay, since he had already exhausted its bequest, and if he remained in America, he would be dependent upon his own resources. The essential decision, then, was whether he wished to remain in the United States. Every intellectual and social drive in Agassiz's personality compelled him to favor America. The Harvard appointment, which made him the only European scientist on the faculty, opened a new and challenging opportunity, not only in New England but in the entire nation. All his experience so far in the New World had been one vast and immensely enjoyable adventure, acted on a stage where he was the sole performer, called before the audience again and again for more demonstrations of his talents. The realities he had encountered in America had exceeded his idealistic anticipations. It was not surprising that twice during the past year Agassiz had experienced actual physical illness, literally working himself into a state of nervous exhaustion from the excitement of the moment. The riches of scientific study stretched before his imagination in profuse splendor, and this promise was heightened by the public acclaim that followed him everywhere. The pleasant society and good fellowship of this wider world of men and affairs was congenial to his warm, friendly spirit. Here in America, Agassiz was approaching the public renown that Humboldt knew in Europe, in an environment that honored him as the world honored his mentor. Always responsive to acclaim and to the entertaining company of influential laymen, Agassiz began to think of himself exactly as his society evaluated him, a symbol of Continental learning and culture. He could begin to enjoy what the intense intellectual drive of his European years had made possible.

The only cloud on this new horizon was the thought of Cécile. The attractive and devoted woman of his German youth was now an invalid, suffering from tuberculosis and living in Freiburg with the three Agassiz children. If Cécile had been homesick for Germany in Switzerland, it was not likely that she would have made the trip to America, even if her illness had not ruled out this possibility. Moreover, acting against the advice of his mother, Agassiz had brought to America the man Cécile could not tolerate in her household—Edward Desor.[47] By so doing, Agassiz had removed himself a step further from a new domestic life with Cécile in America. The only possibility, then, was a return to Europe after the three-year Harvard position had run its course. This could be achieved only by an appointment at Paris or Berlin, as Switzerland would be too small a world after America. Moreover, the political events of February, 1848, nullified a return to Neuchâtel. The wave of revolution that swept over the Continent at this time carried Neuchâtel with it, and the ties of the principality to the Prussian monarchy were broken and, without support, the academy was disbanded. As for a post in France or Germany, Agassiz made no effort to attract such an offer, and in fact behaved as if he considered America his permanent home.[48] Agassiz's thoughts turned inevitably to the future of his children. He might be able to bring them to America, if circumstances permitted and if he realized the success he anticipated.

Having virtually decided to remain in America, Agassiz determined to bring more of Europe to him. He sent for more assistants and for the indispensable tool of his profession—his library. One of the first of the familiar faces to arrive in East Boston was the old family friend and adviser, "Papa" Christinat. The early months of 1848 witnessed the arrival of a host of foreign naturalists and assistants, some called directly by Agassiz, others seeking his hospitality as a result of the political turmoil in Europe. Among them were Arnold Guyot, Leo Lesquereux, and Jules Marcou. All were welcomed, all were put to work in the new scientific factory.

There was much for its director to do before he began his Harvard teaching in the spring of 1848. Another grand tour to educate America in science commenced in the fall of 1847. The favor of the public was important because it meant more money to supply the wants of assistants and to advance scientific pursuits. Every city Agassiz visited vied with its neighbors in the reception it gave him. New Yorkers, having waited nearly a year to hear Agassiz, finally had their opportunity in October and November. Editor Horace Greeley announced the coming event in the *Tribune:* "Never have our citizens enjoyed the opportunity of acquiring so large a measure of knowledge of the laws of nature . . . as these lectures will afford them."[49] Stenographic accounts of the lectures were reprinted in each day's *Tribune* and in response to popular demand were then republished as a pamphlet, with appropriate illustrations.[50]

The introductory remarks of the reporter who recorded the lectures recaptured some of the mood of the audience who heard Agassiz at the hall of the College of Physicians and Surgeons: "Professor Agassiz is personally a man of very striking and prepossessing appearance. He is tall, and formed with as much strength as elegance, with a rather florid complexion and dark hair. In his manner and bearing there is a singular grace and benignity." New Yorkers who could not hear the lectures could read them faithfully reported:

I must first apologize for the deficiency of my language. Happily . . . Natural History has an interest entirely apart from the form in which the subject is presented. . . . I shall do all in my power to make up for the deficiency of . . . language, by the interest directly derived from the subject itself—[Applause].[51]

The lectures were in fact the same addresses New Englanders had heard the year before, and thus the marvelous "plan of creation" of the animal kingdom was extolled once again, with popular presentations of complicated scientific material flowing easily from Agassiz's gifted tongue. Reward for this sort of instruction could hardly be measured in monetary terms, but Gothamites and their near neighbors paid the handsome professor nearly $1,400. Students and physicians at the medical college gave him an extra purse of $250 in silver as a special token of appreciation.[52]

Southerners were not to be outdone. Charleston's inhabitants turned out in great numbers for the Agassiz addresses, and they, too, gave him a gift in addition to the regular fee. Agassiz, responding to such a welcome, heartily enjoyed his first stay in the South, especially the hospitality of the zoologist John Edward Holbrook and his wife, who welcomed him to their plantation. Agassiz was more than politely curious about the character of plantation society; he walked through the fields, watching the slaves at work and observing them carefully. He was perplexed. Were these creatures really men? Or were they another species of man, separate from the white? Such thoughts had been much on his mind, ever since he had seen Negroes for the first time in Philadelphia the year before. "I cannot get rid of the idea that the state of things present in the southern states of the Union might one day be the cause of the ruin of the United States," Agassiz had then observed to his mother.[53] He was, in the future, to return again and again to the

theme of the true nature of the Negro and the character of his relations with the white.

Agassiz was as much at home on Holbrook's veranda as he had
been in the drawing room of John Amory Lowell. He basked in
the admiration of local naturalists such as Lewis R. and Robert W.
Gibbes, Henry W. Ravenel, and Francis S. Holmes. These southerners were no less and possibly even more worshipful of the visiting professor than their northern colleagues had been. Agassiz
treated them as his intellectual equals, examined their collections
of fossils and marine animals with complete absorption, requested
duplicate specimens of as many types as they could spare, praised
the research they had undertaken, and complimented the appearance of the local museum, urging that it be greatly enlarged. All
this earned him the undying admiration and respect of this southern community.[54]

Returning to Boston in January of 1848, the new Harvard professor found that his appointment had been confirmed by the
Overseers of the university and that the city awaited his second
course of Lowell lectures, on a favorite subject, the science of ichthyology. "Agassiz begins a new course of lectures on Fishes next
Tuesday," Longfellow recorded in his journal. "He told me that
he has *une peur terrible* of beginning."[55] There was no cause for
alarm. "I have just come home from Prof. Agassiz's lecture," a
Mr. C. wrote in a letter to the editor of a Boston newspaper. "How
much does the student owe to him who not only teaches the present . . . but carries him forward into all related philosophy and
positive truth? . . . And who was not more than sorry when the
master . . . said 'the hour for this lecture is completed'?"[56]

With such appreciation, Agassiz had no reason for concern regarding popular approval of his efforts to translate specialized
knowledge into terms understandable to all. New projects began
to take shape in his mind, reflecting an awareness of his growing
educational and professional influence. He would organize an
"American Geological Museum" at Harvard. The basis of this museum would be the collection of fossils gathered by James Hall of
Albany. When Agassiz had inspected this collection personally, he
was so obsessed with the desire to have its treasures that he ap-

proached university officials with proposals that they purchase the materials or, if this could not be done, that they bring Hall to Harvard as a professor. Hall would bring his collection with him, and the end would be achieved. This specific project never materialized, but Agassiz never lost his fascination with Hall's materials, or his desire for a university museum.[57]

The museum became all the more necessary for Agassiz as his scientific goals took form and direction early in 1848. He planned to collaborate with those naturalists whose work most appealed to his particular interests. He would write a joint monograph with William C. Redfield of New York, describing all fossil fishes indigenous to the United States. He would do another study with Spencer F. Baird, depicting the natural history of living North American fishes.[58] With the prospect of these contributions to knowledge, he knew he needed official university support in the form of a building to store materials necessary for study and examination and to house the great variety of specimens gathered during his first years in the United States. He sent an appeal to President Everett requesting authority to ask friends in Europe for an exchange of specimens. Moreover, Americans had already made inquiries of him regarding the plans of the university for a museum. It was imperative that the university should act to create the institution that some of Agassiz's friends thought already existed. The Lawrence donation and the organizational scheme for the scientific school contained no provision for a museum, but Agassiz, by speaking of a museum as if it were a reality, effectively created such an institution in the minds of university administrators.[59] As a result, Harvard, acting through treasurer Eliot, purchased an unused bathhouse for four hundred dollars, located at the intersection of the Brighton road and the Charles River, to serve as a storehouse for the growing Agassiz collections. It was not the kind of depository Agassiz had envisioned, but the specimens it contained and the expenditures of the university for their care and increase would both grow with the years.[60]

The bathhouse signified Agassiz's scientific aspirations in Cambridge, but late in April residents of the community knew his social presence as well. The necessity of lecturing twice a week at the university led Agassiz to rent a house on Oxford Street near the Harvard Yard. He thus became the neighbor of Longfellow, Everett, Peirce, Felton, and Davis, men who delighted in the company of a new colleague who could talk of German philosophy and echinoderms with equal authority. These men of Cambridge marveled at the energy that transformed the Oxford Street house, small though it was, into still another scientific factory. The group of Europeans who followed Agassiz like faithful marchers in the cause of science now numbered twenty-two, some living in East Boston—where Agassiz spent as much time as he could spare—some in Cambridge, all busy at tasks designed to give their leader comprehensive knowledge of nature in America. Christinat supervised the domestic affairs of both establishments, assigning work, ruling with a firm hand, and haunting the Boston and Cambridge markets for interesting fish and small game, animals that were first studied and then eaten.

The Oxford Street house became a combination museum, aquarium, botanic garden, and breeding-experiment station; its activities were intensified after September, when the East Boston house was given up and the entire Swiss band quartered in Cambridge. No object, alive or dead, was turned away. The Agassiz zoological garden and its keepers welcomed a tame bear from Maine, snapping turtles sent by an admiring Henry David Thoreau, a collection of snakes from Florida, a group of fishes from Lake Superior. The Harvard naturalist took all that came his way, making use of the dissecting knife and the pickling barrel, or stocking his table with pheasants that had to be eaten to be truly appreciated. Sonrel was soon working in his own lithographic establishment in Boston, his task to reproduce the plates drawn by the artist Burkhardt. Guyot and Lesquereux were of great service in aiding original researches, while the younger scientists were kept busy sorting, classifying, and identifying specimens. Never, it seemed, had New England witnessed such inspired efforts in the cause of natural history.

Agassiz was delighted with all this activity. Even the destruction caused by the Maine bear, who broke into some wine casks and proceeded to reel drunkenly about the house, did not dampen his enthusiasm. He wrote his friend Baird, "All my difficulties arise from having too much on hand. . . . I have before me one hundred species of invertebrated animals spawning and developing. . . . There is work for ten heads and fifteen hands here."[61] Cambridge citizens might have been amused by the antics of these foreigners leading bears by a string and charming snakes, but they also recognized the higher purposes of the energetic student of nature. These purposes were demonstrated graphically when, in the spring of 1848, Agassiz began his first teaching at Harvard.

The small class of special students and college undergraduates enrolled in Agassiz's course benefited from some distinctive educational innovations. Refusing to restrict himself to the time-honored techniques of lecture and recitation, Agassiz enlivened his instruction by encouraging students to visit his laboratory, study specimens by themselves, and acquire direct knowledge of nature by excursions through the countryside and to the seashore. Just as at Neuchâtel, older enthusiasts were not denied Agassiz's instruction. Felton and other Cambridge intellectuals were taken for field trips to points of interest. One of these was recorded by a hack driver:

I drove the queerest lot you ever saw. They chattered like monkeys. They wouldn't keep still. They jumped the fences, tore about the fields, and came back with their hats covered with bugs. I asked their keeper what ailed them; he said they were *naturals*, and judgin' from the way they acted I should say they were.[62]

George Ticknor, a perceptive observer of the Cambridge social scene, characterized the beneficent effects of such activity on the "keeper": "Agassiz continues to flourish. . . . His *bonhomie* seems inexhaustible."[63]

Agassiz had clearly proved to be a much more attractive personality as a scientist than residents of Cambridge and Boston had ever imagined such a scholar could be. Sitting comfortably in Ticknor's parlor, a glass of sherry in his hand, puffing contentedly at a cigar, Agassiz would explain with equal facility the causes of European revolutions or the importance of science in an advanced civilization. In company with Longfellow and Charles Sumner, he would attend a pleasant dinner at Felton's home, and after the meal, scientist, classicist, poet, and politician would very often visit the home of Julia Ward Howe, where more lively conversation on the affairs of the day would take place.

In June of 1848, Agassiz's first American publication appeared. This was the textbook *Principles of Zoology* written with Augustus A. Gould. Gould had been at work on this volume before Agassiz's arrival, but the philosophy of nature set forth in the final product was unmistakably that identified with Agassiz. The textbook, meeting a long-standing need in American secondary and college education, tried to present the most recent information concerning anatomy, paleontology, geology, and embryology and therefore bore testimony to the universality of Agassiz's interests in his European career. It also stood squarely behind the concept of the fixity of species, catastrophes, and the successive creation of flora and fauna by the Deity, all accomplished according to a rigid, Aristotelian plan of creation. For graphic illustration there was a four-color chart as a frontispiece, showing how radiates, mollusks, articulates, and vertebrates had been mutually exclusive zoological categories from the beginning of time, each division identified with particular geological formations.[64] This volume was long one of the most popular of zoological treatises, appearing in sixteen editions during Agassiz's lifetime and being revised and edited by others after him.[65]

The social and educational triumphs of Cambridge were small in comparison with the scientific activities of the summer of 1848. For the first time in his life, Agassiz was able to lead an exploring expedition. The destination was the northern shore of Lake Superior. The purpose was to gather as many zoological specimens as were interesting and important. The method was singular, and characteristic of Agassiz in America. In his view, intellectual activity could be instructive not only to professionals but to all classes of Americans who appreciated nature study. Thus the party of sixteen that left Boston early in June represented a cross-section of Agassiz's American public. Nine Harvard students accompanied their professor. Some were not very robust, but their fathers were assured that the strenuous life of a journey into the interior would make men of them. Two European naturalists, fleeing the turmoil of revolution at home, also came along to serve as instructors and collectors, certain they would find solace in the wilderness of America in the company of their respected colleague. Two New York doctors, eager to learn zoology under Agassiz's tutelage, were also of the party, joined by two admiring and cultivated Bostonians, one of whom would act as chronicler of the expedition. A portable blackboard was an important item of equipment, since at each landfall Agassiz gave impromptu lectures describing the particular geological and zoological features of the environment. The party traveled in three large birch-bark canoes and a Mackinaw boat, "a cross between a dory and a mud-scow," that could be fitted with sails. Accompanied by twelve French Canadians, half-breeds, and Ojibwa Indians, all of whom served as guides, the *bourgeois* and their *voyageurs* were objects of considerable interest to traders and Indians they met along the way. The canoes each carried six travelers, and the boat provided room for ten people and held the collecting apparatus and accumulated specimens. The canoe carrying Agassiz led the way, a large frying-pan lashed to the prow as a figurehead.

There were other traveling scientists abroad in the Lake Superior region this summer. One of them, the geologist Josiah Dwight Whitney, recorded a meeting with the New England band:

Professor Louis Agassiz has a party in the region which is now somewhere on the northern shore of the lake, if not at the bottom of it. They are . . . on a tour of scientific pleasure, and we arrived at the Sault just in time to see them off. We were much amused by their evident *verdancy* in regard to life in the woods. Nobody was captain among the . . . men . . . and when anything had to be done, the only way was to put it to a vote, a precious situation to be in, if overtaken by a squall. . . . I should love to see them in camp and watch their proceedings.[66]

Camp proceedings were eminently instructive, and no mishaps occurred, though certain enthusiasms led to perplexity on the part of friendly Ojibwa Indians.

The professor got a number of fishes. . . . A sturgeon was caught in the river opposite our tent, in a net belonging to one of the Indians, who dispatched him . . . with a fish-spear. Prof Agassiz requested . . . a sketch of this fish, which was some four or five feet long. This took some time, and meanwhile we observed that all the inhabitants of the lodge to which it belonged were assembled and crouching in a row in front of us. We supposed this to be mere curiosity, but one of our men . . . discovered that the whole family had been without food all day, and were waiting to eat the fish as soon as we were done with it. . . . The sketch not being finished, we proposed to them to lunch meanwhile on some of our pork and biscuit, to which they readily agreed.[67]

The journey was truly an amazing experience for Agassiz. His beloved fresh-water fishes appeared in great and fascinating array; gar pike, trout, sturgeon, all were collected and packed for shipment back to Cambridge. It would be instructive to compare North American fauna and flora with the European animals and plants he already knew. As the party skirted the northern lake shore and visited islands and the headwaters of rivers, Agassiz felt more strongly than ever before that a great variety of natural riches awaited the application of his talents. As was the case in his earlier excursion to northern New England and Niagara Falls, he found clear evidence of past glacial action in land areas of New York State and in the shore lines of Lake Erie and Lake Superior, evidence that he would discover time and time again in travels through the northern regions of the country. Students and laymen returned from the journey well versed in the natural lore of the northern lake country and appreciative of the professor for providing them with such an opportunity. Agassiz's reward consisted of eight great casks of fishes, innumerable plants, animals, rocks, minerals, and a store of geological and paleontological information.[68]

Agassiz had at last undertaken a great exploration, not in the company of a Humboldt, but leading a party of his own, with assistants to help his work. Two years later, all the significant results of this journey were published in the volume *Lake Superior*. This was not, however, the kind of book Agassiz might have published as a European. A pleasant and interesting travel account by J. Elliot Cabot accompanied the text; the work had a considerable sale both as a popular book of travel and nature description and as a scientific account. It was well received professionally, particularly by zoologists such as Baird who were interested in fishes and naturalists such as Dana who were interested in geographical distribution and land formation.[69] From England, Charles Darwin wrote a note of thanks to the man only two years his senior:

I have seldom been more deeply gratified than by receiving your most kind present of "Lake Superior." I had heard of it . . . but I confess it was the very great honor of having in my possession a work with your autograph, as a presentation copy, that has given me such lively pleasure. . . . I have begun to read it with uncommon interest, which I see will increase as I go on.[70]

What could not fail to interest Darwin were Agassiz's generalizations regarding the comparative fauna and flora of the Old World and the New and the meaning of their respective geographical distribution.

The purpose of Agassiz's study, whether it related to comparisons of fishes or of plants, was to illustrate one all-embracing principle, that "the geographical distribution of organized beings displays more fully the direct intervention of a Supreme Intelligence in the plan of Creation, than any other adaptation in the physical world."[71] To Agassiz, there was no stronger evidence for "thoughtful adaptation" than "the various combinations of similar, though specifically different assemblages of animals and plants repeated all over the world, under the most uniform and diversified conditions."[72] When he observed striking identities between the vegetation of the northern shores of Lake Superior and the subalpine flora of Europe, or the arctic flora and fauna of both continents, he explained such phenomena by reference to the "laws of botanic geography," the agency of man, or the "uniformity" of the creative plan, with the reservation that intensive specific comparisons remained to be made before a complete identity could be admitted.[73]

Much more at home in ichthyology, he found ample reason to discover striking differences in fresh-water fishes from both regions. Rather than ascribe such distinctions to secondary causes operating to produce variation from a common pair or single center of creation, or admit a "causal connection" between obviously analogous species, Agassiz read the book of nature as he felt the Creator intended it to be interpreted. "The more intimately we trace . . . geographical distribution, the more we are impressed with the conviction that it must be primitive . . . that animals must have originated where they live, and have remained almost precisely within the same limits since they were created."[74] The fundamental justification for separate creation was that "these limits were assigned . . . from the beginning, and . . . the order which prevails through the creation is intentional. . . . It is regulated by the limits marked out on the first day of creation. . . ."[75]

The concept that "physical agents" could act to produce variation and divergence in species was an idea that signified to Agassiz a chaotic world, a world that could never produce change or "call into existence anything that did not exist before."[76] The fact that he observed differences in the comparative ichthyology of America and Europe was proof to him of the ultimate power and authority of mind, intelligence, and thoughtful planning. Moreover, the substitution of intellectual for physical agencies allowed a denial of any relationship between organic forms. The absence of relationship not only reinforced the timeless permanence of the plan; in more pragmatic terms it allowed for the discovery, by Agassiz, of "new" species whenever he observed nature.

American scientists heard these views before they appeared in print. In September, soon after Agassiz returned from Lake Superior, the American Association for the Advancement of Science held its first annual meeting in Philadelphia. The assembled scholars listened to Agassiz deliver twelve separate addresses describing his findings in geology, ichthyology, and botany. Of particular interest were Agassiz's informed analyses of the evidence for glaciation in North America and his excellent descriptions of fishes discovered in Lake Superior.[77]

Agassiz was an important figure at this conclave, but he could not enjoy his achievement. In August, 1848, just after returning from Lake Superior, he learned that his wife Cécile had died in Freiburg a few weeks before, succumbing to pulmonary tuberculosis.[78] Intellectual triumph and social acceptance were small compensations for the grievous sense of personal loss he now knew. Edward Everett understood Agassiz's emotions when the naturalist wrote thanking him for the award of a doctor of laws degree conferred by Harvard at its recent commencement. "There are times in life when such marks of esteem are particularly welcome and contribute to keep up the spirit of devotion which is the soul of science," were words of a man made wretched by sorrow.[79] Even the tremendous success he enjoyed from delivering a series of highly informed lectures on comparative embryology to Lowell Institute audiences in the winter of 1848–49 did not temper Agassiz's unhappiness.[80]

Overwhelmed with feelings of guilt and remorse regarding his life with Cécile, Agassiz could not forget the past because it intruded into the present in a way that could not be ignored. Coming back to Cambridge after the Philadelphia meetings, Agassiz discovered that Edward Desor had been telling all who would listen that his employer had consistently neglected the wants of his family, thus contributing to Cécile's failing health and ultimate death.[81] It was as if the warnings of his wife had been verified by Agassiz's experiences after her death. Friends close to Agassiz both in Switzerland and Boston knew that his domestic life had been unhappy, but such tales as Desor made public were malicious, bitter efforts aimed at tarnishing Agassiz's standing in the community. They revealed, as Cécile, Rose Agassiz, and even Karl Vogt had long ago affirmed, that Desor was entirely disloyal.

Desor had arrived in the United States from Paris in February, 1847. His task in Paris had been to superintend the publication of a manuscript on echinoderms Agassiz had written and to take charge of his employer's financial affairs in Europe. In America, Desor envisioned himself as a scientist just as competent as Agassiz. He had had considerable experience working with Agassiz in glacial research, and although some Americans, like Silliman, properly characterized him as a "scientific amateur," others, like James Hall, Josiah Dwight Whitney, Charles Henry Davis, and Theodore Parker, were impressed with his talent. Desor, aware of Agassiz's success, decided to strike out on his own and win equal recognition as a great European naturalist come to explore America. He was younger than Agassiz and seemed as romantic and dashing to many Bostonians as his employer.[82] He made powerful friends in the community, among them Edward Clarke Cabot, Samuel Cabot, D. Humphreys Storer, Davis, and Parker. During Agassiz's frequent travels, Desor tried to assume command of the East Boston establishment, and he seemed to many the personification of a dedicated scientist.

Agassiz knew different. Desor's allegations concerning his former domestic life were only the last in a series of pernicious actions that Desor had been guilty of in Europe and America. By April, 1848, Agassiz had at last become fully aware of Desor's true nature and had broken with him completely. Such realization should have come much earlier, but Agassiz had persisted in disbelieving what others had urged about Desor until the behavior of his assistant made such disregard impossible. Agassiz's first real understanding of Desor's character came in the fall of 1846 when his mother wrote him that she had received vicious letters from Desor attacking Agassiz's character and complaining of his treatment of Cécile.[83] Considering Cécile's opinion of Desor, such sentiments were indicative only of Desor's irrationality. Next, Agassiz learned that Desor had been dishonest in financial matters, appropriating for his own use money Agassiz had sent to Paris to be applied in settling European debts. In the spring of 1848 Agassiz saw a copy of the published echinoderm monograph for the first time and was utterly astounded to see that Desor's name appeared on the title page as co-author. At the same time Agassiz found that Desor had published under his own name research in the embryology of starfishes Agassiz had personally undertaken in the winter of 1847–48.[84]

By this time Desor's presence had become intolerable to the other Europeans living in the East Boston household, especially "Papa" Christinat and Charles Girard. Desor had arranged for a cousin, Maurice, to join the establishment, and this person, described by Christinat as a "lazy parasite" with a voracious appetite, lived in ease and luxury at Agassiz's expense. Acting on the complaints of Christinat and others, Agassiz determined to put an end to Desor's fantastic assumption of power by using the Maurice issue as a symbol of his distaste for Desor's recent behavior. He insisted that Desor see to it that Maurice leave the house.

Desor refused to do it. Moreover, he threatened Agassiz with public scandal if Agassiz did not withdraw his request, affirming that he would inform important Bostonians about "acts of levity" and improper moral behavior on Agassiz's part. Agassiz was forced to defend his personal and intellectual honor. Having moved to

Cambridge in the meantime, he wrote Desor a bitter letter insisting that this last act and those that had preceded it made it impossible for any further personal relations to exist between them.[85]

Agassiz now suffered needlessly for his failure to rid himself of Desor's influence long before. True to his promise, Desor acted to blacken Agassiz's reputation in Boston. He informed Edward Clarke Cabot that Agassiz had enjoyed an "improper connection" with Jane, a servant in the East Boston home. The story soon spread throughout the community. Professor Agassiz, some said, had studied nature more subjectively than anyone realized. Bostonians were too intrigued by the few details they heard not to wish for more information. They were soon satisfied. Agassiz, it seemed, was not only immoral, he was also dishonest and a plagiarist. By June of 1848 Desor had added new charges to his complaints against Agassiz. The public now learned that Agassiz had appropriated to himself the results of Desor's researches in embryology, had borrowed money from him that was never repaid, and had never compensated him for editorial labors performed in connection with *Principles of Zoology.*

Agassiz now left Boston for Lake Superior. When he returned he found that Desor had been busy spreading stories about Agassiz's unhappy home life with Cécile, alleging that this was the reason for the illness that caused her death. Agassiz realized that he had to act immediately to counter the poison of Desor's charges; he had to vindicate himself in the public mind if he wished to enjoy any authority and dignity in Boston and Cambridge. But he was at a loss to know how to proceed against the man. A lawsuit would have been too public a matter and would have served Desor's purposes by providing a forum for his attacks. Close friends convinced Agassiz that there was a gentlemanly way to settle the matter. The proper procedure would be to convene a private court of honor to evaluate the complaints of each man in one specific regard, that of the disputed embryological research. In this way, a decision in Agassiz's favor would convince everyone whose opinion counted that Desor's assertions were all false. In Boston, men could adjudicate grave disputes equitably without airing their grievances for the delight of a curious public. Agassiz agreed to this procedure. That the affair was a local *cause célèbre* is apparent from a letter Charles Henry Davis wrote Alexander Dallas Bache about recent matters of interest in Boston:

I am not yet able to tell you about the quarrel between Agassiz and Desor—partly because I am not sufficiently well informed. . . . The subject is to be referred to friends. These quarrels, are, as you intimate, truly wonderful. I could like Jacques "moralize" this spectacle for a month to come. " 'Tis passing strange and wondrous pitiful." But as it takes all sorts of men to make a world, so it takes all sorts of dispositions to make a man. Of one thing I feel certain, that in this quarrel . . . [Agassiz] must be right; this is an opinion fire would not melt out of me. . . . The sincere admiration inspired by his character carries me with him without inquiry.[86]

Augustus A. Gould and Edward Clarke Cabot, serving as the private tribunal, found Davis' evaluation substantiated by objective examination. In October, they decided that Agassiz's claims were entirely justified and awarded him title to the embryological research Desor had asserted was his property, material that existed in the form of published and unpublished descriptions and drawings. Desor was given rights to data Agassiz had already acknowledged to be the property of his secretary.[87] Desor was furious at these findings. He refused to accept them, since in his opinion one of the arbitrators, Gould, was incompetent to judge. Late in October, Desor claimed authorship of all the disputed embryology papers and also demanded financial compensation for his work on *Principles of Zoology*. Cornelius Conway Felton and Edward Clarke Cabot were asked to decide this larger question. They confirmed the original judgment and awarded Desor one hundred dollars in compensation for his work on the *Principles*.[88] Desor would not rest. Early in December he made still another public accusation, claiming that he was the real author of *Principles of Zoology*, not Agassiz and Gould. He renewed his slurs on Agassiz's moral character, claimed that his former employer had borrowed money and not repaid it, and had been guilty of damaging Desor's reputation in public!

Bostonians now began to wonder if there might not be some fundamental truth in Desor's charges, since he urged them so vigorously and never seemed at a loss for new complaints against Agassiz. Davis, for example, wrote Desor:

I am very sorry that Mr. Agassiz takes the course that he does. In time to come he will exclaim . . . that he is "the worse for his friends." . . . You have so many and such ardent and faithful friends . . . that you cannot but be very happy, if you will only guard your mind against the intrusions of undue anger.[89]

Agassiz realized that vindication on a more public scale was now vital. He had been willing to pay Desor one hundred dollars and forget the entire sorry business, but when Desor claimed authorship of the *Principles,* Felton, Gould, and other friends advised that no settlement be made until this charge was proved false. The result was the convening of a third panel of arbitrators. Each man picked his own representative, Agassiz naturally choosing his friend John Amory Lowell, Desor asking D. Humphreys Storer to act for him. Lowell and Storer then selected Thomas B. Curtis as the third member of the deliberative body. The task of these arbitrators was to investigate the entire history of the Agassiz-Desor relationship and to reach a decision on the charges and countercharges of each man. Both Agassiz and Desor agreed to abide by the findings of these arbitrators.

These proper men of Boston investigated all matters at issue with an admirable concern for justice to both men. They examined much pertinent evidence and evaluated every allegation. Desor was asked to present the proof he claimed he had to substantiate his charge of immoral behavior; Cabot and Felton were consulted; Christinat, Girard, and even Jane were interviewed. The type specimens described in the echinoderm monograph were sent for to determine whether Agassiz was correct in charging that generic names and identifications he had been responsible for had been changed by Desor in the published account. If Agassiz were proved right in this instance, Desor's action in giving himself credit as an author of this work would be exposed as plagiarism. Letters from Agassiz's mother and Neuchâtel friends were presented as evidence. Maurice was asked for a deposition. Karl Vogt, in Germany, was requested to give his views on Agassiz's relationships with Desor. The entire Swiss-French contingent was asked to give evidence. Neuchâtel and Paris bankers were consulted regarding Agassiz's financial affairs. Boston waited through December, January, and part of February for the results of this exhaustive inquiry.

The outcome firmly established Agassiz's personal and professional character as entirely blameless and beyond reproach. Lowell, Curtis, and Storer agreed unanimously in all their findings, and they detailed their investigative procedure, evidence, and conclusions in a private document of twenty pages. This manuscript was a precise and careful assessment that portrayed Desor as a man of incredible deceit, obsessed with a neurotic compulsion to destroy Agassiz's reputation through any means. The charge of immorality fell under the weight of testimony from Christinat, Gould, and even Maurice himself, who emerged as a person of almost unbelievable duplicity. Apparently this was a family trait. The arbitrators viewed this accusation with utmost gravity. A man who wanted to succeed in Boston and who grieved over the death of his wife could not live under the shadow of such a charge, and the arbitrators understood this. In the interest of social history, their words must be recorded:

M. Desor put abroad reports reflecting upon the moral character of Prof. Agassiz, whom he charged with an improper connection with an Irish girl, by the name of Jane, a servant in the family at East Boston. . . . This charge of incontinency was repeated by Mr. Cabot to Prof. Agassiz, who in the indignation of the moment gave vent to all his subjects of complaint against M. Desor. . . . So far from considering this to be an irrelevant matter, it appeared . . . that it must form the primary object of enquiry. . . . With respect to the charge of improper intercourse with the girl Jane, it was not pretended that either M. Desor himself, nor anyone else was personally cognizant of such a fact. . . . It was said that she had a familiarity of manner towards [Agassiz] otherwise unaccountable; it was said that this had been observed by . . . Dr. Gould. The arbitrators examined Dr. Gould on this point. He said that he had noticed the familiarity and boldness of her manners —but that they were as free and bold towards other men as towards Prof. Agassiz. . . . Mr. Desor averred that Jane was in the habit of going into Prof. Agassiz's room in the evening and staying till a late hour. M. Christinat said that this was not true—that she sometimes went into Prof. Agassiz's room of an evening *when he was absent in Boston* . . . and that he had often known her to go into the chamber of M. Desor of an evening, but that he never drew any unfavorable inference from this circumstance.

Desor had charged that Agassiz had given Jane a gold watch. "He appealed to the referees whether such a gift could be made to a girl in her station except under improper circumstances." The truth of the matter was that Agassiz was grateful for Jane's attention while he was ill, and acting under the advice of Mrs. Talbot, landlady in the East Boston house, he sent to Switzerland for the "shewiest watch for the least money," a trinket that was "worn openly by Jane and everybody knew the circumstances under which it was given." Then there was the matter of the shirts. Agassiz complained to Desor that Maurice would come into his room on any occasion and that once he was trying on some new shirts when Maurice entered without knocking. Desor used this incident to assert that it was Jane, and not Maurice, who had been the person in the room when the shirts were being tried on. Even this charge was investigated with serious caution, and Desor was asked to procure a letter from Maurice as to his version of the event.

Unfortunately, Mr. Maurice, not being blessed with an accurate memory, gives a very different version of the story. It is no longer the trying on of shirts that constituted the indecency; but on his bursting into the room, he found the parties standing by the stove; the front of M. Agassiz's trowsers was in disarray; and Jane afterwards told him that she was sewing on a loosened button! The referees cannot attach the slightest value to such testimony . . . [and] are satisfied not only that this charge is not proved but that it is utterly untrue.

The entire affair convinced Lowell, Storer, and Curtis that nothing had been established to "derogate from the high personal and professional character of Professor Agassiz." They confirmed the award of the earlier arbitration with regard to *Principles of Zoology*, establishing the right of Agassiz and Gould to authorship by letters written by Desor himself. Agassiz's claim that Desor had appropriated the results of his own research was fully justified by the naturalist Storer, who examined the specimens that bore Agassiz's name on their labels yet appeared in the published volume as Desor's identifications. Rather than Agassiz owing Desor money, it was established that the German had used money Agassiz sent to Europe for repayment of debts to finance a trip to Sweden. The hearings also made plain the way in which Desor had come between Agassiz and his wife in Europe, causing Cécile's unhappiness. As for Jane, the arbitrators noted with satisfaction that she had married "a person of some property in Roxbury."[90]

The hearing indicated the nature of Desor's character to influential members of the community. Gould would naturally have nothing more to do with him, and saw to it that colleagues in other parts of the country understood the duplicity of the man.[91] Davis, learning of the full proceedings, was so distressed at having doubted Agassiz even for a moment that he canceled an agreement with Desor to write a joint treatise on oceanography, fearful that he would fall victim to the same kind of controversy that had plagued his friend. He asserted that Desor's behavior toward Agassiz provided him with sufficient moral cause for such an action. Benjamin Peirce moved to have Desor expelled from membership in the American Academy of Arts and Sciences. Smarting from these attacks, Desor in 1852 instituted a lawsuit against Davis for breach of contract. Although the jury agreed that a technical contractual violation had occurred, the judge recognized Davis' claim of moral justification by ruling that only five hundred dollars, half the amount awarded by the jurors, should be paid by Davis to Desor. In a final gesture of ill will, Desor had the entire court proceedings printed in pamphlet form, thus once more making public his former charges against Agassiz, as these had been admitted into the record of the trial. This only gained him the hostility of many scientists and laymen; late in 1852, he returned to Europe, to live the rest of his days in Paris and Switzerland, a broken, sick, and unhappy man.[92]

The Desor episode, instead of detracting from Agassiz's reputation, made him an even more appealing and respected member of Boston society. He himself learned that he had many devoted friends, people who were unwilling to believe his assistant's charges and who offered moral encouragement and support when he was most in need of solace. With his personal and social vindication, Agassiz had once more triumphed over adversity.

Nevertheless, Desor's charges, warped and violent as they were, might be interpreted as further evidence of a fundamental defect in Agassiz's intellectual and emotional character, a defect which had led to controversies with such former friends as Charpentier, Schimper, and Forbes in Switzerland. Taken singly, each incident might be objectively explained from the internal evidence or from the character of the complaining party. Taken together, they suggested an inability to work with others on terms of equality or near equality. Agassiz was too talented to need to appropriate the work of others. Perhaps he reasoned that the mere opportunity to help his efforts would make any man a devoted and loyal assistant or collaborator. Asa Gray sensed Agassiz's need for authority and mastery in matters of science as early as 1848 when he wrote, "A. and I always get on perfectly well—tho' perhaps we might not if we worked in the same field; yet one who understands his impulsive character ought to have little or no difficulty."[93] The questions for the future appeared to be whether he would choose helpers with more wisdom and whether he would be able to enjoy congenial and mutually satisfactory relationships with those who assisted him.

"Agassiz arrived in our city and is . . . entering upon a course of lectures upon the 'Successive Creation of animals and their embryological condition.' . . . He is being lionized at present and daily undergoes the torture of dinner invitations, suppers, etc."[94] As the Philadelphia paleontologist Joseph Leidy described an event of major importance in his city in February of 1849, it was plain that Agassiz had recovered fully from the Desor episode and was ready to take the world as it came and enjoy its riches to the fullest. Charleston, Philadelphia, Washington, and New Haven welcomed him once more, and in each place he visited in this spring of 1849, he luxuriated in the kindness and support of his new American colleagues. Lectures on embryology, tours through the countryside around Charleston, excursions along the coast, lectures at the Smithsonian, all reassured him of the wisdom of his decision to remain in the United States. Wherever he traveled, it was apparent that men like Dana, Henry, and Haldeman had evaluated Desor's true character and supported Agassiz in his time of troubles.

Back in Cambridge, a truly delightful prospect of personal and intellectual distinction opened up before Agassiz in this pleasant spring and summer of 1849. The Feltons invited him to dinner frequently; Peirce and Davis were close companions; Charles Sumner and Harvard's new president, Jared Sparks, found him delightful company. Lowell, Everett, and other Bostonians prominent in the American Academy asked his advice regarding which distinguished foreign scientists to honor with membership. He took an active part in the affairs of the Boston Society of Natural History, delivering papers that reflected the range of his scientific activities in the past two years as he reported on American ichthyology, the ecology of Lake Superior, and the marine life of the Atlantic Coast.

In the larger world of American science Agassiz rapidly assumed the status of a national authority on all things relative to European professional practice. He ordered the most exact of German microscopes for Baird, saw to it that Dana's monograph on zoophytes was known in Europe, and helped American libraries obtain the most recent monographs in Old World natural history. The growing stature of Agassiz in his new country was amply demonstrated at the August, 1849, meetings of the American Association for the Advancement of Science, held in Cambridge. Agassiz literally dominated the formal sessions, in five days presenting twenty-seven separate papers dealing with marine biology, comparisons of European and American botany, insect life, geology, and comparative embryology.[95] One address seemed especially provocative: Agassiz generalized from his earlier work on fossil fishes to describe an entire series of relationships in the animal kingdom, "progressive, embryonic, and prophetic types," as he called them. Using the geological record as his guide, Agassiz affirmed that there were certain intermediate or transitional forms between different classes in past creation. These represented, in microcosm, the thoughtful planning of a beneficent Creator, who had, as it were, given evidence of what He intended for the future by fashioning such forms. Embryology showed a similar series of parallel developments, and in fact the life-history of the entire class as recorded in geological evidence was repeated in the life-history of the embryo. This fact, too, showed marvelous planning and intelligent design and was yet another denial of "causal connection." It also suggested that embryological studies might shed considerable light upon the entire development of the animal kingdom during successive geological periods.[96]

Americans could not fail to be impressed with the talents of such a man. He was already fully identified with the cause of national scholarship. He was able to write James Hall confidently that the New York state legislature had to recognize the necessity for true, patient, slow research and must support Hall accordingly; if not, there would never be an independent science in America. Hall's research in paleontology was recognized all over Europe; it was a pity Americans did not equally appreciate it.

If I were an American I would appeal to my country to shake off this dependence upon European authority for approval of American works. . . . I am not in the habit of appealing to authority . . . but I say that until there are men in America whose authority is acknowledged in matters of science there will be no true *intellectual* independence, however great be . . . political freedom.[97]

Emerson had not expressed the call to intellectual self-reliance more passionately; Agassiz plainly had the potentialities of a cultural hero of first magnitude. The objectivity of the scientist discovering new facts of nature for their sake alone, the admirable wisdom to see the deep spiritual imprint of the Creative Power everywhere in nature's objects, the selfless dedication to instruct American scholars and citizens so that they could see these truths for themselves—these were the qualities of a man so noble that people could only marvel at the national good fortune in his presence in America.

Harvard and its benefactors recognized such qualities at the right moment. It did not matter that Agassiz had not attracted as many students of practical geology as had been anticipated, or that he was not much interested in teaching through formal lectures. If Agassiz complained that the curriculum of the scientific school and the academic organization of the university were not congenial to him because they did not conform to European standards, then these had to be changed, for Agassiz had all Americans as his students. If he insisted that natural history instruction could only be pursued in a museum, with the professor overseeing the independent investigations of students, then his wishes had to be respected. If he demanded more money to preserve and increase his collections, additional room to house them, and the need for more research time, these desires too had to be met. It did not matter that President Sparks had replaced President Everett, or that Abbott Lawrence's original intentions for the school had gone largely unfulfilled in the field of geology. Agassiz's desires were paramount.

When Agassiz inquired of Abbott Lawrence in September of 1849 what his benefactor's plans were about his future, and recited his needs and desires, the results were predictable. Agassiz, who at forty-two felt the pressure of time and the need for assurance that he could remain at Harvard and "settle quietly and permanently" there to spend "those years . . . when a man is capable of the strongest exertions and of doing the best work," was to be granted these wishes.[98] It was the least Harvard and New England could do for a man who, in Lawrence's own estimate, "cannot live without being of essential service in the promotion of scientific knowledge." Lawrence thus proposed, and the Harvard Corporation readily agreed, that Agassiz would now be appointed for a five-year term. Lawrence would add to his original bequest, a sum now reserved for applied science, by guaranteeing Agassiz's salary of fifteen hundred dollars in this period. He could teach whatever courses he and the Corporation thought valuable, or none at all. Instruction in practical geology thus disappeared from the requirements of the original Lawrence donation. It was replaced by Agassiz. Obviously, the plan of creation and the wonders of the organic world as described by Agassiz were more important to wealthy men of Boston than mere mining of earthly riches. This would come, in time, but now Agassiz was to be assured a lasting place in New England culture.[99]

The original role of engineering in the Lawrence Scientific School would be preserved, but the new building constructed for this purpose would have Agassiz's collections, already more than the bathhouse could hold, deposited in its upper story. These col-

lections would benefit from a four-hundred-dollar annual appropriation for their care and increase, the result of an unrestricted
gift recently given Harvard. "It was not without diffidence that the
Corporation offered you so small a sum," wrote Treasurer Eliot,
"fearing . . . you might esteem it inadequate . . . and unworthy of
your acceptance."[100] By June of 1850, such actions had established
Agassiz's role at Harvard. He would be a professor at the scientific
school, teaching a course in geology one term and a course in zoology the next. These were undertaken only out of respect for the
traditional procedures of lecturing and college instruction. The
only way to teach students was in a museum, where they could do
independent, original work on specimens. Agassiz was at the same
time given a large measure of freedom to pursue his own research
and to work for the day when natural history at Harvard would be
studied and taught in a museum he had personally inspired. The
means were readily apparent; the appreciation of men like Lawrence, together with the support of the community and the favor
of the Corporation, would make all things possible. America was
indeed a land of opportunity.

5

NATURALIST TO AMERICA

1850–1857

THE EARLY SPRING OF 1850 was warm and pleasant in Charleston, South Carolina. Agassiz relaxed amid the bloom of early flowers and the companionship of close friends. Charleston provided intellectual activity too, the occasion being a meeting of the American Association for the Advancement of Science, whose members heard the new Harvard professor speak on topics that reflected the diversity of his interests. Life was indeed full and happy for this migratory naturalist of whom there was only one species and one known type specimen. His thoughts, however, were not on the present, but were back in Boston, reflecting on the past and future.

Taking pen in hand Agassiz wrote a confidential request to the ever loyal Augustus A. Gould:

I received some additional proof sheets of *Lake Superior*. . . . Will you allow me to ask a favor of you in reference to it? I wish Miss Cary should have the first copy which can be obtained. Pray speak with the publishers and request that they shall have as early as possible a well bound copy left in Temple Place No. 10. . . . And if by some care she could have the book some days before it is for sale, I should like it the better.[1]

Elizabeth Cabot Cary, an attractive, dark-haired, graceful, intelligent girl of twenty-seven was delighted with this mark of esteem from a man she had come to admire and love. She had lived most of her life in the sedate calm of Temple Place, growing to maturity with her Cabot cousins in enjoyment of flowers, good music, and Brookline parties, and spending her summers at Nahant in the cottage of her maternal grandfather Thomas Handasyd Perkins. Now her future was to be linked with that of Agassiz.

"Lizzie" Cary's sister, Mary Cary Felton, was well aware of the course of events that had led to the engagement of her sister to Agassiz. A Cary family story told how Lizzie's mother, Mary Perkins Cary, had predicted the future as early as October, 1846. Returning from church one Sunday morning, she was said to have asked Lizzie who the distinguished gentleman sitting in John Amory Lowell's pew was, for this was the first man she had seen whom she would like to see Lizzie marry. The Cary girls laughingly informed their mother that this was the great Agassiz, and that he had a European wife. Lizzie, who enjoyed a lecture by George Catlin on the Indians, a series of travel talks on Palestine, or a book on history as much as a party, concert, or theatrical, was curious enough to join the thousands of Bostonians who heard Agassiz laud the "plan of creation" in nature in his Lowell Institute lectures during the winter of 1846–47. Shortly thereafter, Lizzie met the great man face to face, when Agassiz came to Cambridge to meet Edward Everett and other Harvard notables, and to enjoy a supper party at the home of Cornelius and Mary Felton. Lizzie and her sister Caroline had come out from Boston to help entertain the impressive foreigner with songs and piano playing, and if the girls laughed secretly at their mother's prophecy, they were as spellbound as Longfellow, Felton, and Peirce at the tales of Alpine adventures, and the stories of Cuvier, Humboldt, and the Jardin des Plantes that flowed so easily from Agassiz's lips.

Inhabitants of "the Court," as Temple Place was known to its closely knit Cabot, Perkins, Gardiner, and Cary families, soon came to know and admire Agassiz, by far the most distinguished foreigner to visit Boston in many a year. He was a frequent dinner guest at the Cary home, and his correct manners, Continental charm, intellectual attainments, and the fact that he was the protégé of such respected friends as John Amory Lowell and Abbott Lawrence made him eminently acceptable. Agassiz's appeal was strong enough to induce James Elliot Cabot, one of Lizzie's cousins, to accompany him on the Lake Superior exploration and to write the narrative account of the journey that formed a prominent part of the published volume that Lizzie received. When Elliot returned from the north in the summer of 1848, he came back to Temple Place with fascinating accounts of a wonderful summer passed with a man who was as much a warm individual as he was a dedicated scientist. Such reports could not have failed to stimulate Lizzie's interest in Agassiz. Two of Elliot's brothers, Edward Clarke Cabot and Dr. Samuel Cabot, were by this time involved in the controversy between Agassiz and Desor. It was Edward Clarke Cabot to whom Desor had first confided that Agassiz had been guilty of immoral behavior and had neglected the wants of his family in Europe. Like Charles Henry Davis, these men were once somewhat sympathetic to Desor's charges, but like Davis they very soon understood how erroneous their evaluation had been.

Unlike some Cabots, the Carys and the Feltons were loyal to Agassiz from the first. Lizzie's faithful friendship was particularly gratifying at this time, the months from August, 1848, to February, 1849. This was the period when the Desor controversy became a public matter, when Agassiz had to bear the news of Cécile's death and when he knew consequent feelings of guilt and remorse and was anxious over the fate of his children.

Since leaving Neuchâtel early in 1845 with their mother, these youngsters had lived under trying circumstances. Cécile had taken Pauline and Ida to Alexander Braun's Carlsruhe home, with Alex Agassiz remaining in Switzerland to complete his school course. When Alex rejoined the family late in 1847, he came to them in Freiburg, where his mother and sisters now lived in order that the family could enjoy some security through being near Alexander Braun, who had moved to the Baden city to become director of the Botanic Garden and professor of botany at the university. Alex found his mother and sisters living in straitened circumstances in a small apartment. His mother's health grew steadily worse, and her invalid condition meant that Alex, only twelve years old, had to act as head of the household.

Agassiz now learned through his mother that Swiss and German friends were critical of him for allegedly failing to provide for the wants of his family before departing for America. This charge was untrue. Agassiz had made an arrangement with a Swiss banker whereby Cécile and the children would receive a regular allowance, the debt being added to Agassiz's former European obligations. The unfriendly stories were the result of Desor's malevolent rumor-mongering in Europe, a preview of his American activities. Agassiz knew from his mother that Desor had been guilty of such lies, and it is incredible that he nevertheless permitted Desor to come to America at his expense and did not break with him until April, 1848. But having taken this step at last, Agassiz had made it possible for his children to join him in Boston, because he knew from his mother that neither she nor Cécile would allow the children to come while Desor remained in his household.

A new life with his children was all Agassiz could realistically hope to salvage from his former family life. By accepting the Harvard appointment in September, 1847, and thinking and acting as if the United States would be his permanent home, he must have realized that a new life with Cécile would be virtually impossible. Even if her illness had not prevented travel, it was hardly likely she would journey to America. In April, 1848, Agassiz wrote Cécile an affectionate letter expressing his delight at receiving some pencil sketches she had made of the eleven-year-old Ida, the seven-year-old Pauline, and the twelve-year-old Alex. Agassiz's words of love, together with the knowledge that he was at last free from Desor's influence, gave Cécile great joy. In July, just a few days after she read Agassiz's letter, Cécile died. It was now imperative that Agassiz act to make his children's existence secure until he could bring them to America. Agassiz now arranged for Ida and Pauline to live with Rose Agassiz in Switzerland, their grandmother now residing at her family home in Cudrefin. Agassiz asked Alexander Braun to keep Alex with him in Freiburg and supervise the continuation of his education.

Having done these things, in August, 1848, Agassiz could face the unhappy events of the Desor arbitrations with the comforting knowledge that his children were at least temporarily provided for. The final decision of the arbitrators was rendered in February,

1849, and, vastly relieved at the outcome, Agassiz wrote to Switzerland and Germany to arrange for Alex's passage to America. By itself, this action signaled Agassiz's conviction that the United States would be his permanent home. With such anticipations, the possibility of finding a new wife as well as a new career in Boston must have been foremost in his thoughts. Success in America seemed inevitable after the highly triumphant lecture tour to Charleston, Philadelphia, Washington, and New Haven in the spring of 1849. Returning to Cambridge in April, Agassiz once more found himself a very welcome guest in the Felton home.

Mary Felton had given birth to her first child in March, and it was only natural that she ask Lizzie to come to Cambridge and take care of the household and of her two stepdaughters, the children of Felton's first marriage. Lizzie came to live in the Felton house in May, and the spring evenings passed pleasantly for Agassiz, in the company of the young lady who wore her hair "in soft ringlets," whose eyes "looked like the water of a brook running over a bed of autumn leaves,"[2] who talked with him in French, sang softly, and played the piano. Cornelius Conway Felton was quite probably delighted by the blossoming friendship of his Harvard colleague and his Cary relative. He had often thought that Charles Sumner might make an ideal mate for Lizzie and had worked to promote such an alliance. But it was plain that the man who spoke of God's scheme of creation had more appeal for the girl of Temple Place than the intense, brilliant reformer who would soon be a senator of the United States.

Agassiz enjoyed the company of the Felton parlor, where he could talk with men like Longfellow, Ticknor, Holmes, and Peirce, who were more than happy at these opportunities to know Agassiz better. But they found it difficult to get his full attention, because their friend was obviously more interested in the attractive, graceful, gentle, young woman whose conversation was so bright and animated.

In June, 1849, Agassiz interrupted this pleasant life to make a one-day trip to New York to greet his son. Alex had traveled to America in the care of his cousin Dr. Charles Mayor, who had decided to live in the United States. Agassiz, delighted with the appearance and manner of the son he had not seen for three years, called him a "charming boy" and proudly presented him to his friends the Feltons, and his particular friend, Lizzie Cary. Not yet fourteen, Alex hardly knew what to expect in this new environment. With thoughts of Germany and his beloved mother fresh in his mind, he was at best a confused and uncertain lad, his strangeness heightened by a natural shyness and his inability to speak English. The Feltons took the young immigrant into their hearts. Lizzie, only thirteen years his senior, treated him with the love and understanding he had known only from his real mother. "From the time I first saw her at Mr. Felton's house," Alex recalled, "there never was a word of disagreement; she belonged to me and I to her. . . . She learned to know me through and through and placed in me the most unbounded confidence."[3]

The summer of 1849 was important in the lives of both father and son. No sooner was Alex introduced to Cambridge society than he was packed off to nearby Nahant, fifteen miles away from Boston, to spend part of the summer with his father on the peninsula. In Colonel Perkins' stone cottage, the first summer outpost of "Cold Roast Boston," Feltons, Carys, and Agassizes lived in happy companionship. Alex learned to communicate with the other young people in Latin. His father spoke a universal language to Lizzie Cary, as they strolled on the beach in the cool evening. Echinoderms and mollusks were plentiful at Nahant, but Agassiz for once in his life ignored the data of nature. He won the heart of Lizzie Cary just as surely as he had won the admiration and love of Lowell Institute audiences. As for Lizzie, her happiness was unbounded. Her dear Agassiz was, she knew, a lonely man, understood by few, despite his shining public countenance. She resolved, with a determination that spoke well of her ancestry, to know this man fully and make him completely happy. It was a large task, but the rewards were also large, and Lizzie Cary was an exceptional woman.

Agassiz left Nahant with his spirits high. He and his New England friend had an understanding about the future; it was up to him to realize their dreams as soon as possible. Thomas Graves Cary, in keeping with his former occupations as a merchant in the China trade and a lawyer, was a prudent and cautious man. Treasurer of the Hamilton and Appleton Mills in nearby Lowell, and a man of substantial means, he was concerned that his second daughter marry happily and well. Mary Felton had married a professor, and she seemed contented. But her husband was a classical scholar who lived a proper life in a proper Cambridge home. Lizzie was asking permission to marry a Swiss whose only fortune was a salary guaranteed by Abbott Lawrence. The foreign professor had debts still unpaid in Europe, and, so nephew Edward Cabot may have intimated, had a past that bore investigation. Agassiz seemed entirely likable and was obviously devoted to his daughter. It was also true that John Amory Lowell and Thomas B. Curtis, men whose word was a primary recommendation, spoke of Agassiz in the highest terms. Agassiz appeared determined to succeed in America; a man would not bring a young son to a strange land if he did not intend to make it his permanent home. Nevertheless, it seemed best to withhold parental consent to the marriage until investigation and inquiry had shown Agassiz's past to be as blameless and honorable as his present character and behavior.

Agassiz, on his part, acted to make his future secure enough so that the Cary family would be convinced of his intention to remain in America and of the dependability of his financial prospects. Abbott Lawrence was eager to depart for England to assume his duties as minister to the Court of St. James's. He would not leave, however, until he had acted to make more certain the future of the man he had come to admire so much. In September of 1849, he received a letter from Agassiz. "I have determined to remain here because I have strong hopes for the future," were the words Agassiz used to express the confidence Lizzie's love had given him. Agassiz's appeal for financial support to continue his Cambridge association resulted in Lawrence's agreeing to underwrite an additional five-year tenure for the Harvard professorship, even before the original term of appointment had expired Agassiz now had even more reason to feel he could establish an American home.

It was apparent to the social and intellectual elite of Boston and Cambridge in this fall of 1849 that Agassiz and Lizzie Cary were more than good friends. Some speculated that they had in fact become engaged to marry. Agassiz did not want Charles Sumner to learn this news at second hand, and so he wrote him a short note that signified the end of any hopes Sumner might have entertained with regard to Lizzie. "Do you care to learn from myself that I am engaged to be married to Miss Lizzie Cary? I wish you should, and if so, excuse me for not sending you sooner this expression of my particular regard and friendship."[4] Agassiz's engagement was real enough to him and Lizzie, even though the marriage had not yet been sanctioned by Thomas Cary. In writing Sumner, Agassiz anticipated future success in this affair of the heart, in the same spirit as he was able to anticipate intellectual accomplishment. If Sumner was surprised at the news, he did not tell his friend Longfellow, who had to make direct inquiry about the rumor of the day from the wife of a Swiss naturalist who had followed Agassiz to New England. The lady told him that his suspicions were correct, that Agassiz had had difficulty in deciding upon taking such a step but that he had chosen his new wife well, and that his decision was made because he needed someone to look after his house.

If the Swiss community of Boston could rate Lizzie's talents no higher than this, Thomas Graves Cary had a realistic concern for the happiness of his daughter, who would be wife, not housekeeper, to Agassiz. On December 30, 1849, shortly after Lizzie's twenty-eighth birthday, her father wrote Agassiz that he had just received some news from Europe which pleased him very much:

When you expressed a wish that I should sanction your proposals of marriage to my daughter, you referred me, for information of yourself, to the gentlemen here who had investigated your relations with Mr. Desor, and proposed that I should make such other enquiries as I might think proper of the character that you had sustained in the various places where you had resided in Europe. I was so well satisfied with what I heard from the arbitrators, in confirmation of my own judgement in your favor, that I was not disposed . . . to enquire further. Still, it seemed best, in a matter of such deep importance to my child, to omit no precaution; and I wrote, accordingly, not only to our Minister, Mr. [Abbott] Lawrence, whose presuppositions in your favor are so well known that he might be supposed to act under the bias of friendly prejudice, but also to those who would . . . report with the impartial accuracy of men of business and the world. I wrote, therefore, among others, to a banker in London who is connected by marriage with the Belgian Minister, and had therefore the means for extensive enquiry. The first letter that I received brought . . . impressions that every thing was as I could desire. . . . The next letters came with the intelligence that there were points which required investigation. . . .

I have learned today that thorough enquiries, made through these several channels, result in entire confirmation of the favorable impressions that I had received of your character.

I have, therefore, great pleasure in now saying to you, that I cordially consent to your union with my daughter. . . .[5]

Thomas Cary never had cause to regret his decision.

Soon all of Boston and Cambridge had heard of the coming event. Asa Gray wrote Sir William Hooker in England, "Agassiz is really becoming an American. He is engaged to a Boston lady—a friend of ours, Miss Cary."[6] Hooker might well have informed Sir Charles Lyell, a man always interested in Agassiz's welfare. Lyell knew America and Boston well; he also knew Agassiz intimately. Hence his perceptive comment: "Agassiz is about to marry a young lady of good Boston connections. If so, he will be a New Englander for the rest of his life, and will be the founder of a school of zoology . . . of a high order. . . . His enthusiasm is catching, especially when he has good soil to work upon."[7]

The soil of America would yield no finer reward than the love of Lizzie Cary. As Agassiz and others realized, his marriage would be the final bond that united him to America. Mrs. John E. Holbrook of Charleston, like Lyell, was one of the few people able to penetrate the hearty exterior of Agassiz to discern his essential character.

You will be happy, indeed you will, for now you will have no feeling unemployed—and [in] spite of what you told me of the chains in which science binds you I think I have divined in you the strongest capacity for all the domestic affections—One could not caress Alex as you do without a fountain in him, the gush of which should not be wasted on the Ideal. . . .[8]

Mrs. Holbrook would find that Lizzie Cary had perceived the same quality in her husband-to-be. Lizzie knew that Agassiz was a lonely man with deep sources of affection and emotion that had found too few outlets in the past. The man who sparkled in social intercourse and who knew the importance of appearing "decently" before the ladies had always lived a life that was a public perform-ance. He had forced himself to be overly serious and older than his years. This self-discipline had served his purposes well. In Amer-ica, reaping the fruits of accomplishment made possible in part by this conception, Agassiz could relax and enjoy an expanding in-terest in people and their companionship. It was as if he had left the hushed workrooms of the Paris museum, with Cuvier bend-ing over his shoulder, and entered the Boston fishmarket, where every vendor eagerly asked praise for his wares.

Few people knew what the real Agassiz was like; he seldom talked about personal drives or desires in letters or conversation. He would discourse on the "progress of science," but the inner thoughts of its primary advocate remained unknown even to friends like Ticknor, Longfellow, Dana, and Felton who knew him as a man whose *"bonhomie* seems inexhaustible." Lizzie ex-tended her love and thus touched a deeply responsive chord in Agassiz's nature. He had not known the complete devotion of an-other person since the early Neuchâtel years. Lizzie would share his appreciation of nature's grandeur. She had to because she loved Agassiz. She promised her lover, "You will see that I have eyes for its beauty, ears for its melody, and a heart to understand its teach-ing, for there is teaching in these fairest works of God."[9] Lizzie was also completely devoted to Alex, and to Agassiz's public am-bitions, promising him support that "will not fail you in any time of trouble."[10]

It was proper that Agassiz and Lizzie be married in King's Chapel, the Boston church that had long been the spiritual home of the Cary and Perkins families. This was an important event in Boston, and many people who had made Agassiz's American career possible and successful gathered at the Tremont Street church on April 25, 1850. Caroline Cary recorded the scene in her diary: "Lizzie looked lovely, dressed in a green silk, white camel's hair shawl, straw bonnet trimmed with white, [with] feathers on each side. After the ceremony they drove directly out of town."[11] If Agassiz smiled happily as the carriage sped away from Boston, no one could deny that he had good reasons.

"Agassiz . . . has cleaned his house of all loafers, stays at home . . . and is going ahead on his own hook. . . ." This was A. A. Gould's impression of the new environment at the Oxford Street house that welcomed Agassiz and his American bride.[12] Newly renovated under Lizzie's supervision, the house lost its horde of European assistants and friends, only the artist Burkhardt remaining. Some of the Europeans returned home, others remained in Boston, and some, including Girard, Lesquereux, and Guyot, took up independent careers in America.

Elizabeth Agassiz thus began providing a comfortable domestic existence for her husband. Cambridge friends such as Peirce, Felton, and Davis soon discovered that the Agassizes were genial hosts and that the pursuit of science did not preclude good dinners and pleasant entertainment. Visiting Carys, Cabots, and Gardiners from Boston were similarly impressed. Lizzie had found snakes in her bedroom closet, left there by an absent-minded husband, but, while Agassiz still worked in his home, it was plain that Lizzie's desire for an ordered domestic existence impressed on her husband the need to use university buildings for the storage and study of specimens. It was also clear that Lizzie, while entirely in sympathy with Agassiz's intellectual goals, knew the importance of devoting a fixed sum of money to household expenses. There would, it appeared, be a new orientation in Agassiz's pursuit of knowledge. Since it was necessary to provide for a home and family as well as to spend large sums for science, then more money would have to be found to satisfy personal and public drives.

Lectures were always a welcome source of income, but they often became taxing, time-consuming activities that detracted from original investigation. Harvard would have to understand that a research scientist could not live on a salary of fifteen hundred dollars and be expected to teach and carry on specialized studies. Harvard would also have to realize that as teaching and research proceeded, appropriate facilities, such as a museum, were required. By the fall of 1850, these needs had been mildly satisfied by the availability of the upper story of newly constructed Engineer Hall for the deposit of specimens, the grant of four hundred dollars as an annual research appropriation, and a subsidy of an equal amount from the ever generous John Amory Lowell. In the next year, however, Agassiz spent nearly half his total income on additions to the collections, materials that represented both European and American natural history in all its branches.[13]

Agassiz was secure in his domestic existence, even though he might have wished for more money for science; life in Cambridge was more pleasant than any personal existence he had ever known. Lizzie had written to Pauline and Ida, telling them how much she wanted them to come to America to live. In August, 1850, the girls of thirteen and nine arrived from Switzerland, accompanied to New York by Auguste Mayor, Agassiz's cousin. Lizzie became a wise and devoted parent to these girls, teaching them the ways of their new homeland and helping in the difficult process of adjusting to a new society. Under her steady guidance this did not take long. Soon Agassiz found he had two happy, secure American daughters to grace the home so aptly managed by his American wife. Alex Agassiz also cherished the love and affection of his new mother and profited quickly from her beneficent influence. "She was my mother, my sister, my companion and friend, all in one," he recalled in later years.[14] By the fall of 1851 he had entered Harvard College. Two years later he decided on engineering as a profession and undertook a geological field trip through New York State, supervised by his father.

Alex was only one of a number of students who came to appreciate the teaching of Louis Agassiz. In 1850 and 1851 young Americans began arriving in Cambridge, drawn by the reputation of the new professor. Joseph Le Conte and William Jones were the first of these, and they were soon followed by William Stimpson and Henry James Clark. These students were already well educated; three of them had college degrees, and Stimpson was a practiced amateur naturalist. Their teacher was delighted by the opportunity before him. Natural history had to be learned from direct experience. Lectures and constant supervision were old-fashioned techniques. Agassiz had become a master by acting like one, and these young Americans would learn that nothing was so sacred as the dignity of knowledge gained by personal conviction and experience. It was unfortunate that a Jardin was not available as the headquarters for such endeavor, but reasonable substitutes were discovered.

These students were amazed to discover that their professor did not want to know what their previous knowledge or training had been. Nor did he seem very interested in what they were doing. Le Conte, for example, was given a group of one thousand shells and told to separate them into species. He worked at this task for an entire week, rarely seeing Agassiz. When he had finished, he learned from his professor that he had just made a substantial improvement on respected authority in this subject. Students introduced into the realm of discovery in this way would never forget their teacher.[15]

The federal government aided this educational process. Agassiz had found a devoted admirer in Alexander Dallas Bache. In the early part of 1851, Bache provided a Coast Survey steamer to take Agassiz and two of his students on a survey of the Florida reefs. The result was that important information on the process of reef-building was obtained for the government, many new specimens of marine biology were secured for the Cambridge collections, and students received firsthand experience in natural history exploration.

Agassiz made the Florida journey without Lizzie, who took the occasion of their first long separation to write him her innermost thoughts and feelings:

It seems to me incredible that you are truly mine; that you have chosen me for your wife. . . . You say you are homesick—so am I, as much here as you are there, for though this [house] has the name of home, no place is so for me on earth but where you are. . . . I think you must not go away again without me if we can do otherwise without doing wrong.[16]

Lizzie would not live as Cécile had, alone and lonely while her husband traveled in pursuit of knowledge. Only rarely did the future find him studying nature at first hand without the presence of a wife who was at once companion, mother to his children, and intellectual partner. The fact that no children were ever born to Agassiz and Lizzie meant that their lives were almost entirely devoted to the external worlds of science and society.

Agassiz remembered clearly the long months spent in the museums of Europe in search of information. In America, his students would study the material of nature freely and independently, working with a large variety of specimens. Agassiz had by this time developed a program to master American natural history. He would bring the dignity of European scholarship to America but would make knowledge freely available to all who could profit from it. When Le Conte and Jones received the newly established bachelor of science degree from the Lawrence Scientific School in June of 1851, Agassiz had taken a first step in this direction.[17] His words to Haldeman were reflections of his aspirations: "The time has come when American scientific men should aim at establishing their respective standing without reference to the opinion of Europeans. . . . Let us make an effort to be what we can without the assistance of anybody, and let me include myself in the list, if I can be welcome."[18]

American scholars had no need for the learning of Europe when a scholar from the Old World had joined their ranks with such energy and purpose. He would bring standards of professionalism to America and in so doing would strive for a proper public recognition of the dignity and freedom necessary to science and its practitioners. American scientists like Bache, Peirce, Henry, and Dana found such a vision most appealing; it coincided with their own views on the place of science in the national culture. In 1850, Agassiz was chosen president-elect of the American Association for the Advancement of Science, to serve for the year 1851–52. An honor that had gone to Redfield, Henry, and Bache before him, this was a sign of Agassiz's growing command of the institutional forms of science in America.

George Engelmann, St. Louis physician, botanist, and former student with Agassiz at Heidelberg, was a fascinated observer of this process. "Agassiz is a host indeed—a great acquisition for America. . . . By happily combining several eminent qualities [he] has gained a great and controlling influence which I hope will be very beneficial perhaps for generations to come."[19] Agassiz gave ample public demonstration of this role in March of 1851 with an achievement that fitted in perfectly with his public ambitions.

In 1849, James T. Foster, a public school teacher in Albany, had published *Foster's Complete Geological Chart,* an alleged representation of typical American inorganic formations, which was to be used in New York secondary schools. The man most directly affected by this grossly inaccurate chart was James Hall, New York's foremost geologist. Hall showed the offensive document to Agassiz, and both men published their unfavorable opinions in Albany newspapers. Agassiz was convinced that such examples of "quackery" and "charlatanism" had to be exposed by reputable scientists. When Foster's publishers sued Agassiz and Hall for libel, the battle was joined. Agassiz could not understand how the free expression of expert opinion could be put to the test of public evaluation by a jury of laymen. To him, the entire affair was symptomatic of the need for placing science above the level of popular emotion or judgment. He was happy to receive public acclaim for translating science into popular terms, but this was responsible education, while such signs of the times as spiritualism and the claims of amateurs to professional status were symbols of the defects of science in a democratic society. He determined, therefore, to bring the full power of his authority to bear against these tendencies.[20]

Agassiz and Hall enlisted the aid of the nation's leading geologists in their fight. All these men stood ready to support Agassiz in his trial for damages claimed in the amount of twenty thousand dollars. It was arranged to have Agassiz's case tried first. In March of 1851, backed by supporters from all over the country, many of whom had come to Albany to be on hand if needed, Agassiz defended his estimate of the Foster chart in the courtroom of Judge Amasa J. Parker. The aid of other naturalists was unnecessary. Agassiz's testimony was so convincing that defense attorneys were able to secure a quick dismissal of the case, with the result that the suit against Hall was never pressed. The controversy, one followed closely by the profession, firmly established Agassiz's defense of the dignity of science. It was an impressive demonstration of the increasing professionalism of American science.[21]

Agassiz had clearly determined to effect a rapid transformation in the social relationships of science. He wanted to duplicate in

America the intellectual conditions he had known in Europe. He had spurned the possibility of an appointment at the University of Berlin, but he would bring the idea of Berlin to America. In 1852 the French Academy of Sciences awarded him the Prix Cuvier for his *Poissons fossiles,* making him the first recipient of an honor accorded the naturalist whose work best reflected the scientific achievements of the great French paleontologist. Agassiz thought to provide institutions of similar status in America, where intellectual distinction alone would determine the public rank of the scientist. He had expressed these views as early as 1849, when he wrote to the Harvard Corporation: "Science must be in advance of the wants of the times . . . [and] it is a duty for every civilized country to provide . . . for the means of original research. . . . Unhappily, this is neither generally understood nor done to any appropriate extent in this country."[22]

In August of 1851, Bache echoed these sentiments in his address as retiring president of the American Association, when he called for a supra-organization of scientific men under the sponsorship of the federal government for the purpose of guiding "public action in scientific matters."[23] Both Agassiz and Bache understood that such goals depended upon two factors for their realization: the unity of scientists of similar convictions and the support of science by public and private institutions.

The first opportunity to work for these aims came in 1851, when citizens of Albany established the "University of Albany" and asked the aid of Agassiz, Hall, and Peirce in organizing the institution. Here was a ready-made situation and these men moved to gain acceptance for their particular view of progress. They proposed that a truly national "American" university be established, one that would stress graduate instruction, basically in the sciences. Classes would be few, research time plentiful, and salaries high. James Dwight Dana and John P. Norton of Yale, Agassiz, Peirce, Lovering, and Wyman of Harvard, and Hall of Albany were the leading names proposed for the faculty. Although local citizens responded to appeals to make their city the American home of a University of Berlin, moral support was more plentiful than material backing, and the noble vision had faded by 1853. The only tangible result for science was the establishment in 1856 of the Dudley Observatory in Albany, directed by Benjamin Apthorp Gould and managed by a "Scientific Council" comprised of Henry, Bache, and Peirce.[24]

The Albany opportunity, however, had served to bring men of common interests into close alliance. Peirce, Agassiz, and their fellows had proposed an effective controlling influence in American culture by means of a kind of "traveling faculty" staffed by men of recognized ability in the sciences who would teach at the national university while holding appointments at their home institutions. If circumstances were favorable, these men would give themselves entirely to whatever institution provided the freedom and financing they required. Agassiz expressed this ambition in a letter to an Albany educator:

... you have now in your hands the destinies of more than one of the ablest men in the country, who will join you, if everything is done as it may be, or keep aloof until another opportunity is presented. . . . Our number is too small now to allow several universities to be founded. . . . The state which will go for it fairly, will have the lead for years to come, without the possibility of competition.[25]

By 1853 these ideas provided the intellectual and social basis for an informal organization devoted to common goals. Cambridge was its place of origin, its name the Florentine Academy.[26] The founders of this alliance designed to control the institutional forms of science in America were Peirce, Agassiz, Benjamin A. Gould, and Cornelius Felton. Felton, a classicist, took great pleasure in this enterprise because it placed him in the vanguard of the spirit of scientific progress so forcibly represented by his close friend, Agassiz. Coast Survey Superintendent Bache, the Florentines' most influential member in the city of Washington, had the drive, ambition, and professional influence to transform the Florentines into a national alliance by 1855. In keeping with the Italian free spirit, Bache and his Cambridge friends now called themselves the "Scientific Lazzaroni."[27]

Lazzaroni were Neapolitan beggars or idlers. The men who so designated themselves were far from idle. They gathered once each winter to enjoy "one outrageously good dinner together" and to review efforts and plans to secure university appointments for men who had their approval, to defend the operations of the Coast Survey and the Smithsonian Institution, to conclude alliances with men in strategic positions, and to wage war on "old fogeyism," "charlatanism," and "quackery" in their profession. In Peirce's words, this union was "a frigate in which the wise men may seek safety."[28] These wise men were formidable advocates of the "progress of science" as they chose to define the concept. Irresponsible public criticism and popular interest in utilitarian "applications" of science would not be allowed to detract from the primary requirements of science for social recognition and institutional support.[29]

Bache had already assumed command of the public relations of young American professional science, and as "Chief" of the Lazzaroni he had influential lieutenants in Cambridge, Washington, New Haven, Albany, Philadelphia, and New York. James Dwight Dana represented Yale University and the national power of the *American Journal of Science*. Washington and the federal scientific agencies were ably accounted for by Henry and Bache. Philadelphia and the University of Pennsylvania were represented by the physicist John F. Frazer, the City College of New York by the chemist Wolcott Gibbs. Beginning in 1856, Albany and the Dudley Observatory were commanded by Benjamin A. Gould, while Peirce, Agassiz, and Felton cemented the alliance by their authority at Harvard and in the Boston-Cambridge area.[30] The national dominance of the Lazzaroni was demonstrated by the fact that its members held the office of president of the American Association for the Advancement of Science six times between 1848 and 1855.

Outsiders saw these men as the "Cambridge clique" or the "Bache and Company party," whose purposes were to advance personal power, control the American Association, the Smithsonian, and the Coast Survey, and to exert a dominant influence over the entire structure of American science. As one suspicious critic put it, ". . . there is a clique bound to each other's interests and defense at all times and under all circumstances . . . [and] their inter-communication is constant."[31] The fact was that the public view of the Lazzaroni was in the main a correct estimate.

They had high aims and ambitions, and if people confused these public aims with private motives, this was simply a failure to appreciate the Lazzaroni's urge for a self-conscious direction of science in America. As early as 1858, Peirce and Agassiz gave evidence of the magnitude of these aims by urging the establishment of "a great university at Washington the members in which should be naturally self-elected . . . a cult . . . exclusively composed of men who had been selected and elected by each other, because of the preeminence which each was known to have in his specialty."[32] The Lazzaroni would never forget this goal.

Agassiz was a perfect figure to plan Lazzaroni objectives because their aims were his own. Significantly, of all the inner core of Lazzaroni, only one other man, Dana, was a naturalist, and his services were required because of his key position with the *American Journal of Science*. Agassiz's public activity during the 1850's saw him forging lasting ties with men who were administrators and organizers of scientific activity and whose fields of intellectual prominence were in the physical sciences. Being thus the essential authority for natural history in matters relating to Lazzaroni interests and activities, Agassiz began to think of himself as the spokesman for natural history in the nation. Such an estimate was justifiable. By 1853, Agassiz was an editor of the *American Journal of Science* in charge of zoology, a valued adviser to Henry and Bache, the recipient of the Prix Cuvier, a past president of the American Association, and an honored figure in the social and intellectual life of Boston, Cambridge, Washington, and Charleston.

His prestige, however, did not supply the wants of the moment. These needs were financial. The personal attraction of the Albany opportunity was the fact that it had promised a salary of about four thousand dollars a year. Harvard only paid him fifteen hundred and, despite generous sums for research and the preservation of specimens, these monies were hardly sufficient to provide for a family of five in Cambridge and to pay the costs of continual collecting efforts. Thus, when Charleston friends proposed that Agassiz spend his winter vacations from Harvard in their city and lecture on comparative anatomy to the students of the Medical College, he saw the opportunity to collect materials representing local natural history and also to supplement his income by about two thousand dollars annually. This meant he would hold two professorships and teach more than he liked to, but his financial needs made such an arrangement seem the only solution.[33]

Agassiz took his wife and family to Charleston for the winter seasons of 1851 and 1852. Friends provided the family with a pleasant old home situated on Sullivan's Island outside Charleston Harbor. Local visitors were entertained frequently, and the Agassizes were in turn guests of many distinguished families. Casks of marine animals were sent back in a steady stream to Cambridge, a reminder to Harvard colleagues that the Agassiz spirit was still very much among them.

As much as Agassiz found this life congenial, the penalties of such a double professorial existence were physical ones. In the winter of 1852–53, his normally strong constitution was weakened by an attack of typhoid fever in Charleston. He found all work impossible, and doctors advised a long rest and convalescence before he took up his labors again. Agassiz realized that the southern climate was too much of a strain on his health. His eyes had been giving him trouble of late, too, and he resolved to seek a change of scenery and to end his yearly lectures at Charleston.

As a sort of convalescent experience, Agassiz and Lizzie returned to Cambridge by a roundabout route. During the early months of 1853 they traveled through the Lower South, up the Mississippi Valley to Chicago, and overland back to Boston. Agassiz could not resist the pleas made for his appearance on the lecture platform, and hence citizens of Memphis, Mobile, New Orleans, and St. Louis had their first opportunity to appreciate what New Englanders were now familiar with.

The convalescent tour was taking on the appearance of a typical Agassiz adventure. Fortunately for science, the poorly developed southern railroad network did not allow for very direct transport between various cities, so that the actual journey involved travel through much of South Carolina, Georgia, Alabama, Tennessee, Louisiana, Mississippi, Missouri, and Illinois. Consequently, Agassiz had his first opportunity to explore the region drained by the Tennessee, Ohio, Illinois, and Mississippi rivers.

This experience was the inspiration for a most ambitious project that Agassiz conceived. His early interest in the geographical distribution of fresh-water fishes was reawakened. Aided by many local naturalists, men flattered by the opportunity to be of service to the famous professor, Agassiz gathered large quantities of fishes. Upon his return to Cambridge in the early spring of 1853, he hastened to unpack the many barrels of specimens that awaited his attention. True to his expectations, close study revealed that the Mississippi River system contained a great variety of ichthyological data never previously described or analyzed.[34]

Rather than finish projects already begun, Agassiz decided to write a volume depicting not only the ichthyology of the Mississippi River system but the entire natural history of the fresh-water fishes inhabiting all the known inland streams, lakes, and rivers of the United States. The earlier and less ambitious collaborative volume planned with Baird was forgotten. This new undertaking would provide fundamental data relative to the geographical distribution of fishes as well as comparative material of first importance. It would also supply Agassiz with material for determining the nature and extent of variation in species, another problem that seemed to call for intense modern investigation. In all, the proposed *Natural History of the Fishes of the United States* would do for American zoology what *Poissons fossiles* had done for the study of paleontology.[35]

It was obvious to Agassiz that he would have to examine great numbers of fishes native to every section of the country. This seemed an impossible task for any single investigator, but Agassiz had resources that far surpassed those of any other American naturalist. He would bring to bear on this project all the power and influence he enjoyed in America. Agassiz thus began to marshal a collecting force representative of every segment of society that had applauded his past efforts. Amateur naturalists, sportsmen, fishermen, government officials, and professional scientists were organized in a common effort inspired by Agassiz. In July of 1853 he composed a circular defining his objectives. The message outlined

tentative plans for publication, stressed the necessity for thorough-
ness in gathering fishes from all over the United States, and ended
with instructions for shipping these materials to Cambridge.[36]

In the next two years, over six thousand of these circulars were
distributed, many of them containing personal notes from Agassiz
to the recipient. The entire membership of the American Association, the officers of the Coast Survey, the national mailing list of
the Smithsonian Institution, the membership of many local scientific societies, and most teachers of natural history in colleges and
universities received the appeal. More important, perhaps, was the
fact that Agassiz's quest was brought to the attention of countless
local enthusiasts of nature.[37]

After broadcasting his appeal, Agassiz's next effort was to secure the co-operation of individuals who could assist his work in
special ways. By far the most outstanding was Spencer F. Baird, recently appointed assistant secretary of the Smithsonian in charge
of natural history. Baird's aid was essential because the Smithsonian was the depository for all specimens of natural history owned
by the government. Baird, moreover, was a highly competent ichthyologist whose personal collections were valuable. Even more
important was the fact that the 1850's witnessed many government
explorations and surveys of the trans-Mississippi West. Baird, the
primary adviser to all such expeditions, was in a position to urge
the collection of natural history data and thus commanded a great
array of natural history specimens in Washington. It was not at all
certain, however, that a permanent national museum would be
established to house these materials, although Baird had been carefully establishing the nucleus for such an institution through his
comprehensive collecting activity. Agassiz had no official museum
to serve his interests either; but he had always enjoyed excellent
relationships with Henry and Baird, the Smithsonian even having
subsidized some of his early research, and he saw no reason why he
should not have full access to government materials.[38]

Accordingly, letter after letter went from Agassiz to Baird requesting fish specimens. Baird was anxious to help Agassiz in any
way he could, but he was sometimes understandably reluctant to
part with holotypes or original specimens from which a specific

identification had been made. He probably knew too well the sad fate of many collections that had been sent Agassiz for temporary study, specimens that were rarely returned to their original owners. Yet Baird also knew Agassiz's considerable influence with Henry and the great authority residing in Agassiz's name. Agassiz used all the force of this authority to overcome Baird's reticence. Employing a characteristic image, he reminded the younger man that "in Europe" directors of museums were thoroughly appreciative of their public responsibility to scientists and were quite willing to lend specimens in their charge. "When," he asked Baird, "shall the directors . . . of Museums in this country understand the proper use of collections? The Museum of Paris allowed me original specimens of Lamarck for five years, and I had some fossil fishes during nine years in my keeping . . . and nobody ever pressed me for them!"[39]

Such arguments brought results, and it was not long before Agassiz began to receive shipment after shipment of specimens gathered from all over the country by the exploring expeditions of the 1850's. Along with this aid, Agassiz profited from the work of many ardent helpers who gathered fishes for him in the Southwest, the Central Plains, and the Pacific Northwest. Americans of all ages and classes helped his purposes, and he in turn instructed them in the art of establishing local depositories of natural history. Baird suspected that Agassiz was rapidly assuming the functions of a national director of zoological research. His Cambridge colleague was quick to explain the motives that guided him: ". . . we cannot be too careful in shaking out every selfish tendency. You must judge every step of mine in that light. Far from urging anything that might be chiefly useful to my aims I make it . . . a point with all my correspondents to induce them to establish local collections . . . where there is a chance that these may be preserved and made useful."[40] Agassiz's aspiration to improve the public study of natural history was a happy complement to his personal ambitions.

Amateur collectors vied with professionals for the honor of aiding Agassiz. Co-operative fishermen and nature lovers from all over the country sent thousands of specimens of every sort to Cambridge with the hope that something might prove of interest. Such men as J. Wilson Sanders of Memphis wrote Agassiz that "even if a single specimen of a new variety" were discovered in the great batch of fish he had forwarded, "I should consider myself amply rewarded."[41] E. R. Andrews of Marietta, Ohio, was stimulated enough by Agassiz's circular to ask the famous naturalist to send him "some Foreign Works on Fishes" which he might study to learn the scientific character of the creatures he caught in brooks and streams.[42]

Agassiz not only encouraged such people to begin their own collections but conducted what were actually correspondence courses in natural history, patiently answering questions about various fishes and explaining the technical aspects of his studies. Fishermen and enthusiasts of nature were delighted to learn the nomenclature of ichthyology and to discover that their efforts had supplied Agassiz with answers to problems that had been puzzling him for months. Teachers of natural history in isolated regions and young men just learning to appreciate nature study were similarly pleased by the realization that they were contributors to a great research project undertaken by a world-renowned scholar. Many of these people persuaded friends and acquaintances to begin collections and send them to the Harvard professor. Gathering a batch of fishes in a net, sealing them carefully in casks and barrels, and sending them off to Cambridge seemed to have become a favorite pastime of rural young America. "Circulate some of my circulars!" was a constant appeal from Agassiz to all his friends. Bache, Dana, Hall, Engelmann, and even Senator Charles Sumner, who found a rare fish in the market place and rushed it to Agassiz, all aided the great campaign.

The only problem was that all of this cost money. The people of America were delighted to send their fishes to Agassiz, but with them came invoices for casks, barrels, and alcohol. Moreover, the buildings of Harvard were needed for instruction in less romantic but equally important subjects and could not be turned into fish storage houses. Agassiz determined that he would raise a large sum of money. He would also insist that Harvard or its benefactors supply him with a museum building to house the natural history collection that grew larger every day.

Samuel Atkins Eliot had always been a devoted Agassiz admirer. Constant appeals for funds, bills the like of which Eliot had rarely imagined the university would be called upon to pay, and the obvious dedication of the former European naturalist, had transformed this prudent fiscal officer into a man who almost dreaded another Agassiz appeal yet was nevertheless fascinated by the progress of the professor's ambitions. Eliot had determined to resign his office at the end of the 1853 term, but not before he had acted to make the task of his successor easier by furnishing Agassiz with the means of living in the future, or at least for some time to come. Agassiz had expected that the Harvard Corporation would take positive action to establish a permanent museum in 1853. Knowing full well that this was impossible in the present state of university finances, Eliot turned to Abbott Lawrence for the funds necessary to ease the labors of the man brought to Harvard by this philanthropist's earlier generosity. Lawrence, however, was unwilling to do more than guarantee Agassiz's salary for another five-year term. Eliot reported to Agassiz that Lawrence,

as a man of business, goes upon a little different principle from you. . . . He saves what he can and spends slowly and carefully, whereas I believe you spend, on principle, as fast as you can. When a man has so many objects as you have, including nearly the whole animal kingdom, not only of this age, but of all past time, it is not wonderful that he cannot easily find the means to make the vast acquisition.[43]

The plan for a museum building was shelved for the time being, but Eliot's next effort solved the monetary needs of the moment and also provided an essential basis for Agassiz's ultimate goal. When Lawrence proved unreceptive to his plea, Eliot moved to exploit other resources. The collections that were rapidly turning Cambridge into a giant storehouse of natural history suggested in themselves the solution to the problem. With Agassiz's enthusiastic approval, Eliot approached his friends in Boston with a proposal that they contribute to a fund to purchase Agassiz's collections for the university. Accordingly, the summer and fall of 1853 witnessed the first large-scale alliance between Agassiz, Harvard, and the rich men of Massachusetts. Acting under Eliot's urging, men of wealth contributed $10,500 to Harvard. The university added $1,800 to this figure and paid the entire sum to Agassiz, who in turn gave Harvard full title to his natural history collections, materials that represented the vast extent of his labors in Europe and America.[44] Eliot could now resign his position as Harvard's treasurer, convinced he had served Agassiz, science, and his university well.

Harvard now possessed a rich collection, the nucleus of a natural history research museum, but there was only one man who knew what to do with these materials. The university was thus committed to maintaining Agassiz and appropriating funds to preserve and increase its new scientific property. Either a suitable building and proper financing must be provided for the care of these collections or the initial investment of the friends of Agassiz and Harvard would be wasted. Much had been accomplished in gaining the support of such subscribers to the collection fund as Nathaniel Thayer, Samuel and Nathan Appleton, Thomas B. Curtis, Francis Cabot Lowell, John Murray Forbes, and William Sturgis. But if these men believed, with Eliot, that their action was enough to satisfy Agassiz's ultimate ambitions, they were mistaken. Eliot received fair warning of the magnitude of this vision for the future when Agassiz thanked him for his recent efforts with the words: "I want it well understood I have made no bargain in parting with my collection, but only secured permanency for a Museum which it is my earnest desire to render more and more important in devoting all my energies to its increase and scientific arrangement."[45]

This was no idle boast. Agassiz would not allow Harvard to forget his needs and purposes. In less than two years he spent six thousand dollars of his own money in acquiring new collections and preserving the materials now owned by the university. He was helped by annual university appropriations that totaled six hundred dollars. It was only fitting that Harvard recognize such devotion. Early in 1855 Engineer Hall became in truth a "zoological hall," the Kirkland Street edifice having proved too small for the apparatus of both applied science and natural history. Technology gave way to fishes and echinoderms, and the Corporation turned the entire building over to Agassiz's use so that students like Theodore Lyman, James McCrady, and Henry James Clark could now learn the techniques of their craft in a building dedicated solely to natural history. Even this aid was not sufficient, because the building was already too small, in Agassiz's estimate, to allow for display cases and the proper study of specimens. A two-story wooden building was clearly not the limit of Agassiz's ambitions for natural history at Harvard.[46]

In May of 1854, Agassiz's European reputation provided the opportunity to emphasize his present needs and future hopes. A letter came from the naturalist Oswald Heer, offering him a professorship at the newly established Federal Polytechnical School of Zurich. Patriotism and the inducement of a museum directorship were advanced as compelling reasons for his return to Europe. Nothing could have been more remote in Agassiz's thoughts. A short time later, he received a similar proposal from the University of Edinburgh. This too failed to interest him.

Harvard had recently advanced him seventy-five hundred dollars on a mortgage loan and with this he and Lizzie built a spacious two-story house at No. 36 Quincy Street, near the corner of Broadway, facing the College Yard. They moved into their new abode in the fall of 1854. Here Agassiz had his own library, a comfortable study, and neighbors like Peirce, Charles A. Lowell, Henry N. Parker, and Charles Henry Davis. Henry Greenough and Davis had advised the Agassizes on construction details, and Lizzie wrote Davis that their new home would be "our palace, where we shall be King and Queen, and you shall be our Prime Minister."[47] The pleasure of the Quincy Street home was inducement enough to remain at Harvard, but New England had other permanent advantages. Thomas Graves Cary provided Agassiz and Lizzie with a small summer cottage near his own at Nahant. The cottage was shared by the Feltons, and with Longfellow a Nahant neighbor, Agassiz could plan Lazzaroni activities, talk philosophy and history, and gather sea animals to study in the laboratory built onto the house.

But Agassiz felt that Americans should appreciate what he had spurned and make provision accordingly. Abbott Lawrence learned of the Zurich offer immediately. "All my sympathies are now with this country, my best affections are here," Agassiz told his benefactor, at the same time pointing out that he required the equivalent of a Jardin des Plantes in order to be happy.[48] Harvard's new president, Reverend James Walker, a man not yet fully educated regarding the requirements and ambitions of his Swiss faculty member, learned of such desires too. A great American museum was needed at Harvard so that America could "return to the old world a sum of information equal to that imported."[49]

If America presently lacked the museum that would make Agassiz perfectly happy, Oswald Heer did not learn of it. In refusing the Zurich proposition, Agassiz wrote his friend of the great strides being made in American natural science. Learned societies, the activities of the federal government, and the work of men like Joseph Leidy, Wyman, Dana, and Gray were all symbolic of the intellectual progress of America. There were, however, more personal reasons, and Agassiz could be frank with an old friend: "I have now been eight years in America, have learned the advantages of my position here, and have begun undertakings which are not yet brought to a conclusion. I am also aware how wide an influence I already exert upon this land of the future, an influence which gains in extent and intensity with every year." He would remain in America "for an indefinite time, under the conviction that I shall exert a more advantageous and more extensive influence on the progress of science in this country than in Europe."[50]

Apparently his goals were clear to Agassiz, but what exactly were they, and what did people think of him? President Walker thought a museum was Agassiz's main purpose in life. Ohio fishermen thought he was determined to write a comprehensive history of American fishes; the Lazzaroni thought he had joined them in an effort to control the institutional forms of science. Harvard students thought he was dedicated to their intellectual advancement. Bache thought he was writing a report on the Florida reefs. Charles Wilkes believed he was hard at work describing the fishes gathered by the United States Exploring Expedition. Spencer F. Baird knew him as an eclectic scientist interested in any object of natural history. Sir Charles Lyell thought he was writing a masterwork on embryology. Longfellow, Felton, and Holmes saw him as a vibrant companion, anxious to plan with them the organization of a dinner club which would meet at the Parker House one Saturday each month and talk of kings, philosophy, slavery, and science. A former student visited him at Quincy Street and found him a happy domesticated naturalist:

We spent a week with Agassiz, and a more delightful week it is impossible to imagine. The domestic life of Agassiz was indeed charming, for Mrs. Agassiz . . . was not only an affectionate wife but one of the wisest of women. . . . He took a great fancy to my little Sallie . . . then just three years old . . . taught her the names of all his dearest specimens . . . [and] called her "the little Echinoderm." . . . He was continually playing with the child . . . taking her on his back and getting down on his hands and knees and playing horse all around the dining table. . . . At evening tea I first became acquainted with Oliver Wendell Holmes. . . .[51]

The true Agassiz was, of course, everything particular individuals imagined of him. But despite his wide influence and reputation he found it difficult to get down to the serious business of science. Agassiz was worried that he could not concentrate with the same intensity that had once marked his efforts. He had, to be sure, published a number of papers describing his ichthyological and embryological researches of recent years.[52] These, however, were only small samplings of the tremendous amount of material he had before him. He confessed to Mrs. John E. Holbrook that progress was slow and work often frustrating:

. . . in the evening I feel mostly too tired to write. . . . The fact is I am greatly changed in physical energy and the few hours I feel capable of real exertion daily are now constantly devoted to writing. I have discovered that even if I should keep steadily at it, henceforth, I can no longer expect to finish the work I have already cut out, so it is high time I should begin securing the best of it, by putting it into such shape as will save it from loss.[53]

Agassiz's difficulties were not entirely physical; they were of his own intellectual and social manufacture. There was first the strain induced by his unswerving adherence to the concept of the fixity of species. When he had hundreds of fishes spread before him on a work table, these convictions were of such force that even his keen powers of observation and his excellent ability to compare diverse types failed him. He insisted on identifying specimens that seemed even the slightest degree different from one another as separate species rather than as variants. In one analysis alone, for example, he described nine separate "species" of fishes that were in actual fact reducible to four schools of single species.[54] Consequently, whenever Agassiz examined a group of specimens, he increased his intellectual labor by failing to see variation and thus imposing the necessity for labored descriptions of "species" that were in fact varieties. "New" species had to be carefully labeled, identified, and described. Although his fame spread as each carried the symbol *Ag.* after its specific designation, *Ag.* himself was not able to appreciate the really significant aspects of the research that was so enthusiastically supported by Americans. And, strangely enough, he was ruthless in his criticism of "species multipliers," condemning "the race some naturalists are running, for the questionable distinction of being the first to name species."[55] It was no small task, therefore, to find new species everywhere, to keep up to date on the false descriptions of new species that were published by others, to unpack cases of specimens that arrived in Cambridge daily, and to plan activities for the Lazzaroni.

The spring of 1855 found Agassiz resolute in his determination to accomplish something more of lasting value to science. Museums and other dreams and obligations could wait. He had to publish a great work. Conscience may have argued that he had already announced a great work in his appeal for fishes of 1853. But Agassiz's spirit needed the constant refreshment of new ideas representing grander projects than those that had preceded them. Especially enticing was the swallowing-up of one project by another larger, and hence more exciting, enterprise.

Agassiz now planned a publication venture which would be a supreme contribution to the science of the United States. He would write a series of volumes covering the entire sweep of American natural history. In so doing, he would demonstrate to scientists of his adopted land the most advanced descriptive and analytical techniques of European scholarship. This would be a labor of love, repaying the people of America for their past generosity and hospitality. "I have tried to make the most of the opportunities this continent has afforded me. Now I shall be on trial for the manner in which I have availed myself of them."[56] These volumes, he explained to Sir Charles Lyell, "will show that I have not been idle during ten years' silence."[57]

Agassiz would demonstrate, by means of careful and precise analyses, the proper approach to natural history, rather than the hurried and incomplete descriptions that sometimes characterized American zoology. These volumes, moreover, would be instructive to all levels of the American mind. "I expect to see my book read by operatives, by fishermen, by farmers, quite as extensively as by the students in our colleges or by the learned professions. . . ."[58] This scheme, then, was a perfect representation of the man Agassiz had become by 1855. The universality of nature would be captured within the framework of his all-embracing intellect. He would make no money by this venture, the project being undertaken "for no other purpose than to contribute my share towards increasing the love of nature among us."[59] This would be an "American contribution to science," and its analyses of embryology, ichthyology, classification, geographical distribution, and kindred subjects would show Europeans that "there is something doing in this part of the world, for which it may be worth their while to look out."[60] In one great effort, Agassiz would repay Americans for past devotion and in turn provide them with a cultural symbol that would give their nation permanent distinction in world scholarship.

With these purposes, the grandest project Agassiz had yet envisioned, *Contributions to the Natural History of the United States,* was launched with characteristic optimism and enthusiasm. All other activity would be secondary to this ambition. "Here goes the bird! And so I am tied for ten years." Thus Baird learned of Agassiz's intention to do nothing for the next decade but publish ten volumes of *Contributions.*[61]

But Agassiz had somehow to finance the new project, and it would take a larger sum than he had ever obtained for any of his ventures. Once again Americans responded to Agassiz's enthusiasm. Francis Calley Gray, wealthy iron manufacturer and patron of science, was the Bostonian who now came forward to join the popular but select legions formerly led by John Amory Lowell and Abbott Lawrence. When Gray, who had come to admire Agassiz through the years of their membership in the American Academy of Arts and Sciences, learned of the naturalist's desire to publish the results of his American researches, he suggested a scheme whereby adequate financing could be secured. This entailed the co-operation of Little, Brown and Company, whereby the publishing house agreed to produce the volumes if a sufficient number of purchasers were guaranteed. With Gray's help, a "Prospectus" and "Private Circular" were printed, describing Agassiz's purposes and asking for advance subscriptions to the series. A price of twelve dollars per volume was set, payable on delivery to the subscriber. Agassiz and Gray estimated that if 450 advance subscriptions could be secured, it would be possible to begin publication. Each volume would be replete with color plates and illustrations of all sorts, no expense being spared to make these truly admirable contributions to American natural history.[62]

Heartened by the logic of this plan, Agassiz spared no effort in his personal campaign to solicit subscribers. On May 28, 1855, his forty-eighth birthday, he launched a campaign for popular support that proved him a master of modern public relations techniques. If any literate individual in America or Europe failed to hear of Agassiz and the *Contributions* during the years from 1855 to 1857, it was surely an accident. The 1853 campaign for fishes seemed a minor flurry compared to the 1855 project. Agassiz exploited his great circle of personal, professional, and institutional relationships in full measure. Asa Gray, Jared Sparks, James Hall, Joseph Henry, Spencer F. Baird, Charles Sumner, and countless others made sure that the scientific profession and the general public were exposed to Agassiz's appeal. The world-wide system of professional interchange and communication conducted by the Smithsonian Institution was put at Agassiz's disposal, so that announcements of the project could be sent to Europe free of postal charges. In all, nearly ten thousand copies of promotional literature were sent all over the world from Cambridge and Washington.[63]

The results were not unpredictable; they reflected Agassiz's

world reputation. By August of 1855, well over five hundred subscribers had come forward; by November of 1856 their number
had swelled to twenty-one hundred, and the final total exceeded
twenty-five hundred subscribers. This response meant that Agassiz could proceed on an even more grandiose scale than he had
originally anticipated. He determined to make no profit from the
venture and pledged the use of funds promised in the form of advance subscriptions to make his volumes larger, with more text,
more illustrations, and more expensive lithography than originally
planned.

Agassiz was clearly overwhelmed by this demonstration of public esteem and belief in his ambitions and purposes. He wrote
Charles Sumner, ". . . when my subscription list reaches Europe
my friends will not credit their own eyes. I do not think Humboldt himself could obtain in all Europe put together such a subscription for so expensive a work."[64] The implication was plain.
Agassiz was doing for American science what Humboldt had done
for Europe, and America was rewarding its own Humboldt in an
unprecedented fashion. Agassiz could not resist any opportunity to
show Europeans the extent of his influence in America. He wrote
Valenciennes at the Jardin:

I have just had an evidence of what one may do here in the interest
of science. Some six months ago I formed a plan for the publication of
my researches in America. . . . I brought out my prospectus, and I have
to-day seventeen hundred subscribers. What do you say to that for a
work which is to cost six hundred francs a copy, and of which nothing
has as yet appeared? Nor is the list closed yet, for every day I receive
new subscriptions—this very morning one from California! Where will
not the love of science find its niche![65]

Subscriptions to the *Contributions* came from every state in the
Union, as well as from Europe, South America, Canada, and even
the Philippines. Massachusetts understandably provided the largest number, well over one-third, and quite naturally every family
of social prominence was represented in the subscription list. This
document comprised twenty-six printed pages, in itself another
large-scale demonstration of Agassiz's dominant role in the national
culture. Only seventy-eight subscriptions came from Europe, Agas-

siz's former colleagues quite possibly deciding that America could provide well enough for him.[66]

Agassiz's reputation was publicized in a series of newspaper articles written at this time by Cornelius Conway Felton and the historian and former Harvard president Jared Sparks. Together, these men comprised the Agassiz promotional staff, their efforts designed to stimulate interest in the *Contributions* and other Agassiz ventures. Advised by Agassiz, classicist and historian did their work well.

Published in Boston and Washington newspapers in the spring of 1855, these articles furnished material for magazine stories, biographies, and other widely circulated accounts throughout Agassiz's lifetime. If Americans did not already know of Agassiz's past accomplishments, they were amply informed here. The citation from the Academy of Sciences awarding the Prix Cuvier was reprinted. A sketch of Agassiz's European researches and publications followed this. Americans had the honor of Agassiz's presence in their country. He was completely dedicated to this society, even though "tempting offers have been made to him from eminent institutions in Europe in the hope of obtaining his services and the influence of his name." He had determined to remain in America, "resolved to devote his future life and labors to his adopted country." Now, he was engaged in writing the *Contributions,* a work that

will not be less honorable to the country than to its author. It will diffuse a knowledge of American science, and contribute to elevate it in public estimation wherever intelligence and learning are considered as an index of intellectual culture and progressive civilization. A work of this kind must necessarily be expensive . . . yet it can scarcely be doubted that, from patriotic feeling, it will meet with a generous patronage from the intelligent and liberal minded generally, as well as from those who are specially interested in scientific inquiries.[67]

Agassiz was "the man, who, beyond all question will take the foremost place in the world of science, when the venerable Humboldt shall have ceased his labors."[68]

Such adoration did not, however, supply the needs of the moment, even though Agassiz's publication venture had yielded promises of three hundred thousand dollars in advance subscriptions. Working at the first two volumes of the *Contributions,* Agassiz had determined to set up an entirely new system of classification in zoology and to demonstrate the utility of this taxonomic approach by a special monograph on North American turtles. But such aims meant more research before publication could proceed. Research required more specimens, especially turtle eggs, and thus once more financial problems beset Agassiz.

In Berlin, perhaps, a grant from the king of Prussia would have met the needs of the moment. In America, Ralph Waldo Emerson received a note from Agassiz in April of 1855: "It will give me great pleasure to have your daughter attend my school and I feel proud in the confidence you place in me in trusting to my care . . . one who must be so dear to you. I trust this circumstance may lead to a personal acquaintance between us which I regret has not been brought about before."[69]

Agassiz's excellent wife had moved on her own accord to end the financial uncertainty that had seemed to plague her husband whenever a new project was born. She realized that he had spent six thousand dollars since receiving from Harvard the payment for his collections. She knew, too, that such expenditures would increase as long as he needed his personal income to finance his intellectual pursuits. Aided by Pauline and Ida, she planned a means whereby both the household and Agassiz would be guaranteed a regular source of income. The Quincy Street home would house a school for girls of college age. The idea was a sound and practical one, since there were no institutions readily at hand where girls of Boston, Cambridge, and surrounding communities could receive a good, modern education. Tuition would pay for Agassiz's turtles, and general culture would benefit.

As soon as he found out about the scheme planned independently by his wife and daughters, Agassiz entered into the spirit and purpose of the conception with typical enthusiasm. Natural history was a primary requirement in the education of any modern young woman. Had not Massachusetts schoolteachers for years come to lectures by Agassiz and held squirming grasshoppers in their hands as he illustrated salient points of classification? Had not Mary Peabody Mann begged the Harvard Corporation for permission for female normal school students to attend Agassiz's college lectures, only to be refused because, in the words of President Walker, it was "inexpedient" to accede to such a request? Alex, now a handsome youth of twenty just about to graduate from Harvard College, would instruct in German and French. What more could an Ellen Emerson or Lillie Lodge ask for in the way of education than the instruction of a handsome if by now portly father and a shy but equally charming son? Mrs. Agassiz would be in charge of the entire curriculum and would teach English composition and music appreciation.

The resources of Cambridge were far from exhausted. Another Cary-Agassiz-Harvard exponent of higher learning, Cornelius Felton, would be an obliging and jolly teacher of Greek, Latin, and literature. With this faculty, New Englanders knew the education of their daughters would be in good hands. Eliots, Lymans, Shaws, Russells, Appletons, Lodges, and Bigelows all sent their daughters to Quincy Street, as did members of the Harvard faculty such as President Walker and the philosopher Francis Bowen. George Ticknor wrote Agassiz his feelings: "I only wish I were a little girl and were to be sent to Mrs. Agassiz. In time . . . I should come to something."[70]

Agassiz now talked of echinoderms and starfish to daily audiences of young ladies, often joined by parents come to wish them well and unable to leave until they too had heard the professor. Upstairs, Henry James Clark worked valiantly studying the development of turtle eggs under the microscope. Songs of Germany and tales of Homer came floating through the partition, as Professor Felton and Mrs. Agassiz worked at their tasks. Outside, turtles sunned themselves in a specially built inclosure.

The Agassiz School was a success from the beginning. Tuition was $150 per term, and it was such a popular venture that it lasted until 1863, when the money it brought in was no longer needed with such urgency. Thus the efforts of Mrs. Agassiz provided her naturalist husband with an added income of about two thousand dollars a year, funds that were most welcome at a time when the country had to be searched for the best materials to make the *Contributions* truly what subscribers had been led to expect.

Meanwhile, Agassiz worked at a furious pace. He had never

been more engrossed in a project. He sent letters all over the country, penned hastily and with one purpose, to get as many turtles of all kinds, ages, and environments as he could. Baird was once more asked by Agassiz to serve the great cause, and the Smithsonian Institution became a veritable collecting agency for Agassiz, since it was simply impossible for Baird to resist the constant pleadings of his Cambridge colleague. Winthrop Sargent of Natchez, Mississippi, journeyed great distances in the service of Agassiz, and took a trip of over a thousand miles to deliver a collection personally to the Harvard naturalist. A Massachusetts schoolteacher, who knew Agassiz wanted turtle eggs that were not more than three hours old, camped patiently near a pond for three weeks until he found fresh eggs, and then rushed them to Cambridge by employing in succession a horse and buggy, a freight train, and a delivery wagon so that Agassiz would have the eggs on time. With such help, Agassiz worked with the constant goal of finishing the first two volumes of the *Contributions* before his fiftieth birthday, because birthdays were also special times for celebration to him.

On the evening of May 27, 1857, Agassiz had before him the proof-sheets of these first two volumes. He read them slowly, carefully, with a feeling of deep pride of accomplishment. Before him, on his library work table, was a bouquet of fifty different kinds of flowers, placed there by a loving wife and devoted daughters. He paused, the hour being late, to enjoy a quiet glass of wine and an hour of good cheer with the wife whose love had been such a resource to him. Midnight came, and with it the sounds of a Bach chorale sung by a group of his students, gathered outside the house, carrying bouquets of flowers to honor their teacher. As the early hours of May 28 passed and Agassiz listened to his American students serenade him with songs that brought back memories of Heidelberg and Munich, his heart overflowed with happiness and gratitude. Perhaps he remembered the day, ten years before, when he had spent a lonely birthday, uncertain about the future, anxious for the companionship of those he had left behind.

In the afternoon of the day that had begun so pleasantly, Agassiz journeyed to Boston with his Cambridge neighbors Felton, Peirce, Longfellow, and James Russell Lowell to attend the monthly meeting of the Saturday Club at the Parker House. This was an occasion planned in honor of the man who had come to enjoy such eminence in Boston and Cambridge. Some called the Saturday Club "Agassiz's Club," in recognition of the esteem in which its members held him. These men included Richard Henry Dana, Jr., Judge Ebenezer Rockwood Hoar, John Lothrop Motley, Edwin P. Whipple, Horatio Woodman, Emerson, and the Cambridge friends who joined Agassiz this day. Saturday Club meetings saw Agassiz at one end of the table and Longfellow at the other, the naturalist holding forth gaily, the leader of animated conversation, talking of science and politics and progress, finding in Emerson a man whose spirit and ideas he understood and loved. Emerson was equally attracted to Agassiz, listening intently to his words, finding him "a man to be thankful for; always cordial, full of facts, with unsleeping observation, and perfectly communicative. . . . [He] has a brave manliness which can meet a peasant, a mechanic, or a fine gentleman with equal fitness."[71]

Emerson recorded the events of this May afternoon in his journal: "We kept Agassiz's fiftieth birthday at the Club. . . . Agassiz brought the last colored plates to conclude his *Contributions*. . . . The flower of the feast was the reading of three poems, written by our three poets, for the occasion. The first by Longfellow . . . the second by Holmes; the third by Lowell. . . ."[72] Longfellow's tribute, rendered after a hearty dinner made better by much good wine, meant the most to Agassiz because it was a testimonial from a man whose friendship he had cherished for a decade. His words for the occasion brought tears to the eyes of his naturalist friend. Thus Longfellow told how Nature, "the old nurse," had shown the boy Agassiz a story-book "Thy Father has written for thee," inviting him to wander with her

> Into regions yet untrod;
>   And read what is still unread
> In the Manuscripts of God.
>
> And whenever the way seemed long,
>   Or his heart began to fail,
> She would sing a more wonderful song,
>   Or tell a more marvellous tale.[73]

As if this grand birthday party were not a sufficient sign of Agassiz's prominence, wealthy men of Boston soon supplied still another. Publication of the first two volumes of the *Contributions* was delayed until the fall of 1857. The reason was that Agassiz had to have even more illustrations, and these had to be reproduced in lifelike detail, regardless of the cost. Thomas Graves Cary, Nathan Appleton, David Sears, and other Agassiz admirers underwrote this last expense.

These men were thanked in the dedication of the first volume. Above their names appeared those of Ignatius Döllinger and Francis Calley Gray. The Munich scientist had taught him how to study nature professionally; the American businessman had made it possible for this knowledge to appear before the world. Gray had died in December, 1856, and he was unable to witness the result of his aid. Gray and Döllinger were properly joined together with Boston's men of wealth. On one page was represented a significant segment of Agassiz's life since 1827.

It is doubtful whether Boston's cultivated people found much to excite their sustained interest in Agassiz's precise and highly competent analysis of the embryology, physiology, anatomy, geographical distribution, and ecological characteristics of North American turtles. Yet reviews in popular journals were uniformly favorable. They spoke with pride of the great naturalist who was doing so much for American scholarship. Oliver Wendell Holmes wrote glowingly of "Agassiz's Natural History" in the pages of the *Atlantic Monthly,* recently founded by many Saturday Club members.[74] It was also appropriate that Humboldt wrote a letter to George Ticknor, overflowing with praise for the achievement of his beloved Agassiz. Taking his lead from the earlier efforts of Sparks and Felton, Ticknor had the letter translated, saw that it was reprinted in the newspapers, and claimed that "half a million" readers had learned in this way of Agassiz's success.[75]

These books were impressive physically, in keeping with the aspirations of their author. They were beautifully bound, over-sized, and heavy. The second volume contained illustrations in color and black and white, drawn by Clark and Burkhardt, engraved by Sonrel, and conforming to the standards of excellence set by previous Agassiz publications. Moreover, Agassiz's immense knowledge of his subject provided his readers with an encyclopedic survey of the bibliography and intellectual history of zoology that showed a command of the tools and concepts of his profession at once singular and profound. The student of Cuvier and Döllinger also provided his readers with a study in classification and embryology that illustrated the best aspects of his education. The great effort to collect turtles from the coasts, inland waterways, and land areas of North America had yielded magnificent results. A notable feature of these volumes was Agassiz's analysis of the order of Chelonians, wherein the classification, rank, anatomy, physiology, geographical distribution, and habits of life of North American turtles was precisely and fully described. The genera, families, and species of this order were also analyzed at length. Agassiz's insistence upon the importance of embryology was affirmed by the section that detailed this aspect of the natural history of these reptiles. By itself, this description of American turtles was a primary addition to zoological knowledge.[76]

The core of the book, however, was the much heralded "Essay on Classification" that preceded the special studies. This was intended to form the basis for a new evaluation of the facts of nature. The "Essay" did not, however, bear out this intention. Instead, it stressed the timeless permanence of the four grand divisions of the animal kingdom and embraced an approach to classification grounded firmly on the concept of the immutability of species. As such, the "Essay" was a reiterative testimony of faith, on nearly every page, to the wisdom of Cuvier, "the greatest zoologist of all times." What Agassiz had done was to perform two intricate operations in an attempt to construct a valid taxonomic system. He had, first, analyzed all prior attempts at classification, and in so doing had identified that of Cuvier as representing the most valid systematic arrangement because of the profound insight that had characterized types according to a particular and special "plan of structure." This plan was separate and distinct for each of Cuvier's four branches of the animal kingdom: vertebrates, articulates, mollusks, and radiates. But classification, as such, could not be an artificial reconstruction by the systematist superimposed on the reality of nature. A system of classification was only valid to the degree that it reflected the actuality of nature, since systems were "in truth but translations, into human language, of the thoughts of the Creator." The primary task of the naturalist was to discover, in nature itself, the method by which its manifold relations might be properly determined.[77]

Agassiz applied himself to this task and the result was to him a discovery of first magnitude. In examining the facts of zoology as he knew them, the conclusion he reached revealed that "in their general outlines, the primary divisions of Cuvier are true to nature, and never did a naturalist exhibit a clearer and deeper insight into the most general relations of animals than Cuvier. . . ."[78]

Agassiz's effort thus represented a valiant attempt to synthesize the observed phenomena of nature with the findings of Cuvier. This was a truly herculean task. It required that Agassiz build from Cuvier's taxonomy a system of classification similar to that of the French naturalist in all its essentials, yet embracing within its network the findings of zoology made subsequent to the 1820's. Agassiz was thus forced to find fault with systems of classification at variance with those that stressed the ultimate permanence of the four great branches of the animal kingdom and the immutability of species.

The "Essay" contained material that to another would have been highly suggestive of change, variation, and community of descent in the animal kingdom. The chapter on "Relations between Animals and Plants and the Surrounding World," for example, was striking in the modernity of some of its conceptions regarding variation in nature and the importance of studying the habits of animals in relation to their environment.[79]

Agassiz, however, was compelled to demonstrate how the informed naturalist could discover in nature the workings of an all powerful Deity. Every single aspect of natural history thus investigated yielded the same generalization. Embryology, geographical distribution, comparative anatomy, ecology, paleontology, and geology were all specific demonstrations of a general truth. The end of the first chapter of the "Essay" contained a catechism of thirty-one separate "conclusions," just in case the reader had failed to perceive the train of the argument in the preceding pages. Agassiz's resolute effort to unite the facts as he knew them to an ideal system is evidenced by one such conclusion: "The distribution of some types over the most extensive range of the surface of the globe, while others are limited to particular geographical areas . . . exhibits thought, a close control in the distribution of the earth's surface among its inhabitants."[80]

The "Essay" showed conclusively that Agassiz felt he had offered the last word regarding philosophical understanding of the facts of natural history. His words reflected a clear determination to subsume every aspect of the evidence under a rigid categorization of nature into mutually exclusive areas of identity. Branches, classes, orders, families, genera, and species were all ideal and hence permanent categories of the Creative Intelligence, manifested in the earthly reality of individual animals and plants. It was plain that any future evidence Agassiz examined would be studied within the framework of this world view.

The review of these volumes by James Dwight Dana in the *American Journal of Science* was a significant analysis. It demonstrated, in two long articles, the anticipations of American naturalists with regard to the product of Agassiz's labors. Dana shared Agassiz's devotion to the spiritual qualities of natural history and felt much respect for the really useful aspects of the work. He was prompted to affirm, therefore, that Agassiz had raised "science to a higher level than it had before attained, and given a force and direction to thought which will inspire rapid progress towards perfection."[81] But these words were strained efforts at approval. Dana and other American colleagues, like Asa Gray, were hardly as convinced as they had once seemed to be that Agassiz was the ultimate authority on the interpretation of natural history. Dana had expected that Agassiz would deal with pressing questions of the day, such as the nature of variation and geographical distribution, in a new and challenging manner.[82] Instead, he found ideas and explanations really no newer than those contained in Agassiz's lectures of a decade before. The Yale naturalist was hardly convinced, for example, by Agassiz's rigid account of the separate and independent creation of species or by his abandonment of traditional rules for determining genetic affiliation in individuals. Dana wanted more attention paid to the causes of variation. He noted that some naturalists might find it difficult to work realistically with Agassiz's strongly defined divisions and subdivisions of the animal kingdom and had many questions to raise regarding the utility of such a system.[83]

Agassiz gave no indication that he was at all distressed by this mixed professional reaction to his volumes. Had he reflected on the status of his relationships with such colleagues as Gray and Dana, he might have realized that these men had been giving considerable thought to questions of change and variation and that they would have liked to see a much more flexible and open-minded attitude on his part toward such problems. They had expected something more from Agassiz than a mere reiteration on a larger scale of the diagram so confidently displayed as the frontispiece for *Principles of Zoology*.[84]

Agassiz had boasted of what the *Contributions* would accomplish. He had set out to fulfil these aims on a grandiose scale that grew larger and larger as public support came his way. These efforts to promote public and personal goals had assumed forms that some colleagues felt detracted from an ultimate intellectual purpose. Gray had written Engelmann about Agassiz's Cambridge activities: "*I* do not like his plan for keeping school for girls— though he will make money by it—freely. But it will take far more of his time than he supposes and *retard* instead of advancing publication."[85] The fact that the intellectual heart of the *Contributions* did not seem to represent any advance over the ideas Agassiz had espoused a decade earlier appeared to lend substance to these fears. It was true that Agassiz's work demonstrated an immense amount of careful, detailed observation, and many were grateful for this. But facts had to be interpreted, and Agassiz's generalizations were sweeping and dogmatic affirmations that seemed to shut off further inquiry rather than to inspire fundamental questioning.

It was impossible for Agassiz to imagine that he had not advanced intellectually in a decade of labor. He was certain he had made an essential contribution to the world's knowledge. But this certainty was derived from his conviction that the philosophical tools that had served him and his teachers so admirably in the past were more valid than they had ever been. He had done a vast amount of investigation, and new knowledge had confirmed older hypotheses. As his conviction grew stronger, Agassiz became more and more isolated from any understanding of contemporary interpretations of nature grounded on suppositions other than his own. Yet he in no way realized this separation, priding himself on the modernity of his investigations in taxonomy and embryology. His failure to sense any change in the intellectual climate stemmed directly from his American success. A powerful social figure, he now sought approval not from colleagues capable of judging his work but from companions of Saturday Club dinners and Lazzaroni activities. Men like Ticknor and Longfellow considered him a giant of their age, and he now accepted such an evaluation as the yardstick to measure his achievement.

The adoration of a public that had sent him fishes and turtles and subscribed in large numbers to the *Contributions* was only one mark of the influence he now exerted upon the national culture. He enjoyed the friendship of men like Emerson, Motley, Lowell, Prescott, and Felton. He could bring his power to bear in helping to decide major questions of public policy in science. He selected the books on natural history that would fill the shelves of the Boston Public Library. The weight of his reputation was so powerful as to impel editor Horace Greeley to refuse to publish an attack on him by an amateur naturalist and to warn the man that it would be futile to pursue a quarrel against such a personage. Any distinguished European visiting Cambridge was bound to be welcomed and entertained by the famous Harvard professor. New Englanders delighted in their beloved naturalist who had come from Europe and preferred America.

Agassiz cherished the memory of the years since 1850 because their fruits were those he had come to esteem. Emerson, before he knew Agassiz, recorded what must have been a common impression: "I saw in the cars a broad-featured, unctuous man, fat and plenteous as some successful politician, and pretty soon divined it must be the foreign professor who has had so marked a success in all our scientific and social circles, having established unquestionable leadership in them all; and it was Agassiz."[86] Agassiz delighted in his role as a grand figure. He associated the meaning of his life with a wider social existence, a career predicated upon activity in the larger arena of public affairs and popular inspiration. He felt no separation between scholarship and public education, between professional activity and the life of the mind. He could serve equally in all capacities, and the Lazzaroni and the *Contributions* reflected his personal synthesis of science and society.

Eleven years before, he had announced his American arrival to A. A. Gould by affirming that he only wanted to study nature "rather than to sparkle in the world," since he had no "pretension to figure in society in any manner whatsoever." America had transformed the author of *Poissons fossiles* into the man who presided at Saturday Club gatherings and taught schoolteachers how to study grasshoppers. These qualities had been evident, of course, in Neuchâtel, but it was the America of the 1850's that brought them to full flower. He would be at once an original investigator, inspired teacher, and the man who would bring European standards of professionalism to the United States. Such ambitions demanded incredible physical energy. It was understandable that Agassiz's health began to suffer and that he found himself unable to work with the same concentration as in earlier years. But he could allow no opportunity that came his way to go unexplored. Asa Gray characterized one aspect of this urge for national dominance when he wrote a colleague: "I met Agassiz . . . and read to him the remark and offer about the insects of Texas—which he jumped at. . . . He is *carnivorous*."[87] Agassiz had to know all there was to learn about nature in America, and at the same time to direct the organization of science. He never paused to reflect on this course. Everywhere he turned, either in Boston or New Orleans, men like James Russell Lowell or admiring fishermen or mechanics pleaded for his company and instruction. Dana or Gray might have felt that the tangible intellectual contribution offered by Agassiz in the glory of this renown was not representative of his real talent. If such colleagues thought they were too familiar with the conceptions of the "Essay on Classification" and wished for more fundamental analysis, Agassiz knew that Longfellow and Lowell felt he had performed a singular service. Apparently Agassiz had established a permanent pattern for present and future self-evaluation —the opinion of appreciative laymen. He was certain, therefore, from the events of these years, that he had accomplished a true identification of private interest with public welfare. With the two volumes of the *Contributions* as evidence of this feat, he felt eager to do more for the advance of science in the service of American civilization.

6

BUILDING A MUSEUM

1857–1861

THE TWO HANDSOMELY BOUND VOLUMES of Agassiz's *Contributions* were received by over twenty-five hundred Americans in the fall of 1857. To these people the author of their volumes was a man to be admired as the personification of science, a grand figure of nature study. Southern planters, New England merchants, common fishermen, college presidents, all joined in praise for this dedicated yet personable scholar.

"As he strode homeward from his walks in the outer fields or marshes, we eyed him gingerly," a Harvard student recalled, "for who could tell what he might have in his pockets?"[1] Henry Adams, reflecting on a Harvard education that seemed of little value, was sufficiently impressed by the romantic quality of Agassiz's descriptions of organic creation to record of his Cambridge years that "the only teaching that appealed to his imagination was a course of lectures by Louis Agassiz on the Glacial Period and Palaeontology, which had more influence on his curiosity than the rest of his college instruction put together."[2] The daughters of good New

England families felt the same appreciation as their parents and brothers. America, drifting toward disunion, torn by sectional dissension, seemed to agree on one thing. Professor Agassiz was a man to be thankful for.

"He was strikingly handsome, with a dome-like head under flowing black locks, large, dark, mobile eyes set in features strong and comely, and with a well-proportioned frame."[3] This was the man of fifty years as seen by college students and Cambridge friends. His increasing stoutness seemed to reflect his growing dominance in national culture. He was respected by the best people, for his ideas of the spiritual quality that pervaded all material creation fitted in perfectly with the conservatism of men like James Walker as well as with the transcendentalism of Emerson. He wanted to see creation as a whole, by thoroughly understanding its parts. By his own account, Agassiz knew he enjoyed "an influence which gains in extent and intensity every year" and was content that he had turned his back on Europe in favor of the rich soil of America.

As the comfortable Quincy Street house welcomed European scholars, American heroes, and Cambridge friends, its genial cigar-smoking host, surrounded by handsome children and a charming, devoted wife, appeared to want nothing more to make life complete and entirely comfortable. Lizzie might have wished for more money for the household, but she probably realized by now this was an impossible desire, since her husband was likely to continue spending all his available funds for increasing his collections. The girls' school and an occasional lecture tour were his "milkcows," as much as he begrudged the time thus expended. Besides, they were a social responsibility, since knowledge of nature, in the social democracy of Agassiz's public vision, was the property of all.

Subscribers to the *Contributions* who could not entirely understand descriptions of the embryology of turtles or Agassiz's taxonomic analyses were nevertheless impatient to receive the next of the promised annual volumes. Also anxiously awaiting the results of Agassiz's labor were Captain Charles Wilkes, who looked forward to seeing the volume describing the fishes collected by the exploring expedition under his charge, and Coast Survey Superintendent Bache, who was similarly eager for his friend's report on the Florida reefs. Baird and others hoped that forthcoming volumes of the *Contributions* would contain Agassiz's heralded natural history of American fishes. In all, America's naturalist had enough to occupy his energy and intellect for years to come, if he chose to work at these tasks.

People who shared these anticipations did not fully know Agassiz. They may have thought him contented, but he was far from happy. They may have believed him eager only to advance specialized knowledge of nature, but this was but one aspect of his current ambition. Agassiz could no more retire to a solitary contemplation of fishes than he could have returned to Europe. He was as devoted as ever to the "progress of science," to be sure. But such dedication was a public one, and Agassiz had become in essence a public man. A man who, at fifty, felt the power of his authority swelling in the public consciousness with every effort he made, could not rest content or confine activity to one sphere. He felt himself at the center of the decision-making process in matters of science, education, and culture.

There are, of course, no definite lines of demarcation and definition to which a man can point as signposts of his changing role in life. Agassiz was less able to do this than anyone else. In the larger sense, Agassiz's public orientation had been defined from his first entrance on the stage of American science and culture. Now, at the height of his fame and influence, Agassiz could not rest until imagination and ambition discovered yet another public goal worthy of his energy. This could not be any ordinary aim; it had to be a magnificent one because of its creator.

Agassiz knew what this goal would be almost as soon as he accepted his appointment at Harvard. He had informed President Everett in 1848 that "some resolution should be taken in reference to a museum . . . [for] without collections lectures will remain deficient."[4] Whenever the opportunity appeared, and in many instances when the occasion did not warrant it, Agassiz made clear to influential persons that he needed a great museum, with large financial backing and appropriate research facilities, in order to be entirely content. As he grew in power, his appeals became more persistent, his demands more repetitive. Harvard had responded in what seemed ample demonstration of its agreement with Agassiz's urgings. First the Charles River bathhouse, then the upper story of Engineer Hall, then the purchase of the Agassiz collections, then the entire wooden building on Kirkland Street, now no longer serving technology, and by 1855 appropriations for the care and increase of collections that totaled six hundred dollars annually. All of this was merely incidental for Agassiz. In fact, he seemed rather to wish he had known none of these small advantages, since they were but inconsequential suggestions of the grandeur of his ambition.

Agassiz strove to secure a physical symbol of his dominant cultural role. He could not be satisfied with a small wooden building that was not only too small to hold his collections but hardly representative of the Agassiz influence in America. Colleagues believed, in 1855, that their Cambridge friend had sworn to devote the next decade to publishing what he had already studied. He had told them so, repeatedly. Now, once undertaken, this project was replaced by an even greater one.

Agassiz realized his physical endurance was not what it had been ten years ago. His eyes, tired from long examinations at the microscope, often gave him trouble, so much so that he had to stop all work at times until he regained his powers of observation. Friends advised a rest, a change of scenery, less preoccupation with public matters. A trip to Europe was urged for the same reason. But he could not rest until he had founded a great museum in America.

The romantic spirit is only unhappy when there is no grand vision to focus on, or when the means to accomplish ideal ends are absent. Agassiz had his ambition, but where was he to find the sources for its realization? That which he desired was, simply, "the means of building up a Museum which . . . would be as important for science as those founded by John Hunter in London or by Cuvier in the Jardin des Plantes." He thus confessed his ambition to Abbott Lawrence in 1854, affirming that "every year I feel deeper the necessity of leaving no time in securing these opportunities, which would prevent my life passing away without having done what I might do."[5] Lawrence, however, died the following year, without making permanent provision for a great American museum. Lawrence did provide, though, for Agassiz's permanent support at Harvard. His will left the university fifty thousand dollars for the salary of the man whose Cambridge career he had made possible.[6]

Although Lawrence was gone, Agassiz was not left without resources. The museum he envisioned needed more than the backing of a single philanthropist; he planned an institution unparalleled in the history of American higher education. There were, to be sure, "mineralogical cabinets" at both Harvard and Yale, but these, in Agassiz's view, were mere collections of local rocks and assorted curiosities. Of existing American museums, those of the Academy of Natural Sciences of Philadelphia, the Boston Society of Natural History, and the Smithsonian Institution came closest to what Agassiz planned for Harvard. These institutions, however, did not have great sums of money to spend and were not teaching museums. Agassiz imagined that he could establish in Cambridge a world-renowned museum that would train new naturalists by virtue of its faculty and resources. The kind of museum Agassiz wanted would be independent of both Harvard College and the Lawrence Scientific School, a separate institution of the university, a natural history center with a separate faculty, a great building, and the resources of Harvard to draw upon if necessary. This would be an educational innovation of first magnitude. Of all the scientists in America, only Agassiz could have dreamed such a dream with any conviction of success.

Agassiz urged the Corporation again and again to do something in support of his project. During the 1850's, he effectively created a university museum by the expedient of referring to the "museum" in reports to university officers. He wrote potential donors of collections that their materials would be in good hands in his "museum"; he hired assistants to work at preserving and identifying the collections he had gathered. At his own estimate, which does not seem unreasonable, he spent ten thousand dollars of his own funds, from the time Harvard purchased his collections until 1859, in adding materials to his establishment. The people of New England knew that Agassiz had no time to waste making money for himself; they knew that all he earned went to the cause of his science. A will made in 1859 stated that Agassiz's only personal property was his collections and his library, confirming the public image of a selfless man.[7] The group of prominent men serving as the Visiting Committee of the Overseers to the Lawrence Scientific School in 1855, including Edward Everett, Jacob Bigelow, and William Greenough, recognized such devotion by reporting that "the extensive and valuable Museum of Zoology and Geology . . . has only been thus far preserved by the generous sacrifices made by the Professor himself . . . in that cause of science to which he has devoted his life . . . to give to our ancient university the rich results of his labors, and the eminent distinction of his name."[8]

Bostonians read the same sentiments in a newspaper editorial, titled simply "Professor Agassiz," written at this same time by the always loyal Felton:

Mr. Agassiz has formed a large Museum of Natural History, which has been recently purchased and permanently established at Cambridge. If this were supported by an adequate fund, and the continued labors of Professor Agassiz were secured in connection with it, in a few years our University would become the acknowledged center of science in the United States. We hope the time is not far distant when a consummation so devoutly to be wished shall be accomplished.[9]

Both comments bore a striking resemblance to words Agassiz had written President Walker just the year before. Furthermore, the Corporation was reminded continually by Agassiz that the wooden building on Kirkland Street housing both his and Harvard's collections could be destroyed at any time by the smallest "spark of fire." By 1856, Agassiz felt more strongly than ever the need to attain his goal. Turtles, fishes, fossils of all sorts, in fact the entire sweep of American natural history gathered by him made some solution imperative. Cans, boxes, casks, and barrels were piled up in great disorder in the rooms of Engineer Hall. He found it impossible to work under such conditions. The great public appeals of 1853 and 1855 for specimens had made the *Contributions* possible. They had also made a museum a necessity.

Agassiz would not be restricted by a conservative university administration or the lack of capital. He determined to make it

impossible for the good-natured but unimaginative President
Walker to refuse his demands. Agassiz had no patience with the
apathy and lethargy that marked Harvard's spirit in the 1850's.
His European background made him discontented with an intellectual environment centered in a college for undergraduates. Professional scholars, he held, needed a university atmosphere to shape
their efforts. Agassiz wanted command of an independent branch
of such a modern university, a "Museum . . . no longer . . . considered as a mere appendage to the Scientific School or the College,
but . . . in itself the object of the special benevolence and liberality
of our citizens at large."[10] If Harvard would not provide the ideal
home for this ideal institution, then Agassiz would take his talents
and idea to another environment. He was determined to enliven
the Boston and Cambridge community with the excitement of a
grand project and to demand concrete evidence of its faith in him
and in the vision of progress he symbolized.

This was not idle speculation. By late 1856, Agassiz had found
the kind of man he had sought since Lawrence's death. He had
discovered a scholar, philanthropist, and industrialist of liberal
spirit who agreed entirely with his own views regarding the needs
of the age. The man was Francis Calley Gray, who would figure
prominently in Agassiz's career were it only for the aid he had rendered in organizing the popular subscription for the *Contributions*.
Gray, former member of the Harvard Corporation and long a
believer in the necessity of modernizing educational philosophy
and practice, was a perfect successor to Abbott Lawrence. As Agassiz and Gray worked together to bring the *Contributions* into being, the two men had come to know each other intimately and to
recognize a mutual devotion to the cause of higher learning.
Agassiz confessed to Gray the character of his guiding ambition.
Gray decided to make this possible, since he was wholly taken with
Agassiz's dedication to a great public end.

Agassiz had learned much regarding the social relationships of
American philanthropy during the past decade. He knew that men
of good will like Abbott Lawrence needed to be guided in their
generosity. Lawrence had founded a school of science but had
never provided for the primary requirement of the man he brought

there to teach "geology." Moreover, no adequate financing for really advanced graduate education had ever been provided. Agassiz had had to depend upon the enthusiasm of private collectors, the aid of the Smithsonian, the generosity of friends, and his own funds for the means to build up his collections. Even such small aid as Harvard supplied could not provide the necessary means for educating advanced students, young men who, to understand their subject, needed specimens from all over America and Europe in large quantities. As he had long ago informed the Harvard Corporation, the progress of science in America depended upon long-range financial support and a measure of public belief that did not demand immediate results for money expended. When Gray made plain his wholehearted agreement with these principles, Agassiz made equally certain that the money Gray gave would be proffered in conformity to the aims Agassiz envisioned.

In the fall of 1856 Agassiz learned that his ambition had been realized. Gray informed him that he had made provision in his will for the founding of a museum. The name of the institution was to be chosen by Agassiz. At the time of drawing up his will, Gray inquired of his naturalist friend concerning his wishes. Agassiz's reply leaves no doubt about the inspiration for the name, or his intellectual purpose in framing the character of the new museum:

Comparative Physiology has a technical meaning, as definite as Comparative Anatomy, applying only to the study of the functions, as the latter relates to that of the structure of organs, neither of these branches of science considering the broad field of Zoology proper. So if you do not like the title of *Comparative Zoology,* which I think is that likely to prevail for our science . . . or that of *Comparative Zoology, Embryology and Paleontology,* which may be too long, it would be best to go back to the old name of Museum of Natural History. Physicians would at once claim as belonging to the Medical School an appropriation made for a Museum of Comparative Physiology, which I do not think is your intention. I hold still that Comparative Zoology is the best as it includes Zoology in all its branches, embryology and paleontology, covers in fact the whole range of Natural History to the exclusion of Botany and Mineralogy for which there are already separate [branches] in the University.[11]

The title of the institution was a precise expression of Agassiz's long-standing opposition to what he considered to be merely descriptive zoology, the identification and naming of species without reference to their taxonomic relationships, embryological character, anatomical structure, or position in the past and present scheme of creation.

Francis Gray was so convinced of the importance of making possible this Agassiz conception that he did not rest his efforts with making provision for the museum in his will. Soon after drawing up this document, Gray planned to approach his Boston friends with a scheme to raise a private subscription for the museum. In December of 1856, however, this devoted friend of Harvard and Agassiz died, unable to accomplish what he had planned or to see the first published volumes of the *Contributions*.

Agassiz, however, learned immediately of the specific character of Gray's will. He knew that Gray had provided that two years after his death, his nephew William Gray, as executor of the estate, was to give fifty thousand dollars to Harvard College, "or such other institution as you see fit," for the establishment of a "Museum of Comparative Zoology." In January of 1857 it was Agassiz's duty to read an obituary notice of Gray to the membership of the American Academy of Arts and Sciences. A good number of Agassiz's audience on this occasion were Harvard faculty, Overseers, and Corporation members. With these words, he announced the general character of Gray's philanthropy, although he said nothing of its specific form:

He had conceived the plan of a great institution, devoted chiefly to the study of Natural History . . . which should in course of time be for this country what the British Museum and the Jardin des Plantes are for England and France . . . and his will bears testimony to the importance which he attached to the establishment of such an institution.[12]

In making this announcement Agassiz knew that a basic provision of the Gray will instructed William Gray to give the sum specified to Harvard or another institution of his choosing, with only the annual income from the investment of the fifty thousand dollars to be used for the museum. This income, moreover, could not be used for constructing buildings or paying salaries of faculty or staff, its purpose being to provide money for the scientific operations of the institution. Finally, the will provided that the museum was not to be attached to any other department of the college or university but was to be a separate institution under the supervision of an independent faculty responsible only to the Harvard Corporation and the Board of Overseers. Clearly, these provisions, just as the name of the museum itself, were inspired by Agassiz.[13]

By January of 1857, then, Agassiz had created and benefited from a set of circumstances that made a great American museum inevitable. He had impressed Harvard officials with the fact that the collections purchased in 1853 for the university and his own materials gathered since that date were in a state that rendered them almost valueless, because of the inadequate building that housed them and because the treasures could be destroyed at any time by fire. Next, Agassiz and Francis Gray had made it imperative that Harvard and its benefactors create a museum suitable to the Agassiz conception. It was unlikely, of course, that the Gray bequest would go to any institution other than Harvard. Yet the fact that the money thus provided supplied only the means for an institution whose building, land, and salaries would have to come from other sources made it vital that the university and its friends supply these necessities. The Corporation might not have known of the exact conditions of the will, but it knew at least, from Agassiz's own announcement, of Gray's intentions and what might very likely take place in two years. Yet during the period 1857–58 neither Agassiz nor the Corporation referred to the impending Gray philanthropy in their numerous interchanges concerning the establishment of a museum. In fact, these years witnessed an almost comic-opera series of communications between Agassiz and President Walker, with neither man referring to what each knew would soon occur. Agassiz was determined that Harvard and its supporters provide the kind of environment he thought necessary for bringing the Jardin idea to full flower in America. As for the Corporation, it could not of course act in reference to an event of the future. Yet there was the necessity of pleasing Agassiz, because there was always the possibility that William Gray might grant his uncle's money to another institution.

With the Gray bequest, Agassiz was almost certain of success, but he left nothing to chance. The fact that early in 1857 prominent citizens contributed seventeen thousand dollars to be added to the Ward Nicholas Boylston fund, this money to be used for the building of Boylston Hall, where an expanded mineralogical collection, anatomical museum, and chemical laboratory would be housed, made him even more optimistic.[14] He was encouraged enough to write Joseph Henry of the recent events, showing both his knowledge of the Gray bequest and expressing his faith in the future:

My hope for a Great Museum . . . now rests, not only upon the desirableness of the plan, but upon the fact that there are actual steps already taken in that direction, though covering only part of the ground. The Museum of . . . anatomy . . . will lead to a general Zoological Museum . . . an effort to combine all these beginnings appears to me inevitable.[15]

The inevitable was made even more certain by the fortunate events of the fall of 1857, events that were acute reminders to New England of Agassiz's world fame. In September of 1857 he received a letter from Auguste Rouland, minister of public instruction of France, offering him the chair of paleontology at the National Museum of Natural History. He would be paid a salary of two thousand dollars, would enjoy complete freedom to undertake any research he desired, and the scientific and financial resources of the institution would be his to command. He would have his own house, provided by the government, and would need to give only a minimum number of lectures each year. Emperor Napoleon and many friends in Paris urged him to accept.[16]

Agassiz could no more have accepted this position than he could have gone to Zurich a few years earlier. The same reasons dictated his rejection of this offer, but now they were even more substantial. To Rouland he affirmed that he was committed to America for the rest of his days and could not interrupt the work represented by the *Contributions,* even though he recognized the French position as one of the most brilliant posts in the domain of natural science.[17] To Sir Philip Egerton he made plain his view of the American environment: "On one side, my cottage at Nahant by the seashore, the reef of Florida, the vessels of the Coast Survey at my command . . . and on the other, the Jardin des Plantes, with all its accumulated treasures. Rightly considered, the chance of studying nature must prevail over the attractions of the Museum."[18] Elizabeth Agassiz believed that if the French offer had come two and a half years before, when her husband had not yet planned the major work of his career, he would have accepted.[19] She was quite right in thinking that Agassiz, caught up in the excitement of seeing the first two volumes of the *Contributions* through the press, was too involved in American activities to give the Jardin much serious thought. She was less aware of his present determination to bring the Jardin to America.

The only real benefit France could confer on Agassiz was the publicity that came from the fact that he had received and rejected the offer. Friends in Boston and Washington saw to it that Rouland's letter and Agassiz's reply affirming his devotion to the United States were reprinted in the local press.[20] Fellow Lazzaroni and Saturday Club members cheered him for his continued faith in their society, and Agassiz spent the winter vacation of 1858 studying the marine animals and coral reefs of Florida, one of the attractions that outweighed the inducements of Paris.

But Agassiz was to hear more from Paris. Returning from Florida in April of 1858, he found still another letter from Rouland, written in November of the previous year, which somehow did not arrive before he left for the South. Agassiz was now informed that the previous offer had been merely exploratory. The emperor had instructed that the position be left unfilled for two years, in the hope that Agassiz would be able to finish his American labors in that period. He was promised a position superior to any other at the Jardin, with authority to reorganize the entire division of natural history in the Paris institution.[21] Friends in France reported confidentially that there was even more to the offer. Agassiz could expect a total compensation of six thousand dollars a year, a lifetime governmental position as a senator of France, and a newly created post as director of the entire Museum of Natural History. A professorship at the Collège de France would also be his. His authority would be even greater than that enjoyed by Cuvier. It seemed inconceivable to European friends that he would spurn such attractions.[22]

Agassiz once more decided in favor of America. Rouland learned that Agassiz, though highly flattered, had dedicated the next ten years of his life to publishing the *Contributions*. He was, more-over, past the point in life where a man took up a new career in a different environment. His life was forever committed to the future of America. Besides, he felt fatigued after the intellectual exertions of recent years.[23] Charles Martins in Paris heard from his friend Agassiz in more revealing terms:

The work I have undertaken here, and the confidence shown in me by those who have at heart the intellectual development of this country, make my return to Europe impossible. . . . I prefer to build anew here rather than to fight my way in the midst of the coteries of Paris. . . . I like my independence.[24]

Although Martins also read glowing reports about the accomplish-ments of American science, it was plain that Agassiz had decided to use this latest French proposition to test the confidence of those who had been so generous to him in the past.

President James Walker received a letter from Agassiz in April of 1858, the initial effort marking the final stage in the museum campaign. Agassiz wrote at length, including in his letter a copy of Rouland's second offer and his reply rejecting this proposal. Emphasizing what he had just forsaken, Agassiz recounted the his-tory of his public efforts to establish a museum at Harvard. Stating that he had spent "the whole amount of my income from the Uni-versity and occasionally considerably more" on his own and Har-vard's collections during the period from 1853 to the present, he affirmed that he was tired of having to lecture and engage in other occupations in order to gain enough funds to preserve museum materials. He wanted an annual appropriation of fifteen hundred dollars for this purpose and urged that the "museum" would bene-fit by a permanent building to replace its present shabby deposi-tory. Friends of the college should be induced to take the necessary steps to gain this end, and Agassiz was ready to help in any way he could.[25]

If such people were not fully aware of what Agassiz had given up, they had another opportunity to find out. George Ticknor saw to this. Humboldt had written him in praise of the *Contributions*, at the same time reflecting on Agassiz's refusal of the Paris offers. "I never believed," Humboldt wrote, "that this illustrious man . . . would accept the offers . . . made him in Paris. I was sure that gratitude would bind him to a new country, where he finds a field so immense for his research and great means of assistance." Humboldt's letter, together with the new correspondence between Agassiz and Rouland, was given wide publicity.[26] Ticknor's reply to Humboldt revealed how even Agassiz's closest friends shared the public image of the man:

You are quite right. . . . Agassiz will remain in the United States. . . . He is happily married. His social position is as agreeable as we can make it. . . . The field for his labors . . . is new and wide. . . . By remaining here, he not only does well for himself, but for the cause of science, to which he so earnestly and effectively devotes his life.[27]

James Dwight Dana echoed what many believed: "I rejoice to learn that you resist such temptations—it ought to quicken the esteem of the country for you, that even royal munificence cannot lead you to leave us."[28] George Engelmann, writing from Europe, commented on Agassiz's exploiting of the Paris offers by reminding Asa Gray that the botanist had once named a plant *Agassizia suavis* in recognition of the talents of his zoologist colleague.[29]

The Harvard Corporation responded in a way that gave substance to Dana's remarks and Gray's nomenclature. Agassiz had written Walker politely but firmly, determined to force some action from the Corporation in recognition of what he had recently refused. The result was that Agassiz was awarded an appropriation of one hundred dollars a month for the year 1858–59 to preserve and enlarge the natural history collections.[30] Agassiz could now count on sums totaling twenty-two hundred dollars from private and university sources. Together with what he could expect from the Gray fund, fifty-two hundred dollars would be available as a source of support for his "museum," at a time when salaries of Harvard College professors were at best less than half this sum. Agassiz now made certain that influential men were fully aware of his desires and would move to help him when the time came.

On December 15, 1858, just nine days before the Gray bequest was announced to the Corporation, Agassiz took the occasion of his annual report to the Visiting Committee of the Overseers to state his wishes firmly. The time had come for resolute action, and the two years just passed had strengthened his case immeasurably. Addressing such men as Jacob Bigelow, William W. Greenough, James Lawrence, and Ezra Lincoln, Agassiz wrote a moving appeal. But it was unlike any prior statement on the same subject that had come from Agassiz's pen since 1848. He was acting in terms of what was soon to become possible by means of Gray's philanthropy, although he made no reference to the bequest.

Agassiz proposed a program that represented a comprehensive organizational scheme for a Harvard museum. He wanted a fire-proof building, to cost fifty thousand dollars. The museum needed willing workers, and Agassiz offered a proposal that would not burden the finances of the university:

I have thought that a number of Curatorships . . . might . . . be founded by some of our wealthy citizens, which would furnish a small income to students who have already taken their degree, and who, wishing to prosecute further their studies under my direction, might thus come by the means of remaining in Cambridge by assisting in the arrangement and preservation of the collection.

Funds would be required for an artist, for two people who would prepare specimens for study, and for the establishment of "honorary" professorships that would bring distinguished American and European scientists to Harvard to lecture and study at the museum for short periods. All of this, Agassiz admitted, would take a great deal of money, but he was not in the least pessimistic. He told the men of the committee how to proceed. "Let us therefore make an appeal to the wealthy, enlightened, and generous to give us a start." The required financing would provide for the building itself, make possible the establishment of curatorships, and supply the means to open the building to the public. Agassiz underscored his plea by detailing what he had recently given up in declining the Paris positions:

Without any consultation with the authorities of our University, I unconditionally declined these offers, believing that I might under favorable conditions, be able to do more for the advancement of science here than in Europe. . . . I ask nothing for myself—I only hope that sooner or later . . . for a man who is past fifty years of age is no longer young, the means may be secured to enable me to do for the institutions of this country, what I have been so earnestly urged to undertake in the Continent where I was born. . . . Do not say it cannot be done, for you cannot suppose that what exists in England and France, cannot be reached in America.[31]

Having launched what was to become a popular drive for realizing the museum idea, Agassiz awaited what he knew would result in appropriate action in Cambridge. On December 20, 1858, William Gray wrote a letter to the Corporation which was received and recorded at its meeting of December 24. Gray wrote that on December 29, he would act to fulfil the provisions of his uncle's will, contained in a letter of instructions Francis Gray had written at the time the document was drawn up. Harvard was offered the income from fifty thousand dollars for the establishment of a Museum of Comparative Zoology "not to be appended to any other department; but to be under the charge of an Independent Faculty, responsible only to the Corporation and Overseers." This was the initial form of Francis Gray's donation, but in offering it William Gray made the bequest subject to further conditions—provisions that illustrated Agassiz's influence with nephew as well as uncle. "Neither the Collections nor any buildings which may contain the same shall ever be designated by any other name than the Museum of Comparative Zoology at Harvard College" were words that would always reflect Agassiz's particular conception of natural history study. Further conditions specified that the museum faculty should consist of the president of Harvard College and four other members, whose places should be filled when necessary by nomination from the faculty, subject to confirmation by the Corporation. The first faculty, William Gray provided, would consist of President Walker, "Professor Louis Agassiz, Director of the Museum, Dr. Jacob Bigelow, Professor Oliver Wendell Holmes, and Professor Jeffries Wyman."[32]

The Harvard Corporation accepted both the provisions of the Gray will and William Gray's more recent conditions immediately, signifying the establishment of the Museum of Comparative Zoology subject to the approval of the Overseers. A notable feature of the institution formed through the alliance between Agassiz and the Gray family was that the conditions of the bequest served to control the expenditure of future money or property that might come to the museum. The museum would always be independent of other university branches or departments. Its name would remain permanent. The public, however, would always think of the institution thus founded as the "Agassiz Museum." To Agassiz this was a misconception of his fundamental public role, for it was much more vital that the world always remember his particular interpretation of the function and study of natural history. His intellectual authority was more important to Agassiz than a personal, subjective identification with the institution he created. The museum, moreover, would be governed by men entirely in sympathy with Agassiz's aims. Dr. Jacob Bigelow, long associated with Harvard, was president of the American Academy, an accomplished medical man and botanist, and a person of considerable influence in the scientific and cultural life of Boston and Cambridge. Wyman and Holmes were both anatomists and medical men who enjoyed a high rank in the community and in the estimate of their colleagues on the Harvard faculty. With the character of the Gray bequest and the nature of the museum faculty, Agassiz had achieved his great ambition in a way that left him almost unlimited freedom to organize the kind of institution he wanted.

Agassiz's appeal to the Committee of the Overseers was of course perfectly timed. As soon as the Gray bequest had been tendered and accepted by the Corporation, the Overseers of Harvard, men of substantial power and influence in Boston and New England, assumed command of the university's progress in natural history. Early in January, 1859, less than a week after William Gray had made his donation, an ex officio member of the Board of Overseers, Governor Nathan Banks, took the first step in securing public support for the new institution. His annual address to the legislature of Massachusetts contained a strong recommendation for state financial support for the Harvard museum, an appeal that testified to the cumulative effect of years of public pleading on Agassiz's part.[33]

Agassiz and his friends were not content to leave the fate of the new enterprise to the decision of lawmakers. It was fitting that, as Abbott Lawrence had made possible Agassiz's first association with Harvard, his son James played a part in achieving the naturalist's mature ambition for science at the university. On January 18, thirty influential friends of Harvard received a printed invitation from James Lawrence, Jacob Bigelow, and William Greenough, three men of the important Visiting Committee of the Overseers, who, upon receiving Agassiz's appeal, had formed themselves into a subcommittee to realize his requests. These gentlemen invited their friends to "a Meeting . . . at Mr. Lawrence's . . . for the purpose of hearing a statement from Professor AGASSIZ in relation to his Museum."[34]

The outcome of this gathering was exactly what Agassiz hoped it would be. The men of Boston who came to Lawrence's Beacon Street home heard an eloquent appeal from a man with a remarkable talent for making natural history understandable and appealing. Agassiz reaffirmed the arguments and substance of his written proposal to the Visiting Committee. His words had the desired effect, for on January 26 fifteen of the men present at the Lawrence home formed a permanent committee for the purpose of starting a private subscription to build the museum. They printed a circular headed simply "THE AGASSIZ MUSEUM." The document contained a succinct and accurate account of Agassiz's efforts in pursuit of his goal during the past thirteen years. People of wealth were urged to help the cause of the naturalist who, "having brought with him from Europe a reputation as extensive as the civilized world . . . has given himself to the study of its unexplored departments with a degree of zeal and an extent and acuteness of observation which promise to place our knowledge . . . of the whole animal kingdom on an entirely new basis." This dedicated man had resisted

all invitations to leave America for Europe. His collections were
housed in "a temporary wooden building . . . in constant and peculiar danger of being destroyed by fire." The good citizens of
Massachusetts could at least "come forward and share with him
the burdens which he has hitherto so nobly, so courageously and
so faithfully borne *alone*. He only desires to be permitted to do
for the institutions of this country what he has been earnestly
urged to undertake on the Continent of Europe." A magnificent
building, where specimens would be displayed for public edifica-
tion, was the aim of the popular subscription, and recipients of
the circular were informed that representatives of the "Building
Committee" would "call on you at an early day" to receive contributions. Jacob Bigelow and Samuel Hooper headed this group,
whose other members, long an informal association dedicated to
Agassiz's purposes, included Samuel Gray Ward, John Murray
Forbes, Martin Brimmer, Gardner Brewer, Ezra Lincoln, James
Lawrence, and William W. Greenough. Agassiz's former student
Theodore Lyman, now an active professional naturalist, was a key
figure in this fund-raising drive, as was George Ticknor, long a
devoted Agassiz enthusiast.[35]

On January 27 many of the men who had been at the Lawrence
home acted in their capacity as the Visiting Committee of the
Overseers to report to this board the action they had taken to insure the success of the museum. "The most influential quarters
. . . in the community at large" were working diligently to raise
money. Aid from the legislature, a body of "generous and public
spirited individuals . . . to whom no just appeal in behalf of any
really meritorious and worthy enterprise is ever made in vain,"
was confidently anticipated. In the light of the fact that the Massachusetts legislature could hardly have been expected to give money
to Harvard for any other purpose, such optimism was another indication of the remarkable congruence of public and private impulses in the founding of the museum. In keeping with this spirit,
the Overseers concurred in the action of the Corporation and established the museum on the same day they received the report
from the Visiting Committee.[36]

The Overseers, matching the enthusiasm of their Visiting Committee, had the report of this body printed in large quantities as a separate document, and this, together with the circular advertising the cause of the "Agassiz Museum" served as important reading material at many private gatherings held in Boston homes during the winter season. Agassiz visited these meetings, lending personal support to their purpose. People of Boston listened with sympathy to the man whose lectures had charmed them on so many occasions. Now, however, the fees for his instruction were much higher. By March, it seemed certain that the goal of fifty thousand dollars for a museum building would be achieved.

But much more than a building was required. A large sum would be needed to provide permanence for the institution. Agassiz had always been fond of reminding Americans of the example of the University of Berlin, an illustration of the manner in which the Prussian state had supported the cause of higher learning. Now he offered Massachusetts the same opportunity. Acting in accordance with Governor Banks's appeal, the Joint Legislative Committee on Education had been holding hearings on the question of public aid for the museum. In March the committee heard from Agassiz himself, as he stormed the State House in pursuit of his goal.

Agassiz's eloquence on this occasion was a singular demonstration of his power to charm the public when he spoke of the wonders of the natural world. His arguments were unanswerable; they appealed to every facet of public consciousness regarding the role of science in a progressive civilization:

My great object is to have a museum founded here which will equal the great museums of the Old World. We have a continent before us for exploration which has as yet been only skimmed on the surface. . . . My earnest desire . . . is . . . to put our universities on a footing with those of Europe, or even ahead of them; so that there would be the same disposition among European students to come to America . . . that there has always been among our students to avail themselves of the advantages of European universities.

Even though the private subscription seemed certain to assure funds for the museum building, Agassiz saw no reason why the legislature should not also support the project. A great building was the only fitting depository for the collections gathered thus far, materials which would furnish state and university "with what money will not buy for you when I am gone—specimens which will be invaluable because they cannot be procured elsewhere." Affirming that he had spent twenty thousand dollars on these materials since coming to America, Agassiz argued that it was really impossible to place a monetary value on his and Harvard's collections, so great was their intrinsic worth. Yet they were housed in a poor wooden building, and were mostly still undescribed and consequently of little real use. He promised even more treasures, since the Smithsonian Institution would send the new museum great quantities of duplicate specimens from its collections. Agassiz himself would give the museum the personal collection he had made since the university purchase of 1853. Unlimited accomplishments would be possible if legislators would only "satisfy the world that the patronage of a free people may be as effective as that of the most powerful potentate."

Ever the diplomat, Agassiz cited recent advances in applied geology and scientific agriculture which, as museum research progressed, would be of direct economic utility to the state. He spoke in the name of religion, affirming that a museum such as he envisioned would demonstrate for all to see the wisdom and power of the Creator. He argued in the spirit of modern education, promising that students at the museum would enjoy new and unique opportunities to learn the techniques of natural history investigation. He cited the history of natural science since Aristotle, demonstrating that the mind of man had always sought to unravel the mysteries of creation, and he promised that the Harvard museum would epitomize the final strivings for such knowledge.[37]

Massachusetts lawmakers would have been an unusual group had they been able to resist such pleadings. The Committee on Education joined in support of the cause of the day by reporting favorably to the Senate and House on two acts, laws that provided the organic form of state aid for the museum. The museum had already been established by the Harvard Corporation and Board of Overseers. Now the state of Massachusetts effectively created this institution anew. On April 2, 1859, Governor Banks signed into law an act passed by the House and Senate that same day. This legislation granted part of the proceeds from the sale of state-owned lands in Boston's Back Bay to five educational institutions, provided that these grants were matched by private donations of equal amounts. Most important was the stipulation that 20 per cent of land sale proceeds, or a sum not to exceed $100,000, would be given to "such persons as may at the present session of the Legislature be incorporated as the 'Trustees of the Museum of Comparative Zoology.'" None of the sums so granted would be paid until the state itself had received $300,000 from the sale of Back Bay lands.

Having granted money which it had not yet received to a corporation that did not yet exist, the House on April 5 and the Senate on April 6 passed an act signed into law by the governor on the latter date "To Incorporate the Trustees of the Museum of Comparative Zoology." The official character of this body was signified by the fact that the governor of the state was to be its president, and other public officials were designated as trustees. These officers were entirely sympathetic to Agassiz's plans for the museum, so that he could rely upon present and future state support. In addition to Agassiz, other designated trustees were James Walker, Jacob Bigelow, William Gray, James Lawrence, Nathaniel Thayer, Samuel Gray Ward, George Ticknor, and Samuel Hooper. The trustees thus represented the government of Harvard, the state, and the men who comprised the private fund-raising committee. The efforts of this last group were made easier by the provision that the Gray bequest of $50,000 would be considered part of the $100,000 required to be raised by private subscription. The trustees were charged with the responsibility of keeping the museum open to the public free of charge and of facilitating the exchange of specimens with other institutions. The positions of Agassiz and William Gray as trustees could be vacated only by death or resignation, when they would be filled by a vote of the legislature.[38]

The required donations from private sources were certain to be forthcoming, since most of the money had already been promised. There was only the need to raise $50,000, but Agassiz's quest took such a powerful hold on New England that by the end of April the grand sum of $71,125 had been subscribed. The list of donors, like the groups who had purchased Agassiz's collections in 1853 and made possible the publication of the *Contributions* in 1857, included names representative of community wealth and prominence. Of the 136 donors on the subscription list, all of whom contributed over $100, ten gave $2,000 each, including Nathan Appleton, William Sturgis, Samuel Hooper, Theodore Lyman, and John Murray Forbes. The list included the names of nearly every man of means and influence in Boston and Cambridge. The Saturday Club was well represented. So too were the Overseers of Harvard, members of the American Academy, and men of State Street.[39] With over $220,000 donated to his museum in less than six months, Agassiz had chosen well in resisting the blandishments of Paris in favor of the opportunities of America. The fact that the money he had received from the popular subscription was given at a time when New England still felt the effects of the panic of 1857, and that the men who supported him represented the very financial and merchant groups hardest hit by the subsequent depression, was still another indication of the attraction he held for his community.

Agassiz had achieved his ambition as a direct result of personal effort and influence. Fellow scientists rejoiced with him in his accomplishment. From England, Lyell wrote Ticknor, "Sir Philip Egerton told me . . . he must go to Vienna to compare some fossil fish with the recent collection. . . . I expect we shall have to go to Boston instead if one Agassiz has free scope. . . ."[40] From New Haven, Dana sent these words of congratulation: "I believe you are doing your greatest work for American Science in securing this endowment; for its influence will be as long as the ages and as wide as the . . . world."[41]

To make more certain the goal Dana predicted, Agassiz worked at a feverish pace between April and June of 1859. Refusing a sizable gift of money pressed on him by Boston friends in recognition of his efforts, he asked that all money be given to the museum. Harvard University kept pace with all these efforts by donating a tract of land of nearly five acres, north of Kirkland Street between Divinity Avenue to the east and Oxford Street to the west. The trustees appointed Jacob Bigelow as the chairman of their building committee. The architects Henry Greenough and George Snell devoted their services free of charge. Agassiz had long discussed his ideal museum with Greenough, the man who had designed his Quincy Street home. Together, these men planned a brick and granite edifice of two stories, with pine floors, oak beams, and ornate iron staircases.[42]

The task of organizing the internal government and institutional character of the museum was a primary one. In May, 1859, Agassiz, Bigelow, Holmes, Wyman, and President Walker drew up the first "Rules and Regulations" for the scientific operations of the museum in keeping with their authority as its faculty. These were of Agassiz's inspiration. The sphere of faculty authority was defined as encompassing all the scientific interests of the museum. The faculty would be responsible for the arrangement and display of collections, would elect qualified museum officers, and would guarantee the rights of all museum faculty to free and unrestricted use of scientific collections. Collections would be made available to all students of natural history, and could be lent to other museums. The faculty would also provide for a museum library, superintend the preparation and publication of catalogues describing the collections, and supervise all instruction given at the museum. The faculty was also made responsible for the expenditure of the Gray fund income. Finally, a most important regulation provided that no persons connected with the museum would be allowed to make private, personal collections of specimens. While the faculty had charge of museum exhibits, Agassiz made certain that all such displays would be organized in accordance with philosophical principles he held sacred. These were the concepts exemplified in the "Essay on Classification." By this rule, museum exhibits were to reflect the permanence of species that Agassiz believed characterized natural history. They were to demonstrate the four great types of the animal kingdom. Museum displays would exhibit the limits of geographical distribution, the embryological and anatomical bases of classification, and the history of past and present animal forms as demonstrated by paleontology and zoology. If such an intellectual character seemed somewhat rigid as a basis for museum exhibition, this was, after all, only proper compensation for the man who had done so much to bring the institution into being.

On June 2, the trustees and the Harvard Corporation approved a series of "Articles of Agreement" defining the relationship of each body to the museum. The articles recognized that Corporation funds were inadequate to finance the museum, and stated that a suitable building would be constructed on the Harvard land given to the museum. There were then set forth the rules by which "both corporations shall run the museum." The first article specifically recognized the control of museum property by the trustees, although Harvard could give money to the museum in the future if it wished to do so. As director of the museum, Agassiz was to arrange for public visits and exhibitions.

In addition to the title of "director" Agassiz was now made the "curator" or "scientific head" of the institution. The money he received as Lawrence professor of zoology and geology would serve as his salary for duties performed at the museum, since his instructorial responsibilities at the scientific school were now transferred to the museum. The articles provided that the Lawrence professor would be ex officio curator of the museum. The curator was charged with the "entire . . . control of the classification and scientific arrangement" of collections. He was to lecture at the museum to students of natural history in the Lawrence School, and also to teach classes of public school teachers there. The "director" was made administrative head of the museum, the "curator" its scientific chief officer. No formal demarcation between the duties of each office was drawn, except a provision that if they were ever held by different men, then the faculty would specifically define the responsibilities of each person. According to Agassiz, the reason for the specific title of curator and the instructorial duties assigned him in this capacity was to demonstrate "that the rights of the Corporation extended to everything in the Museum pertaining to its educational functions." Agassiz confessed in later years that "all of this will have to be changed the moment I am gone, as no one can hereafter perform the double duties of Curator and Director." Agassiz, however, was confident he could serve in such positions, and he felt that the rules adopted for the museum organization represented radical innovations in comparison to contemporary European and American museum practice, wherein access to collections was often restricted to museum officials and exhibitions catered to popular taste.[43]

"Visited Prof. Agassiz at his house. Found him asleep from pure exhaustion," wrote one of Agassiz's students on June 13, 1859.[44] The next day, Agassiz found his labors of the past months amply rewarded when a small ceremony took place with Mrs. Agassiz breaking the ground for the new museum building. Afterward, trustees, faculty members, and students were entertained at the Agassiz home in celebration of the achievement. In need of rest, his health strained to the breaking point, Agassiz then departed with Lizzie to spend the summer in Europe, to return to his native soil for the first time in thirteen years.

The European vacation proved a delightful experience. Agassiz went to Europe with five thousand dollars to be used for the purchase of natural history materials. He took with him accounts of a grand new museum that was being built for him. He brought an American wife who was clearly devoted to him and perfectly acceptable to Paris and London society. In Switzerland a joyous reception awaited, and Agassiz was proud of the immediate affection shown Lizzie by his mother and family. Lizzie was in turn delighted by her welcome and pleased with her husband's happiness. She reported to her mother that "Louis has rested so completely . . . truly rested for the first time since I have known him, and he shows it already in his appearance. People here seem astonished to find him so young and so unchanged. They all attribute it to my good care of him."[45]

Frenchmen greeted their colleague with strong affection. Any hope that Agassiz might still come to Paris was soon dispelled by his enthusiastic accounts of what he had accomplished in the United States. When Paris scientists heard that Agassiz had been given over $120,000 from private individuals for his museum, they could hardly imagine such munificence. In Paris, Agassiz was awarded the Cross of the Legion of Honor, feted at many banquets, and hailed as the successor to Humboldt, who had recently died. Viewing the collections and exhibit halls of the Paris museum, he experienced a natural pride in the museum he was building in a new land and anticipated that his museum would one day surpass this ancient institution. The reception accorded Agassiz in England matched that of France. Grand dinners were held in London, where he met his old friends Richard Owen, Sir Charles Lyell, Sir Roderick Murchison, and Sir Philip Egerton.

The most pleasing aspect of the journey was the opportunity to purchase Old World treasures of natural history. France, England, and Germany all yielded their scientific riches to the man who seemed to command unlimited amounts of Boston's wealth. Agassiz purchased the fossil specimens collected by his former teacher Heinrich G. Bronn, materials from which he had first learned the rudiments of paleontology. The collections and magnificent library of the Belgian geologist Louis G. de Koninck were also purchased for Cambridge. In less than two months, the director of the future American Jardin had spent the five thousand dollars he had brought with him and had committed his museum to purchases totaling ten thousand more. Unlike some countrymen who came to Europe and paid much for things of little value, Agassiz knew well the value of all he bought. A student reported he had "helped to unpack and arrange five fine specimens . . . which Prof laughingly said cost ten cents each. He said they were worth at least $300 each . . . and told us that he had found out what all the remaining private collections in Europe could be bought for, and said that it would require about $100,000 to make a clean sweep of all that are worth having."[46]

Back in Cambridge in the fall of 1859, Agassiz was rested, refreshed, and ready to expend any energy necessary to make sure the museum was built, staffed, and organized according to his wishes. The following year was a hectic but happy one as Agassiz worked at a near fever pitch of activity, in complete enjoyment of life, doing what was closest to his heart and mind.

In October a long-range operational plan for the museum was established. The main emphasis of Agassiz's scheme was upon securing as large a collection of specimens as money and application would allow. The museum would sponsor "traveling naturalists" who would add to its resources from all over the world. Regional collectors would be subsidized, and a complete system of identification, labeling, and exchange of materials with other institutions would begin. The museum would have primary teaching obligations as well. These educational purposes could proceed apace with research, since fifteen students could work as curators of individual departments in return for the free tuition Agassiz granted them, and at the same time could pursue their studies. A full staff of twenty people was thus organized, and a faculty of eight professors planned for the future.[47]

The Harvard Corporation gave him the use of the old Engineer Hall at a moderate rental, and the building was moved to the museum grounds and became "Zoological Hall," serving as a dormitory for museum students. The Lawrence donation of 1855 yielded more than the fifteen hundred dollars needed for Agassiz's salary, and in December of 1859 the Corporation raised this to two thousand. In January, 1860, Agassiz could look forward to a new period of progress for his ambitions when Cornelius Conway Felton succeeded James Walker as president of the university. Alexander Agassiz, a twice-graduated young man, had received an engineering degree from the Lawrence School in 1857 and spent a year in California as a collector for his father and an assistant with the Coast Survey. He now planned to return to Cambridge "to take up Natural History and go in with father." Alex served as a curator of the museum, its general business manager, and a traveling collector. His salary was generously provided by Theodore Lyman, who also worked as a museum curator. Lizzie also looked after accounts, wrote business letters, and made sure that Bostonians learned of the museum's progress. By 1860, then, the faculty of the Agassiz school for girls ran a university and a museum. In the summer the operations of both institutions could be managed from Nahant, where Feltons and Agassizes lived in congenial if somewhat cramped quarters in the cottage by the sea.

Within the museum itself, the year 1859–60 was marked by intense activity. By November of 1859, the first floor of the building had already been constructed, and Agassiz and his students worked hard moving great quantities of specimens from the former laboratory and storing them in the new building. All winter long this labor was supervised by a happy museum director.

These students were a group of dedicated and willing workers, all recently arrived in Cambridge to learn from Agassiz. Alpheus Hyatt, Frederick Ward Putnam, Edward S. Morse, Addison Emery Verrill, Nathaniel Southgate Shaler, Samuel H. Scudder, and Albert Ordway were the outstanding members of a new student body, most of them college graduates, who were to receive the most unique and valuable education in natural history yet offered in the United States.

These aspiring naturalists were each put in charge of a separate category of specimens; all were expected to identify, label, and arrange the material in their care. Agassiz would sometimes help them in their tasks, whistling like a young boy, a cigar clenched between his teeth, a fish in each hand. "He told us how he once studied a long time on an old pair of leather pantaloons trying to find out their *affinities*, mistaking them for a specimen," Verrill reported.[48] Shaler recounted his first impressions of his teacher, "then fifty-two years old . . . he still had the look of a young man. His face was the most genial and engaging that I had ever seen and his manner captivated me altogether." Agassiz spent much time with his students. "In my room my master became divinely young again," Shaler recalled. "He would lie on the sofa, drink what I had to offer . . . take a pipe and return in mind to his student days, or to his plans for work. . . . I have never known a mind of such exuberance, of such eager contact with large desires. . . ."[49]

"Agassiz's young men" were taught to appreciate the value of books and the authority of renowned naturalists, but were also shown that books were sometimes wrong, and that the investigator had to rely upon his own direct experience of nature. Agassiz would often leave students alone for weeks at a time, with a single specimen, until they knew all there was to know about a fish even though they became sickened by the stench of old alcohol. When a student like Shaler studied his fish intensely and became convinced he had discovered something about the specimen that proved Agassiz's ideas in error, the professor informed him, "my boy, now there are two of us who know that."[50] Agassiz introduced his students to the professional aspects of their science, proposing them for membership in the Boston Society of Natural History. He gave most of them tuition scholarships, and in the case of three particularly deserving ones, Shaler, Hyatt, and Verrill, persuaded trustee Nathaniel Thayer to subsidize a salary for them of about one thousand dollars a year.

Inspired by the experience of teaching in his new museum, Agassiz delivered a series of special courses on paleontology, ichthyology, and general zoology. He continually invited questions from students, and when he came to a subject that a particular student had been investigating, the young man was called to the blackboard to give a special demonstration of his knowledge. Jeffries Wyman was also a professor uniformly admired by students. His lectures were models of precise analysis in comparative anatomy; more than this, his careful, patient habits of work and unassuming attitude endeared him to all. Henry James Clark, a graduate of the Lawrence School and an accomplished embryologist, also taught at the museum, his labors recognized in June, 1860, by a Corporation appointment as assistant professor. Harvard had no money to pay Clark, however, and his salary was provided for by Agassiz. Agassiz followed the same procedure in securing the services of the artist Jacques Burkhardt, hoping that such expenditures would in time be assumed by the museum itself. Fortunately, men like Nathaniel Thayer and Theodore Lyman were always at hand ready to provide financial assistance for their dear friend, who, they realized, was too busy to make money.

Agassiz's students shared his excitement, convinced they were important figures in a historic enterprise. Some of them kept careful diaries and journals, eager to record the marvelous experiences of these days. In January, 1860, when Agassiz called all the students to the museum for an important lecture, Ned Morse and A. E. Verrill both took down his words and compared notes later to see that their recollections were accurate.

Prof wished to make a few remarks to us so we all took chairs and sat down around him. He commenced . . . by saying that we were together to pursue the study of Natural History in its true sense, earnestly and scientifically. That we had an opportunity never offered to anyone before. Students in Natural History hitherto have had lectures now and then and a library to go to, but no specimens to handle. He said that in Europe no one can get access to specimens unless he collects them himself. In the British Museum no one is allowed to take a specimen out of the case unless he is really a naturalist and even Cuvier, when he wished to handle a specimen, was obliged to do it in the presence of one of the Officers of the Museum. He said one is obliged to study through the glass of the closed case. He said that we should have the most unlimited license in the use of specimens; to handle when we pleased, but that we should bear in mind that a specimen which is small or common should not be viewed in the light of being worthless.

He said that a specimen worth one cent played as important a part in the museum as that worth a thousand dollars and we must treat it in the same manner. He wished to impress on us the fact that we had a duty of profound interest to perform. That this museum was unlike any other museum ever founded. That it should be carried on in such a liberal manner as to set an example to others and do away with that illiberal spirit which is seen here and there. And in order to make it great, he depended on us to work with a will and an interest. He spoke of the arrangement of the Museum and what he was going to have. One room, as we enter the building, was to be devoted to the epitome of the Animal Kingdom. In it were to be placed types of the Vertebrates, Articulates, Radiates and Mollusks . . . arranged in such a manner that the prominent features of the Animal Kingdom could be seen at a glance. Then he was going to have a suite of rooms for Vertebrates; another for Mollusks, and so on. One large room was to be devoted to Corals. . . . Just then some of the joiners in another room burst out in roaring laughter at some of their jokes. Prof said, "You perceive the difference in the laugh of a cultivated man and the laugh from one that is not."[51]

Impressed by such a program where they were important workers in a grand cause, students of the museum tried to emulate the energy and dedication of their master. They fixed up their own living quarters in "Zoological Hall," aided by gifts of furniture from Agassiz. They formed their own scientific society, the "Agassiz Zoological Club," which held weekly meetings where members read papers on their fields of particular interest, sometimes joined by Agassiz, who must have been thrilled by this American version of the "little academy" of his Munich days. When not at work, they enjoyed gay beer parties in their rooms, went to theatricals at the Howard Athenaeum, or visited the natural history exhibits at Cutting's Aquarial Gardens in Boston. The nearby Somerville Girls Academy provided an opportunity for interesting diversions, as did the occasions when inquiring female students from Mrs. Agassiz's school came to study at the museum. The young gentlemen were gallant. "Met a lady, one of Prof's students who I have never met before, at our snake cage. Conversed with her for some time," Morse wrote.[52] Ladies were invited into "Zoological Hall" to take tea and see more specimens. Scholars visiting the museum were lodged there. "Prof. Bardwell who had gone to bed was treated in a queer and amusing manner," Verrill boasted. "I carried 420 lbs. of human beings across the room and threw it all pell mell from my back on to his bed in which he had just ensconced himself very quietly, as he imagined—poor fellow."[53]

Students listened to "Prof" Agassiz's personal advice as eagerly as to his words of scientific instruction. He urged them to read history and the classics and to learn to appreciate Shakespeare. He told them to follow strict habits of economy if they wished to succeed as scientists. "He said that the other Professors bought *two hats* a year while he made *one* do." He would not have them be misanthropes, however. They were urged "to associate with cultivated persons even if they are not scientific and particularly . . . [with] cultivated and intelligent young ladies."[54]

Agassiz felt free to confess his attitude toward the world to these students. Verrill reported: "He . . . said to me that people did not know how much he sacrificed for science and remarked that he had had a great many offers of from $100 to $500 per evening to give lectures, but he always declined for the sake of devoting all his time and energy to Science." He impressed them with the responsibilities of his position and how hard he had worked to achieve what he now enjoyed: "He mentioned that he would have to attend a meeting of the Trustees tomorrow and said that they were quite a formidable and important body. . . . He also told me how the Gray Fund was willed and that Mr. Gray was under no obligation to give it to the Museum."[55]

Agassiz used the full extent of his range of European friendships to initiate a world-wide system of scientific interchange. Europeans were begged to sell, donate, or lend their collections for the glory of the Cambridge museum. Arrangements were made with a score of sea captains, explorers, missionaries, and amateur naturalists from every part of the globe to act as agents for the museum. True to his promise to the state legislature, Agassiz managed to have the Smithsonian send many government specimens to Cambridge. With Henry still uncertain about the exact character of the national museum, Agassiz employed the weight of his authority to convince him that valuable government specimens could best be studied at a great research museum, namely, the Cambridge establishment. Baird was not entirely happy with this arrangement, probably agreeing with the sentiments of a Smithsonian assistant that "there is a curious prejudice about the absorption of specimens sent to Cambridge for description."[56] There was not much Baird could do, since Agassiz was so persuasive, promising to preserve and identify all government materials until the time came when Baird had sufficient money and room to use them more effectively in Washington. Henry agreed to give Agassiz almost anything he asked for from the Smithsonian, and consequently the Museum of Comparative Zoology was limited only by the size of its building and the scope of Agassiz's plans.[57]

In California and Central America, Tom Cary, Agassiz's brother-in-law, made many important collections. Government-sponsored expeditions to Alaska and the North Pacific included collectors subsidized in part by the Museum of Comparative Zoology. Never before had an American institution of natural history embarked on such a sweeping campaign for specimens. During 1859 alone, the museum received separate shipments of 435 barrels and 98 cans of materials of every conceivable variety, not including the purchases Agassiz made in Europe. A Paris collection yielded one of the rarest groups of fossil plants then known to paleontology. Oswald Heer, who had just five years before invited Agassiz to Zurich, was now informed that he must come to Cambridge to study his specialty of fossil botany, for here he would find materials unique in the world. Heer came, as did other Europeans and Americans, drawn by the treasures collected by Agassiz.[58]

In the spring of 1860, the museum building was completed. Constructed of "Sandy Bay granite and Fresh Pond Bricks," it was, like many other Agassiz conceptions, a partial representation of a grander design. Agassiz had characteristically planned a much larger edifice, but the urge to spend all he could for new collections forced a compromise with this scheme. His ultimate plan envisioned a main building set off by north and south wings forming a hollow square. The initial building was only two-fifths of the planned north wing, a two-story structure that included a basement and attic. True to his promise to his students, the first story was largely devoted to a "synthetic room," containing a representation, in microcosm, of the entire animal kingdom as defined by Cuvier and Agassiz. A lecture room and work rooms were also on this floor. The second story boasted a massive central hall illustrating a systematic exhibition of vertebrate life-history, while other rooms here demonstrated the natural history of the remaining three branches of the animal kingdom—mollusks, radiates, and articulates. Another room on this floor depicted the geographical distribution of animals in North and South America and Europe. Each of the two main floors was encompassed by a gallery that ran the full length of the exhibit halls. The basement was used for storing specimens. In the attic there was space for a library and a photographic establishment, and closets where animal skeletons were stored.

The fireproofing of the building, an aim close to Agassiz's heart when the public campaign for funds was undertaken, raised an immediate problem. Conscious of its public responsibility, the building committee had planned to expend a considerable sum to make the building entirely fireproof. Agassiz, however, sensing the tremendous opportunity before him to buy important collections, pleaded that a building could be too fireproof. Eventually, the desires of the building committee prevailed, although Agassiz did manage to effect some economies. This, then, was the physical representation of Agassiz's ambitions for the institutional progress of American natural history. Like the building, these motivations were still unrealized in their total conception. Both the museum building and Agassiz's aspirations for natural science would grow with the years.

The year 1860 alone witnessed Agassiz making important strides toward fulfilling his urge to gather an entire spectrum of the data of natural history in the museum. Thus Agassiz could report with pride that the year marked the addition of 91,000 new specimens to the museum, these representing nearly 11,000 new "species." "This is a grand result," he affirmed, "if contrasted with the fact that not less than a century ago when Linnaeus published . . . the whole number of animals known to him from all over the world did not amount to 8,000."[59] The progress of zoology in all the world for the past century was mirrored in the progress of Agassiz's museum. Harvard Overseers, museum trustees, people of Boston, and scientists of America were kept fully informed of this progress. Agassiz, the Overseers, and the trustees of the museum vied with one another in their official reports praising the progress of the times. The report of the Visiting Committee of the Overseers for 1860 was typical:

To the Legislature, Executive, and public and private benefactors who have by the influence of position, the contribution of wealth . . . assisted in the founding and forming of this great Normal School of Natural Science, the present and rising generation owe a deep debt of gratitude, for it promises to be, or rather it already *is*, a scientific treasury, from which the teacher and divine may draw illustrations of beauty and of power, illustrating the *word* by the *work of God*. . . .[60]

Agassiz was never more contented. His entire energy was devoted to building the solid scientific foundation for an institution he looked upon as his sole purpose in life. Although he could write, "I am like a drowning man holding to every branch as he goes along, not to be swamped in the current of pressing and daily increasing work," his activities gave him the deepest pleasure.[61] Asa Gray reported to Engelmann about the effect on Agassiz of these endeavors: "Agassiz is well—now *flourishing*. . . . [He is] very busy with his museum—says he writes no letters."[62] Agassiz did write letters, but his correspondence reflected one compulsion, the acquisition of more and more specimens for his museum.

On November 13, 1860, the Museum of Comparative Zoology was dedicated and opened to the public. It was a beautiful day in Cambridge. The sun shone brightly, evidence of nature's good wishes for her student. "There was the most intellectual lot of men together in the lecture room that I have ever seen," Verrill reported of the group that gathered at the museum to open the ceremonies.[63] American scientists, Massachusetts statesmen, donors of money, members of learned societies, university officers, undergraduates, and students of the museum formed a grand procession as they marched to the First Unitarian Church. There, Governor Banks of the trustees received the keys to the museum from Jacob Bigelow, representing the citizens' building committee. The governor made an appropriate speech. Reverend James Walker delivered a short prayer. President Felton spoke. Governor Banks made a dedication address. Dr. Andrew Preston Peabody gave the benediction.

Everyone praised the museum. Agassiz spoke too. He dwelt only briefly on the present condition of the museum. Instead, he spoke of the past and the future. He took his hearers along the road that had led to the accomplishments of 1859–60 but did not let them rest there. If the donors honored in Agassiz's address thought their task was done, the director dispelled such a notion. The museum so nobly dedicated this day could be the great one of the age, if he had more money. A great design such as this should not and could not be allowed to remain fixed and static in its present condition. New collections, a larger building, more assistants, were basic needs. If the gathered assemblage needed proof of Agassiz's talents, they found it in the museum that day, in the displays shown off to best advantage by students, curators, faculty, and director, all proud of what had been accomplished. Speeches, procession, dedication, and museum exhibits were all reported in accurate detail in the next day's press. The newspapers were especially well informed about the contents and character of the museum, since Agassiz had held a press conference for this purpose a few days before.[64]

Agassiz liked ceremonies and was justifiably gladdened by the one he had just presided over. In two years, he had been the recipient of nearly $241,000 in land, buildings, and the materials that made the museum possible. Since 1853 the grand total of monies given or promised to him and his purposes from *Contribu-*

tions subscribers, Abbott Lawrence, Harvard, private individuals, and the state of Massachusetts totaled $591,000, exclusive of his university salary. Except for the museum land and what by now seemed paltry research grants, all this had cost the university nothing. Agassiz had brought munificent support to the study of natural history at Harvard and in the nation. The attraction of his ambitions for science and culture had proved irresistible, even in times of economic distress and national crisis.

In the years between the Gray donation and the museum dedication, Agassiz had spent over $73,000 on the museum building and the collections. Moreover, he had received important additional assistance from private benefactors for museum salaries. The sale of Back Bay lands had not proceeded as rapidly as anticipated, so that the grant from this source would not be available until August of 1861. Somewhat awestruck by Agassiz's ability to consume ever larger sums of money, and with only $3,000 remaining from the private subscription, the trustees convinced the museum director that his spending would have to be budgeted and curtailed. They insisted that the state grant, when paid, should be invested and its principal conserved so as to provide an annual income which when added to the Gray fund would yield $9,000. Agassiz, however, had already planned to spend $20,000 during 1861, and he was not happy with the trustees' scheme, although it would assure a dependable income for the museum over the years. He agreed to this proposal with reluctance and, having satisfied the trustees, planned a new appeal to private and public sources which would allow him to do what he had intended all along.[65]

The years that marked the realization of the museum idea saw Agassiz deeply involved in public enterprise. Scholarly activity, such as the plan for an annual volume of the *Contributions,* was decidedly secondary, but Agassiz did publish the third volume of this series in 1860. This was a study of the classification, embryology, and zoological character of "acalephs," a class of the phylum coelenterata, whose most well-known representatives were the jellyfish Agassiz loved to gather at Nahant. These primitive forms of marine invertebrates were representatives of that branch or type plan of the animal kingdom known as "radiates" in the taxonomy Agassiz advocated. Within this intellectual framework, the monograph demonstrated the advantages and limitations of Agassiz's comparative zoology. Precise knowledge of one class yielded insights regarding the entire branch, and the representatives of one class, compared with one another, produced an understanding of differences and relationships. There were, however, no affinities conceivable between different branches, so that the method was "comparative" only in a limited sense. The dividing walls of this system of classification prevented any inquiry into derivation, relationship, or transmutation in the total context of animate creation. A monograph following this intellectual persuasion could only describe nature in isolation. Also, it was a professional study of interest to zoologists only. The grand impulse to educate all classes of Americans to appreciate nature was forgotten. The public that had made the research possible was pleased with its contribution to the advance of knowledge, but it learned little from intricate analyses of different kinds of jellyfish. And, by 1860, such a work was of only limited value to specialists because of its theoretical orientation.[66]

The contrast between the promise and performance of the *Contributions* was in important measure reflective of the dichotomy between Agassiz's public drives and intellectual motivations in these years. He had thought it possible to synthesize general and special knowledge in the *Contributions* but succeeded in fulfilling only part of this objective. He had convinced himself that the museum would signify a grandiose merger of scholarship designed to advance the national culture and to instruct the populace. With the plan for the *Contributions* he had assumed a tremendous undertaking, one that had not as yet been realized and that now demanded an even greater expenditure of energy.

The *Contributions* had not received a uniformly favorable reception from Agassiz's fellow naturalists. One reason was that the philosophy of nature espoused in these volumes was too sterile and too traditional. Agassiz had become so certain of his authority that he seemed to some to close his mind to the possible validity of any interpretation of nature other than his own. As he told Verrill,

"The practice of some artists of substituting their own fancies for real nature was the result of mere *laziness* and [he] compared such artists to those naturalists who adopt some pleasing but false theory like that of Darwin instead of investigating the difficult points of science."[67] But Agassiz, inevitably committed to furthering the progress of the museum whose future depended upon his ability to attract large sums of money, was in no position to give himself to "investigating the difficult points of science." He had pledged himself to a continual social involvement at a time when the tendencies of the science he thought of himself as dominating demanded intense intellectual consideration of unresolved questions.

But Agassiz did not feel isolated from the internal progress of his science. Convinced of his scientific and social authority, he was proud in having accomplished what no one before him had done. He had made possible a great research and teaching museum, fulfilling a basic impulse from his student days. One by one, the hopes of youth had been realized, and it was difficult for contemporaries to deny the facts of success. Agassiz had provided, in America, the intellectual environment wherein generations of students could investigate nature with all the resources his command of other people's wealth could supply. This was a supreme accomplishment, and no critic dared evaluate this as a less noble or permanent contribution than the fashioning of new concepts by which to understand nature. Furthermore, since Agassiz believed that his understanding of natural history signified the only true interpretation, he stood, in his own view, in an undisputed position of mastery in both the public and professional world.

Thus Agassiz saw the years from 1857 to 1861 as highly rewarding ones. He had not only achieved a great ambition, but had managed to visit his mother once more, to enjoy the plaudits of his European colleagues, and to train his son Alexander and other young men in the profession of natural history. He could even look forward to becoming a grandfather. In the fall of 1860 Alex had married Anna Russell, a former student at the school for girls, and the sister of the wife of his honored close friend Theodore Lyman. When Pauline Agassiz married the wealthy and prominent Quincy A. Shaw shortly thereafter, the family ties of

Agassiz to Boston and America became even stronger. New England had provided the Agassizes with material and spiritual comfort in large measure. Agassiz and his family were happy in the world of the Saturday Club, State Street, Quincy Street, Divinity Avenue, and Nahant.

Nor did Agassiz feel that his labors as a scholar were at an end. People remarked how young and energetic he looked and acted, with the stimulus of the museum providing a constant source of spiritual strength. He could not work as much as he liked to, his eyes continued to trouble him, and he wrote often of being "taken away any minute" and not being able to finish the projects that propelled him ever forward. There was, of course, much to do. Agassiz had for his use the greatest number and variety of natural history materials ever assembled in the United States, housed in an institution which might conceivably outrank all others if the people who heard his dedication address heeded his pleas. If any naturalist possessed both the knowledge and the resources which would enable him to determine the true meaning of natural history, it seemed to many that Louis Agassiz was that man.

7

AGASSIZ, DARWIN, AND TRANSMUTATION

1859-1861

IN THE SPRING OF 1860 neither clouds of civil war nor the need for museum funds could dim Agassiz's optimism. His was an occupation "for which my whole training has fitted me . . . the organization of a great museum," and he wrote Tom Cary of his delight:

I have never been more troubled by the pleasantest kind of pressure than I am now. Imagine yourself unpacking and arranging about one hundred thousand specimens acquired for the Museum within the last six months. . . . Sanguine as I have always been of rapid progress, the reality exceeds all my anticipations.[1]

The museum was the institutional side of Agassiz's life; he lived also in a world that was "full of fun and frolic" and the companionship of Longfellow, Samuel Gridley Howe, James Russell Lowell, and Emerson. There were pleasant summer excursions with fellow Saturday Club members to the "philosopher's camp" in the Adirondack Mountains, where Howe and Lowell hunted and fished while their naturalist-friend busied himself preserving the specimens they brought him. Longfellow was Agassiz's greatest admirer in this convivial society, even though he once declined to go on a hunting trip with Agassiz because Emerson was coming with a gun and he was sure "somebody will be shot." Back in Boston, there were times like the club meeting where William Cullen Bryant was a special guest, and friends insisted that Agassiz preside over a discussion of Shakespeare. He agreed with little persuasion, and one of those who heard him recalled his opening remarks:

Many years ago, when I was a young man, I was introduced to an estimable lady in Paris, who . . . said to me, that she "wondered how a man of sense could spend his days in dissecting a fish." I replied, "Madam, if I could live by a brook which had plenty of gudgeons, I should ask nothing better than to spend my life there." But since I have been in this country, I have become acquainted with a Club in which I meet men of various talents in languages, literature, law, poetry, [and] the conduct of affairs . . . and I confess that I have enlarged my views of life, and besides a brook full of gudgeons, I should like to meet once a month, such a society of friends.[2]

But Agassiz could not enjoy club meetings or Adirondack excursions in full measure. In the midst of a cosmopolitan existence that marked him as a commanding personage of his age, he was forced to return to a world he had almost forgotten. This was the arena of the professional interchange of ideas, a strange environment to Agassiz because he believed in the certainty of his own authority in the realm of science and because he had never found it necessary or profitable to debate his convictions. In November of 1859, as he happily opened boxes of specimens from all over the world, Agassiz received a letter from an English naturalist, a man he had met in 1840 and whose work on coral reefs he had recognized as important. The scientist who wrote him was Charles Darwin, a man only two years his junior, but one whose education, training, and view of nature were sharply distinct from his own. The purpose of the letter was significant.

I have ventured to send you a copy of my Book . . . on the origin of species. As the conclusions at which I have arrived on several points differ so widely from yours, I have thought (should you at any time read my volume) that you might think I had sent it . . . out of a spirit of defiance or bravado; but I assure you that I act under a wholly different frame of mind. I hope that you will at least give me credit, however erroneous you may think my conclusion, for having carefully endeavored to arrive at the truth.[3]

Agassiz was familiar with the factual evidence advanced by Darwin to support conclusions about change, variation, and modification through descent. He was not familiar with the logic that bolstered the Darwinian argument. These ideas were foreign to his view of the world, a view he had confidently espoused for many years. But the publication of the *Origin of Species* made it necessary for Agassiz to examine these ideas and to estimate their significance. Americans expected their interpreter of natural history to inform them of the validity of such views; as an authority in national and international science Agassiz regarded it as an obligation to assess Darwin's work.

Agassiz's public involvements during the past decade, however, and the firmness of his philosophical and scientific convictions illequipped him for such an intellectual role. While he would never admit this deficiency, it had become increasingly clear to men like Gray and Dana that the scholar they had deified in 1846 was less able in 1859. His omniscient philosophy of nature prevented him from sympathetically reviewing the hypothesis of derivation. The world of inquiry had been moving forward, and the work of Darwin, Gray, Joseph Dalton Hooker, and others seemed to provide the kind of insights Agassiz's reaffirmation of Cuvier's natural philosophy could no longer furnish.

Agassiz had consistently warned against the "folly" of the "development theory." To heed the ideas of men like Lamarck or Robert Chambers, the then anonymous author of the *Vestiges of the Natural History of Creation,* was unscientific. For Agassiz the only theoretical outlook worthy of attention was an attitude of mind that interpreted the physical world as the product of an original and continuous intervention by the Supreme Being. Within this framework the naturalist could go forward to examine the facts of nature, ever adding to man's understanding. "Theories" that presupposed change as the result of "physical agents" were as false as they were fanciful; they were the "curse of science." His reactions to Darwin's *Origin* were both understandable and predictable.

Agassiz studied the small green volume with mounting annoyance. The ideas it contained were plainly no different from the notions of development he had long since rejected. "This is truly monstrous!" he wrote in the margin next to a passage he found especially offensive. Darwin had departed from the methods of scientific inquiry so well exemplified in his earlier studies; he had contributed nothing new to the understanding of nature. "What is the great difference between supposing that God makes variable species or that he makes laws by which species vary?" Agassiz asked himself in another marginal note. He refused to give Darwin credit for originality of conception. "What does [all] this prove except an ideal unity holding all parts of one plan together?"[4]

Agassiz's expression of faith in special creationism signified an articulate reaffirmation of an important world view of pre-Darwinian biology. He could thus number such distinguished savants as Cuvier among his intellectual predecessors, and most contemporaries of the 1840's joined him in rejecting concepts of development. Thomas Henry Huxley and Asa Gray shared Agassiz's opposition to the *Vestiges* when the book appeared, but by the 1850's these men showed an increasing awareness of the value of concepts relating to change, modification, and genetic affiliation in nature because of their admiration for the work of Darwin and Joseph Hooker. Huxley's assertion in 1859 that special creationism "is like all the modifications of 'final causation' a barren virgin," was suggestive of the growing dissatisfaction with views such as those held by Agassiz.[5] Agassiz, however, never swerved from dedication to his philosophy of creation.

During the decade of the 1850's, therefore, Agassiz was confronted with a self-imposed task: defending the thesis of separate, successive, and special creation. This was a difficult effort at a time when naturalists were becoming increasingly doubtful of immutability and when his own capacities were limited by the pressures of popularization and other varieties of public activity. In other words, in a crucial decade of scientific investigation and debate Agassiz was the captive of his own authoritative position and relatively isolated from the major professional tendencies of his science.

Agassiz's dedication to the principle of special creationism was strikingly apparent in his interpretation of the character and significance of human origins. His interest in this subject, quickened by his first contact with Negroes in America, had important influences upon American thought regarding the equality of human races in the troubled years before the Civil War. Agassiz's position demonstrated a determination to oppose concepts of "development" from any possible vantage point and widened the intellectual gulf separating him from colleagues like Gray and Dana.

Lecturing at Neuchâtel in 1845, Agassiz had announced his belief that there existed specific "zoological provinces" in nature, notable for distinct flora and fauna and for particular human inhabitants. While the lower animals of these zones were all distinct and separate species, created where they were found and confined to the boundaries in which they lived, the races of man thus circumscribed were "one and the same species capable of ranging over the surface of the globe."[6] Almost as soon as he arrived in the United States, Agassiz, delivering this same lecture, told Lowell Institute audiences that "varieties" of men were confined to specific zoological provinces and that, while they were all of one species, Negroes had a distinct origin from whites and their ancestry could not be traced to the sons of Noah.

This allegation was Agassiz's first effort to refine his special creationist philosophy. By insisting that human beings had originated in more than one place, he was interpreting the history of mankind according to the same logic which he applied to the origin of plants and animals. He went against the teachings of revealed religion by denying a common center of origin for man and by asserting that the existence of non-Caucasian races proceeded from a miraculous intervention other than that recorded in Genesis. The fact that such ideas were advanced not specifically against the Bible but from a dedicated compulsion to expose the fallacies of any concept of the "development" of life from lower to higher forms had led such men as Gray to applaud his reverence for religion and characterize his interpretation as an "original and fundamental" rejection of materialism.[7]

After observing Negroes for the first time during a visit to Philadelphia late in 1846, Agassiz confided to his mother:

I hardly dare to tell you the painful impression I received, so much are the feelings they [Negroes] gave me contrary to all our ideas of the brotherhood of man and unique origin of our species. But truth before all. The more pity I felt at the sight of this degraded and degenerate race, the more . . . impossible it becomes for me to repress the feeling that they are not of the same blood as we are.[8]

When he visited Charleston and the plantations of South Carolina in 1847, southern friends and scientists pressed him for his professional opinion concerning the nature of the Negro. Southerners, determined to defend the slave system, were anxious to discover "scientific" bases for a belief in the inequality of man. Agassiz, just arrived in America, was hardly aware of the use to which his ideas might be put by defenders of slavery. Reflecting his exposure to southern culture and his feelings of strangeness regarding the Negro, Agassiz's opinion on this question now differed considerably from what he had said in Boston. In the North, although maintaining that the Negro and the white were of different origin, he had nevertheless insisted that men were all of the same species. Addressing the exclusive Literary Club of Charleston, however, Agassiz affirmed not only that Negroes and whites were of distinct origins but also that Negroes were quite probably a physiologically and anatomically distinct species. Statements such as these reinforced southern arguments in defense of slavery and established Agassiz as a popular figure in the South. They also showed the extent of Agassiz's belief in special creationism.[9]

Agassiz thus involved himself in the growing American debate on the origin and nature of man. The controversy over the unity or plurality of the human race had been of interest to Americans for many years. In 1839, the accomplished Philadelphia physician and anatomist Samuel George Morton had published *Crania Americana*, a volume based on the study of some nine hundred human skulls native to the Americas, specimens that Agassiz felt were alone enough to justify his visit to America when he first saw them in 1846. Morton's work contained the assertion that there were innate differences in human types inhabiting the continent. Such divergencies were "aboriginal," present from the beginning, and were not to be explained as due to varied environmental conditions. On the basis of this and other publications over the next decade, Morton argued from findings in physical anthropology and archeology that mankind had not originated from a common center and that both human and animal history had been marked by a plurality of origin and creation.[10]

Morton's contentions were given cultural significance in the South by his former student, Dr. Josiah Clark Nott, of Mobile, Alabama. An ardent champion of southern institutions, Nott was motivated by a sort of missionary zeal to convince the general public and the scientific profession of the plural origin and consequent mental and physical distinctiveness of human types. Like Morton, Nott was convinced that the intellectual labors of the pluralist school were findings of "science" and as such sacred; if they contradicted biblical accounts, as some clearly believed, then this was but another chapter in the age-old struggle between the forces of progress against hoary tradition.

In 1844 Nott published *Two Lectures on the Natural History of the Caucasian and Negro Races,* a work which appeared at a time when the southern proslavery argument was becoming more militant. The book offered a rationalization for the ideology of inequality by citing a host of "scientific facts" to buttress the concept of the difference of origin of the two races.[11] Nott's argument received additional support from Morton, who published two essays on the important subject of hybridity in 1847. These analyses overcame a significant deficiency in the pluralist system of belief. Since the time of Buffon, it had been a common assumption among naturalists that if two animal forms could produce fertile offspring, they were of the same species. By the same token, the sterility of hybrids resulting from crosses was taken to be a definite sign of distinct species. Morton's argument was grounded on the assertion that there were many examples, which he cited, of fertile hybrids produced through crossing two known and distinct species. In effect, Morton redefined species by negating a previously accepted criterion, the absence of interfertility. Pluralists could now argue that successful fertile crosses between human types did not necessarily contradict the assertion that man was of more than one species. Significantly, the problem of the sterility of hybrids was of great moment to Darwin, too, but for the opposite reason. Darwin was constrained to show that the sterility of hybrids was not to be taken as a sign of permanency of species. On this question then, both special creationists like Agassiz and Morton and transmutationists like Darwin were equally concerned with removing the designation of fertility or sterility as marks for determining specific identity.[12]

Agassiz's views on the plural origin of mankind were opposed by fundamentalist theologians and others who objected to any reinterpretation of the biblical account of creation. The theological outcry distressed Agassiz. His position was simply that the naturalist had to describe the facts as he knew them, without reference to religious conviction or prejudice. Such a view gained him the sympathy of many fellow scientists who, while not necessarily sharing his opinions on plural origins, nevertheless stoutly defended his right to hold them.

In March of 1850 Agassiz made a significant effort publicly to defend his ideas in the forum of liberal religious opinion. He published an article in the *Christian Examiner,* a leading Boston Unitarian journal that reflected the theology of Harvard and the cosmopolitan religious culture of liberal New England. Here Agassiz affirmed that an unorthodox interpretation of human origins did not preclude a truly devout attitude on the part of the scientist. Demonstrating his real purpose in advocating the plural origin and distinct character of human beings, he claimed that modern science showed that a belief in a common center of origin for all species was the greatest obstacle to the intelligent study of the geographical distribution of animals. He asserted that Genesis referred only to those animals placed near Adam and Eve by the Creator, that there had been at least a dozen separate, successive creations, and that it was consequently impossible to believe with literal interpreters of the Bible that all animals had been created in a single place. There were in nature specific and limited zoological provinces, each containing identifiable animal and human types. These forms "were primitively created all over the world, within the districts which they were naturally to inhabit for a certain time." It was therefore impossible that animals had been created in pairs; it was much more scientific to believe that creation had occurred before the time of Genesis and that it had occurred in many different parts of the world, at many different times, giving rise to separate and distinct animal forms. The only species that Agassiz believed to be not limited to a specific zoological province was man. Man was "a special case" which he would discuss at another time.[13]

Also in March, 1850, Agassiz had his first opportunity to state his views before a professional audience. The occasion was the meeting of the American Association for the Advancement of Science in Charleston, which took place just as the unity-plurality controversy was nearing its most explosive stage. Morton and his opponents were exchanging heated arguments in the public prints, and Nott's writings were a continual source of irritation to fundamentalists. The assemblage looked forward to a definite pronouncement from Agassiz. They were not disappointed.

The sessions of March 15 began with a paper by Nott entitled "An Examination of the Physical History of the Jews, in Its Bearings on the Question of the Unity of Races." A condensation of views Nott had expressed many times before, the paper presented the thesis that a "race," such as Nott considered the Jews to be, remained "pure" when its vitality was not marred by intermixture with other "species" of races. After Nott's paper had been read, Agassiz informed the assembled naturalists that he would clarify his own views on the origin and nature of mankind. All men, he affirmed, enjoyed a spiritual and moral *unity* by virtue of their relationship to the Creative Power. "Viewed zoologically," however, "the several races of men were well marked and distinct." Reviewing his prior designation of zoological provinces, Agassiz now stated that both men and animals inhabited these specific regions, "and this fact . . . tends to prove that the differences . . . between the races were also primitive, and that these races did not originate from a common centre, nor from a single pair." As proof for this contention, Agassiz cited differences between the Caucasian and the Negro, affirming that these distinctions were permanent, as marked in ancient times as they were at present.[14] Since the meetings of the Association were public and the proceedings published, Agassiz had come closer to affirming a belief in the separate and special creation of mankind than ever before. Nott was appreciative of the weight of Agassiz's authority. He wrote Morton: ". . . with Agassiz in the war the battle is ours. This was an immense accession for we shall not only have his name, but the timid will come out from their hiding places."[15]

To make his Charleston statements available to the general public and to establish his fundamental respect for religion, Agassiz published his views once again in the *Christian Examiner,* in July, 1850. The article reflected the extent to which Agassiz had been disturbed by theological dissent. The Bible, he held, was not a textbook of natural history and could not be treated as such by a mind seeking only to discover the truth. Science had a right to investigate these questions without reference either to politics or to religion. Mankind was still one brotherhood in a spiritual sense, Agassiz repeated, and noting that he had been charged with supporting slavery by virtue of his ideas, he affirmed that his studies were concerned with other types of men in addition to the Negro, and that "politicians" could do what they would with the facts— the business of science was simply to discover them.

Agassiz proceeded to examine the evidence. His principal contention was that the geographical distribution of men and animals was the result of creative design and intelligence and not the product of "physical agents" or forces. Neither men nor animals could have originated in single pairs, but rather their creation took place in definite, separated locales. This view, he insisted, actually magnified the power of the Creator in comparison to the traditional interpretation, because it gave Him responsibility for the production of a great diversity in nature, time and time again. Significantly, Agassiz avoided the central question, refusing to state categorically, as he had in public utterances before, that man was of more than one species. He insisted that it was really not very important whether human groups were termed "races," "varieties," or "species"; what was important was a recognition of the fundamental differences between them. In addition to physical peculiarities, Negroes were by nature "submissive," "obsequious," and "imitative"; it was thus mere "mock philanthropy" to consider them equal to whites. Africans, for instance, had been in contact with whites for thousands of years, yet were still averse to civilized influences. White relations with colored peoples would be conducted more intelligently if the fundamental differences between human types were realized and understood. "We hope these remarks will not be considered as attacks upon the Mosaic record," said Agassiz. "We have felt keenly the injustice and unkindness of the charges that have so represented some of our former remarks. We would also disclaim any connection . . . with the political condition of Negroes."[16]

But while defending the freedom of science, Agassiz had done much to supply justification for southern racist doctrines. His claims to objectivity, like his pleading in behalf of special creationism, were vitiated by the conclusions he drew from the scantiest of data and without adequate understanding of physical anthropology or ethnology. Also, he did little to advance actual knowledge of geographical distribution. While his concept of "zoological provinces" was a useful insight, his defense of special creationism did not permit him to see that such provinces might in fact be ecological boundaries that encouraged the transmutation of species. His pronouncements did much to influence popular opinion in favor of pluralism.[17]

In January of 1851, Agassiz emphasized the logic leading to his increasingly dogmatic affirmation of the pluralist position. Writing once more in the *Christian Examiner,* his remarks, titled "Contemplations of God in the Cosmos," had a decidedly theological flavor. Agassiz had never been comfortable with formal theology. Long ago in Neuchâtel he had disputed the assertions of Protestant theologians regarding the conflict between science and the Bible. His recent excursion into pluralist philosophy confirmed his dislike for a theology that restricted freedom of scientific inquiry. Never faithful in church attendance, he had made a comfortable adjustment to the Boston Unitarianism of his wife's family which allowed him the latitude he demanded of religion. His New England friends admired his courage in debating against dogmatic interpretations of the Scriptures, even though some of them were made uncomfortable by the implications of his argument for the equality of man. But this was Agassiz speaking, and Agassiz was a symbol of scientific authority.

In this *Christian Examiner* article, Agassiz made plain to his public the choice that confronted them: either they accepted a belief in the origin of man from a common center and a single pair and hence at the same time embraced the doctrine of development, or else they accepted his belief in the plurality of human species, itself a consistent extension of the principle of separate, successive creation in all natural history. The facts of the case should be examined on the grounds of science, and science alone. And Agassiz's postulates suggested that all life had been created at specific times, again and again, and had not developed from lower to higher forms. The concept of development was accordingly untrue, and special creation remained the only valid interpretation. Agassiz warned that science alone could serve as a reliable guide, but he interpreted science broadly: "The recognition in the animal creation of specific thoughts excludes . . . the idea of a natural development from law, and acknowledges a personal, intelligent God." There was relationship in animal life, to be sure, "but it is not a connection of successive generation, but of intelligent thoughts of the Creator."[18] There had always been continual and direct creative intervention producing successive changes in the organic and inorganic worlds. Species were created by the divine will, and there was no evidence whatever to demonstrate the transformation of one species into another. Animals and plants had been created wherever they were found, and in whatever state they presently enjoyed: men in nations; land animals on land; sea creatures in marine habitats. Life was as diverse and many-hued in former times as in the present; the Creator, for whatever reasons, wiped out one vista and started anew, with models different from, though not progressively superior to, those cast aside. The scheme of creation, then, was a drama whose author could only have appeared to rational critics as a capricious, often purposeless, often callous designer, whose handiwork as a whole bore no relation to its successive acts, and whose actors were never quite sure whether they would live to enjoy the roles assigned to them.[19]

Agassiz had another opportunity in 1854 to express his opposition to transmutation. Heartened by the growing cultural status of pluralism gained through the efforts of Morton and Agassiz, Nott collaborated with George R. Gliddon, an ardent advocate of racial separatism, in writing *Types of Mankind*. The volume was offered as the most modern evidence of the inequality of races, enjoyed a wide popular sale, and was very favorably received by defenders of slavery. The volume contained an essay by Agassiz, his most complete statement on the relationship of geographic distribution to the question of human origins. There were, he affirmed, eight primary human "types"—the Caucasian, Arctic, Mongol, American Indian, Negro, Hottentot, Malayan, and Australian. Each of these groups inhabited specific zoological provinces, zones that were in turn characterized by particular flora and fauna. Such types of human beings were not only distinct from one another; their differences were stamped on them from the beginning. Again Agassiz urged that unless one accepted the unthinkable assumption of development from a primary stock, the concept of the original diversity of mankind, predetermined by the Creator and exemplified in the laws of geographical distribution, was the only alternative. The doctrine of the unity of mankind was consequently "contrary to all the results of modern science." If further proof were required, an impressive map and chart showing the correlations between human and animal zones of creation were included.[20]

Three years later the authors of *Types of Mankind* presented the public with another volume, *Indigenous Races of the Earth*, written with the same intent. *Indigenous Races* represented the culmination of a decade of pluralist efforts to achieve acceptance and respectability. A questionable scientific presentation benefited from Agassiz's authority once more, for he contributed a prefatory essay. He reminded naturalists that the question of the origin of man embraced the entire problem of natural origins. Insisting on the truth of the pluralist position in stronger language than he had ever used before, Agassiz advanced arguments that typified both the range of his interests and the power of special creationism in his thinking. Recent anatomical investigations of monkeys by Richard Owen were cited with approval. Owen had identified three distinct species, and with a logic acceptable only to adherents of special creationism, Agassiz argued that since anthropoids who looked so much alike were not of the same species, then human racial types showing a clear physical distinctness from one another might also reasonably be regarded as comprising different species. Agassiz even ventured into the area of comparative linguistics to support his case. Were the languages of mankind any proof of common origin? Linguistic affiliation between types of mankind might suggest community of origin to some, but to Agassiz this was no proof at all, since plainly distinct animal species shared the power to communicate with one another.[21]

Within ten years Agassiz had provided racial supremacists with primary arguments, had clashed with religionists on the veracity of the Bible, and had made his public aware of the heresies latent in any interpretation of nature based on concepts of development. That Agassiz permitted his reputation to support doctrines of social and racial inequality was indeed tragic. He was not, to be sure, a defender of the institution of slavery. Yet he had many friends in the South and could not entirely free himself from the idea, imposed by his science, that one race was physiologically, genetically, and culturally inferior to the other. It was plain that his judgments were colored by preconceived notions regarding the diversity of species. Although he pleaded that "politicians" were to blame for interpretations placed on his scientific opinions, Agassiz was all too conscious of the weight of his authority not to understand that any pronouncements by him on the subject were bound to be used to serve social ideologies.

The effect of Agassiz's efforts on fellow scientists, however, was not what he hoped. Sir Charles Lyell, hardly a complete convert to evolution in 1860, observed:

I must confess that Agassiz . . . drove me far over into Darwin's camp . . . for when he attributed the original of every race of man to an independent starting point, or act of creation, and not satisfied with that, created whole "nations" at a time, every individual out of "earth, air, and water" as Hooker styles it, the miracles really became to me so much in the way of S. Antonio of Padua . . . that I could not help thinking Lamarck must be right, for the rejection of his system led to such license. . . .[22]

In 1863, Lyell aptly characterized Agassiz's motivation in a paraphrase of the reasoning that compelled his American friend to embrace such ideas:

Were we to admit a unity of origin of such strongly marked varieties as the Negro and European . . . how shall we resist the arguments of the transmutationist, who contends that all closely allied species of animals and plants have . . . sprung from a common parentage . . . ? Where are we to stop, unless we take our stand . . . on the independent creation of distinct human races . . . ?[23]

Lyell also observed that Agassiz's theory

has at least the merit of being consistent with itself, and relieves the opponents of transmutation from the dilemma of explaining why, if so great a divergence from a parent type as that of the white man and negro can take place, a like modifiability should not be able, in the course of ages, to go a step farther, and give rise to differences of specific value.[24]

Agassiz's inability to grant evolutionism plausible status as an explanation of nature was not unique. Many scientists, like Richard Owen and Karl Ernst von Baer, were also opposed to doctrines of development. Lyell and Huxley were reluctant to accept concepts similar to those announced in the *Vestiges of Creation*, and if these two men, who did not hold Agassiz's metaphysical preconceptions, failed to see the truth of evolution before Darwin, it is hardly reasonable to expect Agassiz, whose intellectual history reflected a firm a priori view of the world, to have reacted otherwise.[25]

Agassiz's opposition to transmutation concepts was understandable, but the dogmatism of his attitude was less comprehensible to his American colleagues. Gray and Dana reasoned that when older concepts were challenged, scholars should show an intellectual objectivity and a willingness to grant alternative explanations if they were rational and plausible. Gray, therefore, could write Dana late in 1856 that the times called for intensive investigation by

a number of totally independent naturalists, of widely different pursuits and antecedents, to . . . work towards a common centre, but each to work perfectly independently. Such men as Darwin, Dr. Hooker, De Candolle, Agassiz and myself, most of them with no theory they are bound to support,—ought only to bring out some good results.[26]

Agassiz was one man bound to support a particular theory, and, as Gray was coming to understand with increasing annoyance, this commitment made his fellow naturalist less and less useful to the theoretical progress of natural history in the United States.

In 1846, Gray, Dana, and other colleagues had looked upon Agassiz as a dominant figure, and their admiration was understandable. A decade later, the quality of natural history investigation had improved considerably, partly as a result of the developing institutional progress of science, partly as a result of dedicated Lazzaroni activity, partly in consequence of the coming of age of American higher learning. A museum at Cambridge such as Agassiz founded in 1859, for example, could hardly have been established a decade earlier. These social and cultural factors were complemented and paralleled by the intellectual distinction stemming from the work of Dana in zoology and geology, Gray in botany, Wyman in anatomy, Joseph Leidy in paleontology, Hall in geology and paleontology, and William B. Rogers in geology. Naturalists such as these were doing much to give science in America a professional ranking of the highest order. Gray had an international reputation, had been in correspondence with Darwin since 1855, and was highly respected by Joseph Dalton Hooker and Sir Charles Lyell. Dana, too, was very highly regarded in Europe as well as in the United States and was a very competent teacher of natural history at Yale.

In the public mind, however, the efforts of these men had been overshadowed by the dominance of Agassiz. Those who subscribed to the *Contributions* and helped build the museum at Harvard were hardly as aware of the scholarly contributions of Wyman, Dana, or Gray as they were of Agassiz's authority. He gained support for natural history, while others worked steadily to advance its internal progress. Wyman's efforts are a case in point: through the years he worked in a dedicated fashion to establish a Museum of Comparative Anatomy at Harvard and by 1858 Boylston Hall housed his important collections and exhibits. Yet a "museum" at Harvard in the public mind was a term connected with Agassiz and the edifice on Divinity Avenue; though Wyman was on the faculty of this museum and Agassiz's students benefited greatly from his teaching, the populace associated the study of natural history with Agassiz's name and ambitions.

Agassiz himself had come to enjoy the plaudits of the multitude to such a degree that he seemed to care less and less for the feelings or opinions of his scientific colleagues. Never one to share his authority happily with others, the years of his American success had made him more and more certain of his position. This feeling of intellectual and public mastery led him to enlarge his vision of educating the public in the complexities of natural history in an ever broader scale of activity. Such popular efforts, as they encompassed building a museum or publishing the *Contributions,* were understandable and responsible enterprises, activities that enhanced an appreciation of natural history in the public mind. But when the problem of defending his philosophy of nature came to be of increasing importance, his arguments were primarily addressed to laymen and not to his colleagues. Only twice, for example, in the entire decade of his involvement on the question of the unity or plurality of mankind, did he advance his views in the professional forum. The result was that his interpretation of nature was accepted by non-scientists as the latest expression of scientific truth. But men like Gray and Dana, by 1857 avowed believers in the unity of mankind, were repelled by his public pronouncements.

Gray, by virtue of his close friendship with the English botanist Joseph Dalton Hooker, was in an excellent position to understand the sources of the evolution idea. Hooker's researches in botany were leading him closer and closer to a fundamental understanding of the mechanisms of evolution. He was, moreover, a close friend of Darwin, and his insights benefited tremendously from the stimulus of that naturalist's ideas. As early as the spring of 1854, Gray arranged for the publication in the *American Journal of Science* of the introductory section to Hooker's recently published study of New Zealand flora. The article included analyses by Gray concerning the significance of such research. Hooker's work was significant precisely because it was fundamentally opposed to Agassiz's ideas. Hooker did not deny the traditional belief in the permanency of species, but he departed from Agassiz's view on fundamental points relative to the origin and distribution of species. Using his knowledge of New Zealand flora as a guide, Hooker asserted that the geographical distribution of species was quite probably the result of origin from a common pair in a single center of creation, that distribution was caused by physical agencies, and had occurred over wide areas and produced variation from the original parents. Although Gray raised questions about Hooker's viewpoint, it was apparent he was basically sympathetic to an interpretation opposed to what Agassiz was currently asserting regarding the plurality of animal and human origins.[27]

Gray's developing divergence from Agassiz's view of nature was strengthened by Hooker's approval. Shortly before the excerpt from Hooker's work was published, the English botanist wrote his American friend of his reaction to Gray's accounts of Agassiz's ideas:

I have long been aware of Agassiz's heresies. His opinions are too extreme for respect and hence are mere prejudices. They are further contradicted by facts. Lyell and I have talked him over by the hour. . . . I always think Agassiz an extraordinarily clever fellow and a treasure too as a scientific man, but there are many people whom personally we like and men of science too, but whose views on individual points are best left alone. Giving too much attention, even to oppose, the startling views of such people rather encourages them, and there is an inherent love of getting fame *at any price,* i.e. *getting notoriety,* amongst these French, Swiss, and Italians that leads them to commit themselves on such questions. . . . We have too many clever people in the world, too few sound ones.[28]

Gray's disenchantment with Agassiz as an interpreter of nature reached Darwin through Hooker, and Darwin was very pleased to learn that an American naturalist of Gray's caliber was unimpressed with special creationism. He wrote Hooker in March of 1854:

It is delightful to hear all that he [Gray] says on Agassiz: how very singular it is that so *eminently* clever a man, with such *immense* knowledge . . . should write as he does. Lyell told me that he was so delighted with one of his (Agassiz)['s] lectures . . . that he went to him afterwards and told him, "that it was so delightful, that he could not help all the time wishing it was true." I seldom see a Zoological paper from North America, without observing the impress of Agassiz's doctrines— another proof . . . of how great a man he is.[29]

Gray neither heeded Hooker's advice about opposition to Agassiz nor gave Darwin much time to reflect on the imprint of Agassiz's doctrines on American zoology. In the fall of 1856 and early in

1857 Gray published a statistical analysis of the range and geographical distribution of American flora, papers that Darwin thought "admirable" and "of great importance to my notions." Clearly, Gray was approaching the entire species problem as it related to geographical distribution with an insight and attitude that was to draw him intellectually farther and farther from Agassiz.[30]

It was in this climate that Agassiz published the "Essay on Classification" in the first volume of the *Contributions* and, in Lyell's words, "nailed his colors to the mast." If Gray, Dana, or other naturalists in Europe or America thought that the passage of time or opportunity for reflection had dimmed Agassiz's devotion to special creationism, the "Essay" showed them mistaken. Agassiz had too much to do and had lived too long in the glare of public enterprise to have the time or the inclination for private reflection and assessment. Consequently, every crucial question currently demanding new insights found Agassiz content to rest upon the solidity of the landmarks that had served him so well in the past. He continued to emphasize the idea that every observed fact of ancient and contemporary natural history demonstrated thoughtful planning and an intelligence reflecting

premeditation, power, wisdom, greatness, omniscience, providence. . . . All these facts proclaim aloud the One God, whom man may know, adore and love; and Natural History must, in good time, become the analysis of the thoughts of the Creator of the Universe, as manifested in the animal and vegetable kingdoms, as well as in the inorganic world.[31]

Here, then, was a truly remarkable demonstration of the power of thought over matter. Wherever Agassiz's searching intellect turned in the scheme of nature, empirical study yielded confirmation of a priori truths. Despite its impressive documentation, the fact that the "Essay on Classification" was more of a treatise in natural theology than natural history was a disappointment to Gray and Dana. His colleagues did not have to be convinced of the power of the Deity; they were both men of devout and deep faith in the existence of a first cause. What they wanted from Agassiz was the kind of empirical analysis they knew he was capable of doing, investigation that would shed new light on the questions of the nature of species and the meaning of geographical distribution. Instead, it seemed that beliefs Agassiz did not share were not worthy of discussion, so convinced was he of his own overpowering authority.

Agassiz's authoritarianism was never more apparent to his colleagues than during the controversy with James Dwight Dana. Of all American naturalists, Dana was closest to Agassiz, perhaps because of the theistic view of nature that they shared, perhaps because Agassiz recognized in Dana a man sufficiently removed from his own surroundings to be no threat.

The Dana quarrel began in 1853 when Jules Marcou, a French geologist who had visited America under Agassiz's auspices, published a geological map and description of the country written from his observations as a member of various government exploring expeditions. Geological maps were a touchy enterprise in the America of these years, and when Dana sent the map to James Hall for review in the *American Journal of Science,* he received a bitter condemnation of the work from the Albany geologist.[32] Since Marcou's volume was dedicated to Agassiz and the friendship of the two men was well known, Dana sent the Hall review to Cambridge, asking Agassiz whether it should be published. Agassiz urged publication, feeling that Hall's comments did not detract from the general excellence of Marcou's effort. Moreover, he urged that controversy was good for American science. Hall's review was accordingly published.[33]

Marcou, ignoring the criticism of Hall and other Americans, published a European edition of the map in 1855. It fell to a young and talented American field geologist, William P. Blake, to review this version, also in the *American Journal of Science.* Blake cited his own work and that of geologists such as Hall, David Dale Owen, Josiah Dwight Whitney, and Sir William Logan to refute Marcou's findings. To Blake, the most objectionable aspect of the volume was Marcou's clear disdain for the work of American geologists.[34]

In 1858 Marcou published still another and considerably larger description of American geology, including the map that had been criticized earlier. Once more it was Dana's responsibility to make known the deficiencies of the volume, and this time he undertook the task himself. Noting the unfair presentation of the work of American investigators, Dana, like the two reviewers before him, could not tolerate Marcou's incorrect identifications of many fossils and geological formations. The usually reserved naturalist lamented the publication of such a volume because it was detrimental to "the progress of geological science to have false views about some 500,000 square miles of territory . . . spread widely abroad. . . ." Dana summed up a common feeling when he concluded that "if any American geologist had mapped our strata on such data as have satisfied [Marcou] . . . he would have been deemed young in science, with much yet to learn before he could have a sober hearing."[35]

Agassiz quite properly had no relation to the issue that had been met squarely by competent American geologists. However, reading Dana's harsh words about Marcou, Agassiz wrote a sharp letter to his New Haven friend. He included a reply to Dana's review which he had written and which he wished to appear in the next issue of the *American Journal of Science*. Since the reply contained a frank admission from Agassiz that he had not read the work in question, but could defend Marcou nevertheless on the basis of past scholarship, Dana merely asked his friend to read Marcou before committing his views to print. Highly indignant, Agassiz informed Dana that if his defense of Marcou did not appear he would demand that his name be removed from the list of editors of the *Journal*. Dana acquiesced, and the next issue carried Agassiz's reply to Dana's review.[36]

Agassiz's defense of Marcou, prefaced by his testimony of actual ignorance of the work in dispute, seemed sufficient justification for the loss of esteem he had suffered in the eyes of some co-workers. A weak and inconclusive attempt at rebuttal, the defense repeated Marcou's annoying practice of belittling the work of American geologists. Dana, Hall, William Barton Rogers, and other competent investigators contended that fossils from certain Lake Superior and Rocky Mountain formations were of the Cretaceous period, while Marcou insisted they were of the Jurassic period. Agassiz supported Marcou, even though he had not examined the disputed evidence. Dana followed Agassiz's statements with a short rebuttal, pointing out the obviously erroneous identification of the so-called Jurassic fossils, materials which were later revealed to be European and not American in origin. Dana also pointed out the patent impossibility of a reviewer appraising a book he had not read and called attention to Agassiz's summary treatment of the work of American colleagues.[37]

Marcou was not yet ready or able to concede. When he read Dana's remarks, he published a pamphlet in Zurich, attacking the views of American geologists regarding the Jurassic fossils. What was worse, Marcou had somehow gained access to the private correspondence of Dana and Agassiz relating to the review of his book. Citing this knowledge, he published the false charge that Dana had refused to print Agassiz's defense without exercising editorial censorship. The fact that Marcou also made public Agassiz's threat to resign as a *Journal* editor did not redound to the credit of the Harvard professor. Dana was sick at heart when he read Marcou's latest piece. He wrote of his dismay to Agassiz, recounting the actual facts of their discussions and asking that his colleague publicly correct Marcou's charges.[38]

Significantly, Dana turned to Asa Gray who, like himself, was becoming more and more disenchanted with Agassiz. He asked Gray's advice on how to deal with the situation, sending the botanist a copy of the letter he had written to Agassiz. Affirming that his only purpose had been to get Agassiz to read the offensive Marcou volume before writing about it, Dana voiced what was for him a sharp expression of his sense of ill treatment: "If Agassiz prefers to befriend Marcou . . . to dealing fairly with me, there is a lack of *something* in him. I suppose there was no intentional wrong. But such double dealing is sure to betray itself and prove a double failure."[39]

The July, 1859, issue of the *Journal* carried a statement from its editors concerning the Dana-Agassiz relationship. Presumably written by those editors not directly involved in the affair—Wolcott Gibbs, Gray, and Waldo I. Burnett—this document reprinted Marcou's charges and denied the threat of censorship. The editors made it clear that Agassiz had in fact threatened to resign from the *Journal* and also detailed Dana's effort to have Agassiz read the book before writing about it. They asserted that, had Agassiz done this, he would not have written as he did. Readers were informed that "there has been no interruption of the cordial intercourse that has always subsisted" between Dana and Agassiz, and that if Agassiz had not had to leave for Europe as soon as Marcou's pamphlet reached America, he would have undoubtedly made a public statement denying Marcou's charges of editorial censorship.[40]

The entire affair reflected little credit on Agassiz. Dana could rightly feel suspicious toward his friend, and despite the fact that they continued to enjoy cordial personal relationships, the old flavor of confidence and warm fellowship tended to diminish. Apparently, Agassiz had been guided by personal feeling at the expense of intellectual candor. Marcou had been his confidant and confessor in the sorry days of the Desor episode, and he undoubtedly felt a strong bond of loyalty to the Frenchman for this reason. Also, Agassiz evidently felt that a scientist educated in Europe was superior in knowledge to any American investigator. By defending Marcou, however, Agassiz contradicted his long-standing demands for the dignity of native scholarship in the face of European competition. He had, moreover, been grossly unfair to his old friend James Hall, a man who had spared no effort or expense to provide him with materials for study. He had also slighted the valuable work of the geologist William Barton Rogers, a man whose work was rapidly placing him near the top of his profession.

Apart from illuminating the degree to which Agassiz's authoritarianism influenced the character of his professional relationships, the real significance of the Marcou controversy was its demonstration of the national pride and self-confidence American scientists had come to feel by 1859. Americans of this generation were fully able to evaluate the worth of Charles Darwin's investigations. They were quite prepared to question the authority of Agassiz and to resist any effort to interpret Darwin's work in a manner that implied a prior decision as to its validity. Darwin would receive a full and fair hearing in America if men like Gray, Rogers, Wyman, and Dana had their way. His efforts would be judged with the independence and objectivity exhibited in examining the pretensions of a Marcou or in protesting against the inadequacy of certain aspects of Agassiz's system of classification.

The intellectual scene in 1859, like the work of Darwin himself, had been prepared and conditioned by preceding events and experiences. One American naturalist had determined to challenge Agassiz's rigid interpretation of nature. This man was Asa Gray. Only three years younger than Agassiz, Gray differed from his Harvard colleague in both personality and mental outlook. Gray did not particularly enjoy popular lecturing, and while he counted many notable Bostonians among his relatives and friends, he was much more at home in the company of fellow scientists. In the days when the popular image of natural history in New England was synonymous with Agassiz's name, Gray steadily improved the resources of the botanic garden in his charge, began a conservatory and exhibit of botanical specimens in Harvard Hall, and worked to establish a university herbarium, a goal finally realized in 1864.

Intellectually, two divergent attitudes separated Agassiz and Gray. The botanist looked upon the domain of nature as essentially an open universe for scientific investigation. The more Gray advanced in his special studies of systematic botany and related problems of the geographical distribution of plants, the less satisfied was he that a fixed and final scheme of creation such as that which Agassiz stood for, could explain the data of experience. The more knowledge Gray gained, the more convinced he became that naturalists had to keep an open mind regarding any valid hypotheses that might explain the evidence. This attitude motivated Gray to view Darwin's ideas as worthy of the most serious consideration. Agassiz's conceptions, on the other hand, appeared sterile, retarding further investigation.

In June, 1858, Gray gave clear evidence of his divergence from Agassiz. He wrote R. W. Church in reference to the labors of his Harvard colleague:

If I ever find time I am greatly disposed to write some day upon the principles of classification,—the ground in nature for classification, the nature and distribution and probable origin of species,—knotty points, upon which I incline to differ decidedly from Agassiz, and considerably from the common notions. Some of the more immediate and best-established deductions I hope to bring out in a paper . . . containing the results of a comparison of the flora of Japan . . . with our own of the United States. . . .[41]

The winter of 1858–59 found Gray prepared to expose what he now felt to be Agassiz's relatively useless assumptions, using the data derived from his comparative studies of the botany of America and Japan. This was not a task to be undertaken casually. It was one thing to report in confidential letters to Hooker and Darwin the seemingly absurd character of special creationism as defined and defended by Agassiz. It was quite another to challenge Agassiz in public debate. Hardly a convinced evolutionist, Gray wanted to clear the air for a full and fair discussion of the concept he knew Darwin had already formulated. To do this it was necessary to demonstrate how the facts of nature were open to more than the single interpretation Agassiz placed upon them.

In December Gray presented his research results, data that revealed a sharp contrast with Agassiz's views on the multiple and separate creation of species, to a meeting of the Cambridge Scientific Club. "Agassiz took it *very well* indeed," Gray reported of this encounter to John Torrey.[42] Gray may have been disappointed in not stimulating more excited rebuttal from Agassiz, but his Harvard colleague had other things on his mind. He was concerned with Gray's ideas, to be sure, yet the focus of his attention was not Asa Gray, but William Gray, who was just at this time preparing to announce the museum bequest to the Harvard Corporation. When, in January, the museum became a reality, Agassiz could pay more attention to Asa Gray's arguments. Having succeeded with the body of Cambridge scientists, Gray prepared to go before a larger intellectual public, the members of the American Academy of Arts and Sciences. His purpose, simply, was to "knock out the underpinning of Agassiz's theories about species and their origin—[and] show . . . the high probability of *single* and *local* creation of species, turning some of Agassiz's own guns against him."[43]

If this was war, Agassiz was both prepared and unprepared for combat. He had just published a separate edition of the *Essay on Classification* in England, at the urging of "friends in whose opinion I have great confidence." Certainly, men like Adam Sedgwick and Richard Owen were pleased to have such a succinct and fervent statement of special creationism. Owen regarded the ideas of the *Essay* as "the most important contribution to the right progress of zoological science in all parts of the world where progress permits its cultivation."[44] Darwin, however, evaluating the effort from a different frame of reference, found himself "disappointed" with the scientific testament of Cuvier's last and most devoted disciple. Agassiz could reflect with pride on the *Essay*, which for him dispensed for all time with any ordering of nature that did not presuppose "an intelligent and intelligible connection between the facts of nature . . . [and] the existence of a thinking God. . . ."[45] He also knew that at the very time Academy members were convening in Charles Greeley Loring's home to hear Gray, many of these same men and their friends were planning how best to raise great sums of money for his museum.

On January 11, 1859, Gray delivered an analysis of his Japanese botany research, material soon to be published in the Academy *Memoirs*.[46] Gray's analysis was based on a comparative study of eastern North American and Japanese flora. From the facts at hand it was evident to Gray that there were certain remarkable similarities between the botany of America and that of Japan. How was this relationship to be accounted for? Had these species originated simultaneously in areas where they were now found, with no genetic relationships between them and no possibility of community of origin? This, he informed his hearers, was Agassiz's view. What was the inherent weakness in such an interpretation? Gray did not mince words. The deficiency of Agassiz's view of nature—one that had impressed Gray in 1847—was simply that "it offers no *scientific* explanation of the present distribution of species over the globe; but simply supersedes explanation, by affirming, that as things now are, so they were at the beginning; whereas the facts of the case . . . appear to demand from science something more than a direct reference of the phenomena as they are to the Divine will."[47] And so the student of Japanese botany informed the cultivated laymen of Boston and their scientific friends that the author of *Poissons fossiles* and the man who had elaborated the glacial theory was, for the present at least, more of a theologian than a scientist, the data of natural history having somehow transcended the limits of Agassiz's intellectual capacity. Instead of outworn metaphysics, Gray suggested the need for a new logic and a more rational conceptual framework. It appeared to him that "the idea of the descent of all conspecific individuals from a common stock is so natural, and so inevitably suggested from common observation, that it must be first tried upon the problem; and if the trial be satisfactory, its adoption would follow as a matter of course."[48]

This was the heart of the issue that divided Agassiz from the changing world of natural history. The facts of nature were known equally to him and to those who opposed him. He might not have been familiar with the botany of Asia as compared to that of America, but as he pointed out to Gray in rebuttal, he knew from the zoology of turtles that there were many resemblances between Testudinata typical of each area. Politely and confidently, Agassiz acknowledged "the correctness of the views ascribed to him by Dr. Gray," although this was more of a gesture of good manners than anything else, since he was certain, not only from the compulsions of metaphysics but from the facts of science, that any idea of derivation and descent from a single pair was an incorrect assumption. How would he explain the clear similarity between North American and Japanese flora? He took no refuge in ultimate truths, except to affirm that there had occurred a "primitive adaptation" of forms to similar physical conditions, such conditions remaining unchanged since these types were first introduced. Could the simultaneous appearance of similar forms in different parts of the world be explained by reference to "physical" events or changes affecting dispersal from a common region of origin? Decidedly not. The idea of a unified origin of species was negated by "the warfare which so many species wage upon others." The struggle for existence and the survival of the fittest, then, became Agassiz's explanation for the categorical denial of community of descent.[49]

Gray was not to be deterred by such an argument. Not wishing to avoid direct confrontation, Agassiz had urged that special Academy meetings be held during the next few months for "scientific discussion." Armed with data from geology and paleobotany supplied by Dana, Gray looked forward to the February meeting of the Academy as a further opportunity to expose the fallacies of Agassiz's position. "Your argument is a good one," Dana had assured him. He felt both the Agassiz and Gray positions to be somewhat extreme but was prepared to admit that Gray had made out "a strong case for your side."[50]

In February, Gray once again argued in favor of the single origin of plant and animal species. He cited the similarity between widely separated Japanese and American contemporary flora as proof of genetic derivation and relationship. He accounted for the present distribution of these similar plant forms as the result of glacial action in former times, a physical agency that had operated to divide such flora into the two large but geographically separate groupings in which they now existed. The chance to refute Agassiz with an explanation based on Agassiz's own discoveries led Gray to exaggerate his case. He affirmed the existence of a second postglacial unity of these two floras, with physical agencies once again accounting for their dispersal into present locations. The logical motivation for such an argument was to contradict, by inference, Agassiz's constant assertion that flora and fauna were created anew after the glacial epoch and were hence not derived from prior life. The same compulsion led Gray to cite evidence from paleobotany showing derivation and relationship between plants living before and after the Ice Age. All this was "turning some of Agassiz's own guns against him" with a vengeance, but more was yet to come. Illustrating his developing intellectual affinity with men like Lyell and Darwin, Gray took sharp issue with Agassiz's statement that the similarity of forms existing in different areas was due to "adaptation" from primitive times to the present. Gray's inquiring mind seemed shocked by this dogmatic assumption. Agassiz's position went "to the extreme of implying that the present state of things so strictly represents the primitive condition as to exclude second causes, and to deny that physical influences, known to have been in operation, should have produced their natural effects in former times as well as now."[51] This was exactly Agassiz's position, and he saw no reason to rationalize or apologize for it. Physical "agents," "forces," or "conditions" implied material and mechanistic chaos, and the man who read and reread Aristotle every few years and found him "charming" could only see blind chance as destructive of thought, intelligence, planning, and order in the natural world.

Gray, however, felt confident at being able to bring to bear upon Agassiz's special creationist theology a view of the universe that seemed to him equally devout, yet entirely conformable to the idea of change as affected by secondary causation. The Deity was as familiar and comfortable a center of belief to Gray as to Agassiz. It all depended on the definition of power. How much force, and to what purpose, was to be ascribed to divine cause? If absolute power might be said to corrupt absolutely, the all-embracing superintendence of every moment and facet of creation was to Gray a poverty-stricken view of God's mission in the natural world. If Agassiz could argue from the facts of geology and paleontology to deny descent from empirical as well as metaphysical grounds, Gray, who had pleaded for a scientific discussion of the case, was equally equipped with ultimate explanations, if these were needed for the business at hand. This was urgent work indeed—to show Agassiz incorrect not only with regard to geographical distribution but also with regard to the meaning of supernatural distribution.

Admitting that it was useful for science for different men to approach a problem from different viewpoints, Gray pleaded for the idea that the single creation of species from a common center, rather than the concept of separate, independent creation in many regions, not only was a natural view but also explained the actual facts of experience. More than this, while he was not given to a priori reasoning, he could not help agreeing with P. L. M. de Maupertuis that "it is inconsistent with our idea of Divine Wisdom to suppose that God would use more power than was necessary to accomplish a given end."[52]

Agassiz was fully prepared to counter this argument, both "scientifically" and from a priori grounds, since Gray had left both avenues of rebuttal conveniently open to him. Speaking as a paleontologist, Agassiz quite properly doubted the alleged similarity boasted of by Gray regarding certain fossil plants of the Tertiary period and modern flora. Even Hooker chided his botanist friend that "Agassiz had you on the hip," in this matter, and agreed with Agassiz's words that "if Professor Gray were to exercise the same critical judgment upon the fossil Flora which he does with reference to the existing Flora, he would find differences between the species of the two epochs. . . ."[53] To Agassiz, there could be no genetic relation between pre- and post-glacial flora and fauna, since special creationism would admit of no derivation or community of descent. Here, a scientific explanation seemed the least scientific one according to the canons of a nascent Darwinian like Gray, but Hooker had the best observation on the problem: "The conclusions to be drawn from the facts are hence good if you grant creation by variation, but if you believe in special creations they are comparatively worthless."[54] Facts were good for one set of conclusions if one looked at them with the eyes of a Gray, for another if one saw them in the intellectual coloration, equally protective to be sure, of an Agassiz.

The March meeting of the Academy witnessed Gray trying to demonstrate by means of the mechanics of geographical dispersion and separation the similarities and differences observable among flora. Once more, Agassiz's accursed "physical" causes were discussed, as the villains, or heroes from Gray's viewpoint, of the entire affair. As for Agassiz, he listened to all this in silence, only asserting, by reference to a familiar intellectual bailiwick, that "*even* the fishes of different islands of the Sandwich group differ from each other, as has been proved by the specimens in his possession."[55] It was this kind of reasoning that led Lyell, still secure in a dual skepticism toward both evolution and special creationism, to observe: "Agassiz helped Darwin . . . by . . . not hesitating to call in the creative power to make new species out of nothing whenever the slightest difficulty occurs of making out how a variety got to some distant part of the globe."[56]

In September, 1859, Gray made his views available to a wider audience by publishing a portion of his Japanese botany studies in the *American Journal of Science*. By this time, too, reactions from fellow naturalists had come to Gray, some praising, some criticizing his position. Elias Durand of Philadelphia reported that both he and the paleontologist Joseph Leidy could not side with Gray, but Moses Ashley Curtis told his botanist friend, "I am always suspicious of Agassiz. He has an enormous amount of facts —he is incomparable in the discovery of facts—but I am becoming continually more dissatisfied with him as a *generalizer*.[57]

If these debates resolved nothing, they showed that American scientists were familiar with the major conceptions of the evolution idea before Darwin published. The interchange was also testimony to Gray's rank in the councils of international science; Hooker and Darwin, among others, felt it a primary necessity to inform the Harvard botanist of the progress of their thinking and were in turn enlightened by his research. There was, however, another side of the coin. One reason why the academicians and laymen of Boston were so well informed on major aspects of the new biology was that Agassiz had spent so much time and effort contradicting these ideas. Before 1859, Agassiz had argued with almost every major assumption of the forthcoming Darwinian analysis. As Gray knew and Agassiz indicated by his protestations, the world was prepared for a revival of the "development" theory. But this would be in a form that, as Gray predicted, would obviate many of the older arguments against it. In Agassiz's view, every old argument was just as valid as ever; Darwin's work supplied no new mechanism or interpretation but was simply a rehash of Lamarck, Oken, and the *Vestiges*. It was hardly worth the bother, it seemed, for the director of the Harvard museum to refute the arguments again, but bother he must, because his colleagues would not let the matter rest.

Agassiz's cosmic philosophy shaped his entire reaction to the evolution idea. His definition of the relation of natural history to transcendental conceptions was that such conceptions were basic to understanding and were supported by evidence. Thus he could assert:

There is a system in nature . . . to which the different [classification] systems of authors are successive approximations. . . . This growing coincidence between our systems and that of nature shows ... the identity of the operations of the human and the Divine intellect; especially when it is remembered to what an extraordinary degree many *a priori* conceptions, relating to nature, have in the end proved to agree with reality, in spite of every objection at first offered to them by empiric observers.[58]

An attitude such as this made Agassiz appear to his critics an exponent of a traditional idealism whose German education in the spirit of *Naturphilosophie* prevented him from admitting the validity of an objective interpretation of nature based on observable, secondary phenomena. This was an understandable reaction to Agassiz. There was an unbroken thread connecting his mental outlook with a view of nature stretching back to Plato, a view intellectually close to a concept of being in which the immaterial world was considered the essence of reality. Exemplifying this intellectual tradition, Agassiz saw natural history as the earthly representation of spirit, and thought of the Creative Power as having engineered a timeless, all-encompassing plan for the universe. This scheme of creation was rational, because nature past and present illustrated the creative intention. All facts could be subsumed under this master plan that had been fashioned in the beginning, and all apparent change explained as indicative of a predictable, fixed order in the universe. Species, the individual units of identity in nature, were types of thought reflecting an ideal, immaterial inspiration. The same was true of the larger taxonomic categories—genera, families, orders, branches, and kingdoms. All such categories had no real existence in nature. Reality could be discovered only in the character of the individual animals and plants that had inhabited and were now inhabiting the material world. The individual fossil or living form represented on earth the categories of divine thought ranging from species to kingdom and ultimately symbolized a complete identity with the highest concept of being, God.

For Agassiz there was only one method by which an insight could be gained into this creative process, and that was the method of the natural scientist. The naturalist had an understanding vastly superior to the theologian; it was his expert knowledge of the data of the material world that could provide continual and ever more impressive verification of the power and grandeur implicit in the plan of creation. The fact that Agassiz thought of himself as possessing this ability provided him with the intellectual drive to achieve superior knowledge. It was this life role, moreover, that prevented a simple espousal of traditional idealism. Without constant empirical study, Agassiz would have been deprived of a basis for offering the world new demonstrations of the work of the Creative Power, such as the Ice Age. In drawing a spiritual lesson from his study, Agassiz had to create "species" that did not exist, because he could not admit variation and had to interpret the glacial epoch as another event in a long chain of divinely inspired catastrophes. It was this intellectual quality that made Agassiz such a formidable and perplexing opponent for men like Darwin and Gray. He was quite capable of making the most admirable scientific discoveries reflecting complete devotion to scientific method, but he would then interpret the data through the medium of what seemed to be the most absurd metaphysics. Faced with this kind of mentality, Darwin and his defenders understandably labeled Agassiz the advocate of an outworn idealism.

The tragedy of Agassiz's relationship to Darwin's ideas was that, in a crucial decade of transformation in natural history interpretation, he had given too little thought to justifying his own viewpoint. When Agassiz finally published an integrated statement of his philosophy in 1857, the "Essay on Classification" represented ideas that had little value for his times.

This publication demonstrated, however, that Agassiz was by this time entirely certain that the teachings of *Naturphilosophie* were incompatible with special creationism. He therefore equated this concept with the false notion that "all animals formed but one simple, continuous series," an idea that could readily "become the foundation of a system of the philosophy of nature which suggests all animals as [being] the different degrees of development of a few primitive types."[59] It was but a short step from such a view to one that interpreted animal forms as sharing a unity of origin and genetic derivation, illustrating the transformation of one form into another through modification from "physical" causes. Unable to tolerate this idea, Agassiz found it necessary to abjure what he felt were these larger tendencies of *Naturphilosophie,* all the while retaining the mental attitude once derived from its idealism, the ability to interpret the data of experience as significant of a meaning above and beyond experience.

Naturphilosophie seemed a threat to Agassiz's special creationism primarily because it assumed a continuity in organic creation. Agassiz and his honored master Cuvier, on the other hand, deeply believed that the creative plan was so ordered as to illustrate discontinuity and the independence of natural categories. Thus catastrophes had operated to break the thread of natural history on many occasions. Moreover, since species and the larger units of identity were symbolic of divine intelligence, they were immutable and could never be said to illustrate material connection with each other. Individuals representing the divine plan were created independently and separately. This discontinuous view of creation gave the Deity much more power than believers in "development" were ever able to allow. Multiple and new creations were symbolic of the discontinuity ordained by the creator.[60]

Agassiz did believe, however, in one particular concept of continuity and development. Indebted to his German education from Döllinger, he affirmed that change was to be discerned in the life-history of the individual form, namely, the ontogenetic transformations revealed by embryology. The development of the individual from egg to adult signified, to Agassiz, a progressive, unfolding evolution along a path predetermined by the potentiality of the original egg and ending in a fixed form that was the permanent character of the individual. Change and development were in this view transitory stages in the achievement of permanence. Schelling employed this concept to demonstrate the existence of a supreme being who could ordain the potentiality of highest perfection from the beginning. Agassiz drew similar comfort from embryology, synthesizing empiricism and idealism by insisting that the naturalist had to observe the development of the egg under the microscope to experience demonstrations of absolute power. Understandably, Agassiz insisted that embryology provided "the most trustworthy standard to determine relative rank among animals."[61] This science was the necessary basis for all classification, since study of individual development revealed how the animal conformed to the essence of its type. Individual growth reflected an unfolding of the higher categories of identity,. and by studying a single fish Agassiz could see the entire scale of being from species to branch in the animal kingdom.

Embryology thus illustrated the entire history of life. Agassiz, therefore, could never understand why the evolution concept of Darwin required such a great amount of time to accomplish change in species or types when he could observe change and evolution that occurred rapidly in the individual. If such change was so sudden in the history of life from egg to adult, it was incomprehensible why great periods were required to effect changes in classes, orders, or types. To Agassiz change was dynamic and catastrophic in embryology, just as it was in geology. In each instance, sudden change resulted in preordained, final purpose.

Agassiz could not understand the evolutionary process because he confused two different kinds of evolution. He made the common error of his time of equating the history of the individual—ontogeny—with the history of the type or race—phylogeny. Agassiz believed that the various phases of embryological development or ontogeny were in fact determined by the inherent race history that each individual form contained within its germ as a kind of preview of things to come. Thus the embryology of the animal revealed in successive stages the predetermined scale of categories to which it belonged—species, genus, family, and so on.

Agassiz was consequently very impressed with the "biogenetic law," that ontogeny or individual development is a recapitulation of phylogeny or racial history, the history of the type being the cause of the history of the individual. His student Joseph Le Conte claimed that Agassiz had discovered this "law."[62] This was an unfounded assertion, because the concept had been known since the late eighteenth century, and Agassiz had learned it from his teacher Tiedemann.[63] Agassiz's specific contribution to the recapitulation concept was empirical. In his own words, "I have shown that there is a correspondence between the succession of Fishes in geological times and the different stages of growth in their egg, that is all."[64]

Analysts such as Le Conte and others claimed that Agassiz's association with the recapitulation idea made him a notable forerunner of Darwin. Nothing could be further from the truth. Agassiz's interpretation of the facts of embryology was a cosmic one:

The leading thought which runs through the succession of all organized beings in past ages is manifested again in new combinations in the phases of development of the living representatives of these different types. It exhibits everywhere the working of the same creative Mind, through all times, and upon the surface of the whole globe.[65]

Moreover, Agassiz emphatically contradicted the wider uses of the recapitulation concept by men of his generation, an interpretation that viewed the separate examples of ontogeny as proof of a long history of causally connected phylogenetic transformations in an ascending scale of development from lower to higher forms beginning with the earliest ancestor and ending with contemporary creation.

Agassiz insisted, therefore, that embryology showed a recapitulation of phylogeny only in the repetition of the natural history of the particular and separate type-plan to which the individual belonged. In so doing he reflected his disapproval of the assumptions of *Naturphilosophie*, that there was an ascending and unbroken scale of development from lower to higher forms. He was explicit on this point:

It has been maintained . . . that the higher animals pass during their development through all the phases characteristic of the inferior classes. Put in this form, no statement can be further from the truth; and yet there are decided relations, *within certain limits,* between the embryonic stages of growth of higher animals and the permanent characters of others of an inferior grade. . . . As eggs, in their primitive condition, animals do not differ one from the other; but as soon as the embryo has begun to show any characteristic features, it presents such peculiarities as distinguish its branch. It cannot, therefore, be said that any animal passes through the phases of development *which are not included within the limits of its own branch.* No Vertebrate is, or resembles at any time, an Articulate; no Articulate a Mollusk. . . . Whatever correlations between the young of higher animals and the perfect condition of inferior ones may be traced, they are always limited to representatives of the same branch. . . . No higher animal passes through phases of development recalling all the lower types of the animal kingdom.[66]

Agassiz's interpretation of the recapitulation idea had consequences for the concept of evolution. From the first, Agassiz was much more radical in regard to recapitulation than the embryologist Karl Ernst von Baer. Agassiz believed that ontogeny was a recapitulation of adult ancestral forms, while Von Baer would grant only that recapitulation was limited to a repetition of young or intermediate forms in the life-history of ancestors and that the individual deviated from these resemblances in a progressive fashion during its growth. In 1859 Darwin cited Agassiz's concept of adult recapitulation and Agassiz's belief that this process of repetition in the individual signified the history of the race. For Darwin, this concept "accords well with the theory of natural selection," and he hoped it would be proved in the future. Subsequently, Darwin accepted the Agassiz view without qualification.[67]

Agassiz's view of recapitulation as a direct repetition of final adult forms was erroneous. Darwin's acceptance of it had unfortunate results for the later history of the evolution doctrine. Von Baer's view, on the other hand, laid the groundwork for the modern science of embryology by stressing the fact of individual development from egg to adult, and the very limited recapitulation of younger forms in such development. Had Darwin followed Von Baer and not Agassiz, modern embryology would not have had to rescue Von Baer's interpretations from the obscurity in which they were placed by the triumph of Darwinism and by the ideas of such subsequent advocates of the Agassiz position as Ernst Haeckel. Von Baer, of course, opposed evolution from idealistic presuppositions, and vacillated a good deal in his own relationship to Darwinism. Nevertheless, when modern embryologists who were intellectually equipped to separate Von Baer the idealist from Von Baer the embryologist perceived the value of his view of recapitulation, they could employ it as a means of understanding phylogeny as the result of individual ontogeny in particular periods of natural history.[68]

To call Agassiz a precursor of Darwin on the basis of Darwin's ill-considered use of an erroneous Agassiz conception is a vast mistake. In fact, when Von Baer criticized Darwin for his use of the recapitulation concept, he was in effect criticizing Agassiz. Agassiz was wrong on recapitulation, and Darwin made the same error. Darwin made other errors too, but despite gaps in his knowledge, despite ignorance of the mechanism of heredity, and despite Agassiz, Darwin was right. He was right because the evolution idea did not require the recapitulation theory for its general validity. Darwin, after all, understood phylogeny, and Agassiz did not.

Regardless of the erroneous Agassiz belief that individual development was determined by previous ancestral history, it is most nearly accurate to say that the history of types and races is the result of separate, modified, individual transformations. Ontogeny "causes" phylogeny in the large sense, rather than the reverse of this process, as Agassiz believed. Phylogeny, moreover, is best understood through knowledge of the history of life. Organic development occurs through the introduction and preservation of new and useful variations and the consequent influence of such transformations on the character of subsequent populations.[69]

In Von Baer's criticisms, Darwin paid a heavy price for his use of Agassiz's interpretation of recapitulation. To make matters worse, Darwin did not realize that Agassiz had expressed strong reservations about the very recapitulation idea he advocated and Darwin used. Agassiz criticized recapitulation, moreover, before 1859, and his criticism was both empirical and idealistic.

Agassiz did so because of a growing realization that the concept was useful to advocates of the development hypothesis. Recapitulation, sometimes put forward as proof of a long, continuous sweep of natural history with types and races transformed into more advanced types, was a view of phylogeny Agassiz could never accept. Consequently, he cast doubt upon such continuity, taking issue with the logical extension of an idea he had advocated by citing evidence that demonstrated that ontogeny did not always recapitulate phylogeny in direct repetition, since many characters appeared in the individual in a sequence different from that in which they had appeared in the history of the type.[70] Agassiz joined Von Baer both before and after 1859 in opposing concepts of development with the weapons of idealism. For Agassiz, the reality of the plan of creation was threatened by a historical view of the evolution of types and races; permanence of type was also threatened by a concept of transmutation made possible through the agency of physical processes. Hence recapitulation, to Agassiz, had to prove thought and premeditation.

Embryology provided Agassiz with empirical demonstrations of cosmic assumptions in natural history. He sought the same satisfaction from paleontology. If it was possible to see the correspondence of ontogeny and phylogeny as a demonstration of thoughtful planning in the past, it was equally possible to define such preordained ordering as a characteristic of the future. He thus identified the existence of what he called "prophetic types" in paleontology, intermediate forms like sauroid fishes, that possessed a combination of both ichthyic characters and the reptilian characters of animals that came after them in geological time. These forms were defined as "a prophecy in . . . earlier times of an order of things not [yet] possible with the . . . combinations then prevailing in the animal kingdom." Prophetic types were, therefore, "types which are frequently prominent among the representatives of past ages," combining in their particular structure "peculiarities which at later periods are only observed separately in different distinct types." These forms that "lean towards other combinations fully realized in a later period" afforded the naturalist with "the most unexpected evidence that the plan of the whole creation had been maturely considered long before it was executed."[71] Whereas "prophetic types" might seem to suggest to transmutationists the fossil and living evidence of transformation from one type to another, to Agassiz evidence from paleontology provided a remarkable demonstration of the operations of mind in nature.

The ultimate significance of these viewpoints of philosophy and science was that Agassiz had encountered and evaluated many of the basic foundations of the evolution idea before the *Origin of Species* was published. Agassiz's involvement was not at all unusual, in the light of the prehistory of the evolution concept and his own awareness of the factual data from which Darwin constructed his synthesis. But awareness did not mean understanding. Familiar as he was with the idea of development, Agassiz was neither able nor prepared to understand its logic and its mechanisms. From the viewpoint of intellectual interchange concerning evolution Agassiz had already made up his mind, and his reaction was predictable.

In January, 1860, when the American discussion concerning Darwinian evolution began through a series of debates in the learned societies of Boston, the ground had been carefully prepared beforehand. The January issue of the *American Journal of Science* contained the first part of Joseph Hooker's introduction to his volume on the flora of Tasmania. Once again it was Gray who made the data available to Americans. Hooker's argument was based on a belief in the origin of species from a common center, their subsequent dispersion to different parts of the world, and their variation from the parent stock. Hooker's research, said Gray, represented, as did the work of Darwin and Wallace, "an improved and truly scientific form" of the familiar development hypothesis. He urged scientists to take note of these ideas, the better "to take part in the interesting discussion which they will not fail to call forth."[72]

Gray prepared to help in this assessment. The *Origin of Species* had already met with an uncompromising American opponent, he was quick to inform Hooker:

Agassiz, when I saw him last, had read but part of it. He says it is *poor —very poor!!* (entre nous). The fact [is] he is very much annoyed by it . . . and I do not wonder. . . . To bring all ideal system[s] within the domain of science, and give good physical or natural explanations of all his capital points, is as bad as to have Forbes take the glacier materials . . . and give scientific explanation[s] of all phenomena. Tell Darwin all this. . . . As I have promised, he and you shall have fair play here. . . .[73]

Gray thus drew the lines of dispute. Sure that the past had caught up with Agassiz, he was confident of his own ability to serve as the American interpreter of the Darwinian position. He was not without important aid. William Barton Rogers was also prepared, intellectually and psychologically, to help in exposing the conceptual weaknesses of Agassiz.

As for Agassiz's assessment of the Darwinian impact in America, he told Verrill he "was surprised to learn that Prof. Gray inclined to that theory."[74] Agassiz was of course not surprised at all; what was bothersome was that Gray and others seemed determined to use Darwin to explode the authority of Agassiz. Verrill and his fellow students were curious and excited by the chance to hear their master defend his convictions. It was one thing to take down his every word regarding the way they should conduct themselves at the museum; it was quite another experience to hear Agassiz as he engaged in fundamental debate with other naturalists. Museum students listened avidly to the Darwinian debates at the Boston Society, of which many of them were members, and were eager to learn what occurred in the discussions at the American Academy. If their professor lost the debate, his authority in other areas might also be subject to dispute.

When Agassiz made his first formal statement on Darwin's work to an audience at the American Academy in January, the American evolution debate was underway. But the debate was predestined to resolve nothing, because Agassiz and Gray had argued about most of the basic questions in 1859 and the assumptions of each man had been irreconcilable. Agassiz, moreover, had decided against Darwin beforehand; he had opposed concepts of development for nearly thirty years. The only real issues at hand were whether the discussion of evolution would find Agassiz as able to dispute Darwin, Gray, and Rogers as he had been able to refute the ideas of the *Vestiges* and whether he could convince American naturalists to ignore what he felt to be a rehash of ideas already discredited.

Agassiz's first pronouncements at the American Academy proved him as firm as ever in his denial of evolution. He told the January meeting that there were no genetic relationships between shells of the Tertiary period and those of modern times. William B. Rogers questioned this assertion, and later lent his support to Darwin from the facts of geology and geographical distribution.[75] But, in March, Agassiz took the offensive by telling the assembled academicians that "varieties, properly so called, have no existence, at least in the animal kingdom." Rising to the excitement of the moment, Benjamin Peirce called for special Academy meetings to be held during the next three months for further discussion.[76] At the first of these, on March 27, Agassiz affirmed once more that there was no such thing as random variation in the wild state of nature. As for Darwin's employment of the data of animal variation in a domesticated state, Agassiz discounted such artificial selection as a valid argument. True to his understanding of evolution as individual development that conformed to a predetermined plan, he pointed out that variation was simply a stage of growth or cycle of development in the individual. He saw variation in nature wherever he turned, but such change was only transitory. Since individuals conformed to the permanent idea they represented, these ideas could never change and it was ridiculous to talk of variation in species. Agassiz emphasized this argument time and time again. If he could not understand the phenomena of variation, relying on the ideal concept of species as the unit of identity, Darwin overemphasized the assertion that "species" were the mere invention of the naturalist.[77]

Asa Gray now entered the fray as Darwin's defender. His task was to gain a full and fair appraisal of the *Origin of Species* by reporting Darwin's major arguments and preventing Agassiz from obscuring the disputation by taking refuge in the familiar and unassailable logic of special creationism. Gray tried to show, in April, that his knowledge of botany proved that variations did in fact exist in nature, that variations were preserved by the processes of natural selection, and that to deny variation was in effect to deny the principle of inheritance. Variation under domestication supplied excellent evidence of this truth, Gray held, applauding the impressive amount of evidence Darwin had supplied to substantiate the fact of variation. Agassiz did not play a major role in the effort to counter Gray's argument. His position was defended, however, by the Harvard philosopher Francis Bowen and by John Amory Lowell. Neither Bowen nor Lowell could defend special creationism with the logic of an Agassiz, and Gray had little trouble in exposing the fallacies in their reasoning. Lowell had helped make Agassiz's journey to America possible, had upheld him in the Desor controversy, and had provided him with money for research and travel. He was of no help in the affair of the moment, however, and it was a significant commentary on Agassiz's American years that a Boston philanthropist served as a defender of the ideas Agassiz had learned from Cuvier and had first set down in the *Poissons fossiles.*[78]

In May, Gray continued the attack against "Agassiz, Bowen & Company," predicting the character of his future defense of Darwin by asserting that "no thoughtful theistic philosopher . . . should be justified in charging that a theory of the diversification of species through variation and natural selection was incompatible with final causes or purpose. . . ."[79] Agassiz could never grant such an assertion, nor was Darwin ever very impressed with what would become, in Gray's defense of evolution, a restatement of a traditional argument for design in nature. Gray may have been right, but if final cause was to be the issue of the day, Agassiz had much more talent in interpreting the operations of thought in nature than did Gray. A man who could have the Deity create entire populations and many more catastrophes than the Bible reported was more than a match for an American rationalist such as Gray.

In terms of science, however, the American Academy discussions revealed that Agassiz was arguing for a concept of nature that Gray believed could not be defended logically or empirically. While Gray was willing to grant that the concept he argued for should be employed only as a provisional hypothesis, Agassiz presented a fixed and irrevocable position.

In March of 1860 Gray published a dispassionate appraisal of the *Origin of Species* in the *American Journal of Science* in which he contrasted Agassiz's position with Darwin's. He wanted to do full justice to Agassiz, but he had to write with care as his colleague was "childishly apt to be offended" by criticism, yet his ideas seemed to Gray an example of "science run mad."[80] In Agassiz's view, the data of animate creation were

regarded . . . as ultimate facts, or in their ultimate aspect, and interpreted theologically; under [Darwin's view] . . . as complex facts, to be analyzed and interpreted scientifically. The one naturalist, perhaps too largely assuming the scientifically unexplained to be inexplicable, views the phenomena only in their supposed relation to the Divine mind. The other, naturally expecting many of these phenomena to be resolvable under investigation, views them in their relations to one another, and endeavors to explain them as far as he can . . . through natural causes.[81]

Gray would not admit that Darwin's ideas precluded any theological view of creation. "Mr. Darwin . . . we trust implies that all was done wisely, in the largest sense designedly, and by an intelligent first cause."[82] Although he cautioned that Darwin had not yet offered satisfactory explanations for many aspects of natural history, and that his concept was far from established, Gray was surprised that Agassiz, "a scientific man," should reject the doctrine "on the broad assumption that a material connection between the members of a series of organized beings is inconsistent with the idea of their being intellectually connected with one another through the Deity . . . as indicating and realizing a preconceived plan."[83] Darwin, then, in Gray's view, could be just as theistic as Agassiz, depending on the degree of interference granted the Creator in the broad scheme of natural history.

The entire assessment was an accurate presentation of the Agassiz picture of the universe. Gray exposed its major contradictions and used some of its own suppositions to argue in favor of the Darwinian view. Darwin thought the effort "admirable" and complimented Gray as a brilliant legal dialectician. How would Agassiz respond? Gray reported to Hooker:

Tell Darwin that Agassiz has *again failed* to provide his promised criticism on Darwin for [the] Jour[nal] after promising it over and over in a few days and bespeaking room for it. [He has] failed because [of] the poor stuff—as everybody calls it—he has been pouring out at the Academy. I do not wonder that he hesitates to commit himself to print. I really think his mind has deteriorated within a few years.[84]

The "poor stuff" of the Academy debates had already been published.

William Barton Rogers would have been the first to agree with Gray's estimate, although "everybody" was a rather sweeping evaluation. It was Rogers who played the role of devil's advocate in the meetings of the Boston Society of Natural History. There, from February, 1860, through April, Agassiz presented the classic defense of special creationism. Some of his students took delight in asking pertinent or impertinent questions designed to pit Agassiz and Rogers against each other. Their assistance was hardly needed. Agassiz refused to grant Darwin's ideas any credit for originality.

Mr. Darwin he acknowledged to be one of the best naturalists of England. . . . His works on the coral reefs, on the cirripeds, and his narrative of the voyage of the Beagle, show him to be a skillful and well prepared naturalist; but this great knowledge and experience had, in the present instance, been brought to the support . . . of an ingenious but fanciful theory.[85]

As Hooker suggested, the remainder of Agassiz's arguments followed logically. Refuse to grant the central thesis, and the facts could support an opposite conclusion as well. If transmutation implied a development from lower to higher forms, how explain the fact that certain animals had remained unchanged over long periods of time or that fossils from the earliest geological deposits were as diversified as contemporary animals? Rogers' counter argument that species which had remained unchanged were apparently suited to their environment did not matter. Agassiz would not depart from special creationism, affirming that no fossils had ever been discovered that were identical with forms of a later geological period. "He was prepared to show the occurrence of at least forty-eight successive periods of change" in earth history. How could Gray suppose that design through the agency of transmutation could ever match such miraculous creative power? Rogers even attempted to turn Agassiz's own research against him, citing the very existence of "prophetic types" as proof that certain forms had been inferior to those which succeeded them in geological time. Agassiz was not deterred. Prophetic types, as any reader of the "Essay on Classification" knew, contained, like the egg of a living animal, the "idea" of higher development, but this was an intellectual conception, not to be confused with physical relationship or gradation.[86]

Time after time Agassiz countered the Darwinian generalizations and disputed evidence that supported them offered by Rogers. In March, for example, Agassiz read a prepared paper to show that fossils existed in the same zoological provinces as he claimed characterized contemporary forms. As with living fauna and flora, there had been no connection between types of one geological "zone" and another. Rogers replied by marshaling his extensive knowledge of American geology in favor of the concept of evolution in extinct forms, citing instance after instance where fossils of obviously inferior organization had been discovered in earlier strata than those of more advanced character. Agassiz would not yield an inch. He not only refused to admit any gradation in fossil forms but insisted that time would vindicate the correctness of his views.[87] Whether the facts under discussion related to geology, paleontology, zoology, anatomy, or the past and present geographical distribution of flora and fauna, made scant difference. Such singlemindedness appeared to his students sometimes as masterful, sometimes as unconvincing, but all were impressed with his devotion, a devotion that prevented him from admitting a single fact that seemed damaging.

Prof. Agassiz . . . stated that he knew no such thing as a variety in the animal kingdom, except such as are stages of growth, within the limits of species. . . . In 6,000 fishes, he had not seen a variety except in coloration. . . . So that he would start with the propositions that animals do not vary, and that species remain within the limits of their type. . . .[88]

If New England naturalists felt that Agassiz had come off second best in all such argumentation, they had yet to read his rebuttal of Gray's review. Like Gray, they awaited this analysis impatiently. Agassiz's review of Darwin's book finally appeared in July, 1860. Agassiz dismissed Darwin with dispatch. The arguments of the *Origin* "have not made the slightest impression on my mind," he informed readers of the *American Journal of Science*, "nor modified in any way the views I have already propounded. . . ."[89] With Darwin's intellectual framework in mind, he asked: "If species do not exist at all, as the supporters of the transmutation theory maintain, how can they vary? And if individuals alone exist, how can differences which may be observed among them prove the variability of species?" Neither Darwin nor Gray could see the logical pertinence of this argument, but to Agassiz it was most certainly a telling blow, trapping Darwin in a dilemma of his own making. By affirming the reality of species, Agassiz had once again pointed to Darwin's summary dismissal of the species concept as an invention of the naturalist. But Darwin and his defenders were so convinced that everything Agassiz said represented an entirely outmoded viewpoint that they rejected his perfectly proper question concerning the creation of new species.[90] Agassiz was pleased to be able to point to errors, inconsistencies, and insufficiency of data in Darwin's book. He was, for example, eager to point to the inadequacy of the geological record and unable to admit the inferences Darwin had advanced to buttress his conception. Darwinism was no more than the latest example of false reasoning, "a scientific mistake, untrue in its facts, unscientific in its method, and mischievous in its tendency."[91]

The long-awaited reply to Darwin was greeted with mixed reactions on both sides of the Atlantic. Gray was not deeply stirred by Agassiz's reasoning, and Hooker wrote: "I am perfectly amazed at Agassiz['s] article. . . . I must confess the whole Darwinian controversy has given me a very low opinion of my fellow Naturalists." Leo Lesquereux thought Agassiz had made some telling points against Darwin, but notwithstanding, "The more is said the less we know about the matter." Gray rather proudly reported to Darwin that "a very clever friend" had been converted to the evolution doctrine by reading Agassiz's review.[92]

Gray and Darwin realized that the weight of Agassiz's opinion was still a significant barrier to any popular understanding of evolution. Consequently, Gray took upon himself the task of facilitating the public acceptance of Darwin. He would expound the originality of the idea and absolve Darwin from charges of atheism and materialism. More deliberately, he would review the arguments against Darwinism that Agassiz had raised during the preceding months. Gray's analyses were addressed both to educated laymen and to scientists. He argued, for example, that a conception of God as the first cause behind adaptation and variation was no less theistic than to suppose that a supreme power had created an immutable world. A planned universe could just as readily operate under the secondary laws of natural selection as under the seemingly illogical dictates of Agassiz's Deity.[93] Gray, moreover, published a series of three separate analyses in the *Atlantic Monthly* in the summer and fall of 1860 to explain the major assumptions and evidences in favor of the Darwinian thesis and to answer the arguments of Agassiz and others. Like Rogers, Gray pointed to Agassiz's own ideas as suggestive of the Darwinian view. Point by point, Agassiz's arguments were taken up, examined, and found wanting.[94] The last of Gray's analyses appeared in October, 1860, the same month in which the third volume of the *Contributions* was published. Here, Agassiz reprinted with no changes the entire review he had written of Darwin, so that his opposition had an even wider circulation.[95]

Simultaneously, another dispute between Agassiz and Gray occurred at the Academy. Did variation really operate in nature? Agassiz had examined "a great abundance of divergent forms, which without an acquaintance with the connecting ones, and large opportunities for comparison, might be taken for distinct species, but he found that they all passed insensibly into each other." For Gray to assert that it was these very intermediate forms that "far from disproving the existence of varieties, would seem to furnish the best possible proof that these were varieties" was fruitless.[96] In January, 1861, Agassiz even disputed President Felton's analysis of the common derivation of languages. Having already made use of data from ethnology and physical anthropology, Agassiz was not above generalizing about linguistic science. He informed members of the Academy that despite Felton's analysis, he was certain that later languages were not derived from earlier ones; languages, like species of plants and races of men, were created separately and bore no relationship to one another.[97]

After a year of argument, Agassiz had not retreated from his positions of earlier days regarding the development concept. He had, if anything, become more dogmatic. Agassiz felt his opposition to Darwin to be quite consistent with his prior viewpoint. He had, however, undertaken more than a defense of special creationism in the evolution debates. He had entered the lists against Gray and Rogers in an effort to show American scientists that they should pay no more heed to the *Origin of Species* than they had to the *Vestiges*. While Agassiz was entirely satisfied with the way he had set forth the philosophy of special creationism, his colleagues were not convinced that Darwin could be classed with former propo-

nents of development. Gray and Rogers, by their effort to report
the Darwinian argument, had convinced their fellow naturalists
that evolution had to be evaluated as a potentially valid hypothesis
and assessed in an open-minded fashion. Agassiz had defended his
position, but had failed in his effort to dispense with Darwin. In
fact, of all the professional naturalists of America, Agassiz had
emerged as the sole opponent publicly committed to oppose the
new doctrine. If the actual intellectual content of the evolution
debates resolved nothing, Agassiz had nevertheless been defeated
strategically; naturalists were even more anxious than before to
discuss, evaluate, and test the evolution idea.

While men like Wyman, Rogers, and Gray were hardly convinced evolutionists, their open-minded approach to the issue of
the day was in sharp contrast to that of Agassiz. In Europe, at least,
Agassiz would have had the comfort of intellectual companionship:
Richard Owen, Murchison, Buckland, Sedgwick, Élie de Beaumont, and others who opposed Darwin. In America, to the intellectual eminence of which he felt he had contributed, he had to
submit to opposition in terms which negated his philosophy. The
early Darwinian debates demonstrated the maturity of American
natural science as well as of its practitioners, a development that
made the United States a different intellectual community from
the land Agassiz had known in the past.

The proof of this maturity was the accomplishment of scholars
like Gray, Wyman, and Rogers. The arguments over evolution
demonstrated the degree of intellectual isolation that had come to
mark Agassiz's professional relationships. Virtually alone in his
rejection of Darwin, it was apparent that in the years from 1857
to 1861 Agassiz felt much closer to men like Longfellow than he
did to Gray and Dana. The museum and its supporters, at least,
were a refuge from the argumentation provoked by the Darwinian
hypothesis, disputes that in their personal overtones were in sharp
contrast to the behavior of legislators and Corporation members
still captivated by the Agassiz charm. Whether Agassiz ever really
had the time to study a single group of specimens gathered for him
through Boston's generosity is doubtful. Even if he had done so,
it is unlikely that his study would have led him to accept Darwin's
argument. For, as he relished reminding his friends and critics,

Darwin's work was contrary to modern science as he had defined it. Was he entitled still, in 1861, to such self-assumed authority? In April of this year the legislature responded to Agassiz's pleas concerning the needs of the museum and granted a special appropriation of $20,000 to the institution. In August, friends of the museum gave Agassiz over $3,000 with which to start a library for his beloved establishment. In December, the Royal Society of London awarded him the coveted Copley Medal in general recognition of his intellectual labors extending back to 1829. Murchison informed him that he now belonged to the select group that included men like D. F. J. Arago, Humboldt, and Lyell. As if to make proper response to such honors, Massachusetts added $3,500 to its special appropriation at this time.

With all this in his favor, the Darwinian debates and the consequent loss of intellectual stature could hardly seem an indictment of Agassiz's aims and ambitions. Significantly, the third volume of the *Contributions* received its major professional review not from Dana or Gray but from Agassiz's devoted museum assistant and trustee, Theodore Lyman. Young men of the Lyman caliber might honor his teachings as his colleagues were apparently unwilling to do. With the initial argument over evolution ended by early 1861, American scientists were well aware that Agassiz had not proved his case, but some expected that he still might be of singular service to science. As for Agassiz, he could now look forward to identifying the museum with the struggle over evolution, confident that eventually he would fill its halls with enough material to prove once and for all the falsity of this "fanciful doctrine." He had not in any sense given up the field of battle to Darwin and his American supporters; but the character of his American success suggested that there were other methods by which to oppose evolution. The museum existed to serve America. This was Agassiz's mission, too. What better course than a public campaign to inform the American people of the fallacies of evolution?

Such were the thoughts taking shape in Agassiz's mind as he rested his case against Darwin. He had tried to show his colleagues the folly of the new idea. If they chose to reject his pleadings, it was their intellectual error. If they talked smugly about new discoveries in botany and domestic breeding, the museum could furnish evidence to refute all such alleged knowledge. Agassiz had never undertaken a project or an enterprise and tasted failure. Having ventured into the field against Darwin's American adherents, he had not been victorious. But the museum had been a long time in building, too, and Agassiz was philosophical enough to know that since he was on the side of right, intellectual virtue as he defined it would triumph. Victory might be his if he redefined the lines and locale of battle. If he worked steadily at building up the museum and returned to the pleasant lecture experiences of earlier days (where questions from the audience were rare), he might still know the sense of accomplishment he needed to feel with regard to the evolution question.

Moreover, these were times of crisis that called for the highest level of popular culture and awareness. Two months after the outbreak of the Civil War, Agassiz defined the essence of his public view to Lyman:

It is something worth living for to be here and to see the uprising at the North. . . . Somebody asked me . . . whether I did not regret not having gone to Europe and thus avoided these troubles. My answer was that the possibility of sharing in the present movement was in itself reward enough for whatever I may have done for my adopted country . . . and . . . the chance of working successfully in the Museum, so that when our troubles are over it may appear that Civil War even cannot cripple the onward progress of science in the new world, is a great one. . . . In difficult times to show that the intellectual interests of the community are taken care of with as great solicitude as in ordinary times, that would be worthy of a great nation. If only half a million could now be turned over to scientific purposes . . . the whole world would look with admiration upon America.[98]

The course of the future was thus mapped out, and if Agassiz could instruct Americans as to the true meaning of natural history, at the same time as he advanced the stature of their cultural institutions, his goals should find only unanimous support and approval from grateful colleagues and a grateful Union. To Agassiz, the impulses of his intellect were in perfect tune with the pulse of America. Darwinian heresies and the southern rebellion would both be met by a staunch opponent.

8

THE TRIALS OF A
PUBLIC MAN

1861–1866

In December of 1860, Agassiz wrote Charles Wilkes that he was "comparatively at leisure" and therefore wished to begin work once more on his report describing the fishes collected by the United States Exploring Expedition of 1838–42. The letter reported the naturalist's satisfaction at having published the third volume of the *Contributions,* organized the Museum of Comparative Zoology, and secured the assistance of highly talented students, but it did not mention the recent debate over the concept of organic evolution.[1]

Agassiz never wrote the report about the Wilkes Expedition fishes. His failure to do so illustrated once more the dispersal of his talents and energies into more grandiose and dramatic fields of public endeavor. The years of the Civil War and its aftermath marked a period in Agassiz's life when he was at the height of his public fame, his influence extending into politics, diplomacy, and educational administration. While some of his fellow scientists had by 1861 lost their once high regard for Agassiz's intellectual capabilities, Agassiz himself still happily shared the popular view —that he was the outstanding naturalist of the nation, perhaps of the world. That he felt no contradiction between his career as a scientist and his life as a public figure is a revealing indication of Agassiz's conception of his own role in life.

Agassiz's "leisure" was devoted, first, to proving once and for all the fundamental truth of his interpretation of natural history. Being in charge of one of the largest collections of natural history materials ever gathered, he believed he could produce data which would demonstrate that species were permanent creations and had not evolved from simple to more complex forms. Compelled by this objective, the museum director worked without stint to build his institution into a leading research laboratory.

Early in 1861, his students Frederick Ward Putnam, Addison Emery Verrill, and Albert Ordway were sent to Washington, where they busily sorted and packed Smithsonian Institution specimens for shipment to Cambridge. "Prof gave us money to go . . . and instructions," wrote Verrill; "he told us to get as many corals, birds, bird's eggs and mammals of all kinds and to gain the good will of Prof Henry as much as possible by talking with him. He told us many other things."[2] Baird, who had charge of the materials, was not overly pleased at the arrangements, but he was powerless to act in view of Joseph Henry's great friendship for Agassiz. Some time before, Baird, at Agassiz's urging, had instituted an international system of specimen exchanges with the great museums of Europe. Now, in an ironic turn of events, Agassiz told the assistant secretary that valuable objects should be kept in the United States, as European museums had sufficient collections and their naturalists should not be allowed to rob America of her great treasures.

In the museum itself, activity continued at an accelerated pace. Collectors sent in shipment after shipment of materials from all over the world, ranging from turtle eggs of Australia to photographs of European racial types and "one head of a North American Indian, in alcohol." Alexander Agassiz, educated by his father and by now a graduate of the Lawrence School with a degree in natural history, was a competent administrator, supervising the world-wide network of scientific exchanges and striving to keep the accounts of the museum in order. Despite the receipt of generous state and private contributions totaling nearly forty thousand dollars during 1861 and 1862, this was no easy task. "Father cannot make up his mind to draw his horns in and has laid out all sorts of financial plans which I am doing my best to knock in the head," Alex confided to Theodore Lyman. "When I urge objections, he simply says I don't know anything about it and that he gets in that way money to carry on the Museum. It is pretty bad to fight not only against him but against all the dodges he invents to keep the Finances up."[3] Fortunately, Lyman was completely devoted to Agassiz's purposes; this former student could always be counted on to provide generous funds to bridge the gap between Agassiz's ambitions and his actual resources.

In these dark days of national strife and turmoil, Agassiz advocated the Union cause with an uncompromising patriotism. He made application for American citizenship, feeling that the least he could do was to affirm in this way his allegiance to his adopted land. He wrote letters to friends in England, chiding them for the policy of their country toward the United States. He became outspoken on questions of social democracy, complaining that too many people were deprived of the right to vote because of various property qualifications in certain states. He wished for a national program of education in citizenship to instruct the "lower classes" in the values of democracy.

Agassiz was not, however, a complete equalitarian. He complained to friends that the "philanthropy" of the past had resulted in such "socialistic" schemes as Brook Farm, and he hoped that this same reformist spirit would not now result in a complete eradication of social and intellectual distinctions. He was asked by Samuel Gridley Howe of the government-sponsored American Freedmen's Inquiry Commission for his opinions on the social and political adjustment of the Negro in the South once emancipation had been achieved. In his response, the old convictions reasserted themselves. The Negro could never be the physical equal of the white. Moreover, there were permanent barriers to social and economic equality. Negroes would very likely take over the govern-

ment of the reconstructed southern states, but intelligent north-
erners could meet the pressing question of the day only by resolv-
ing to grant the freed slaves equality before the law. This equality,
however, should be gradually tendered, since the social advances
of white men had been achieved only through a lengthy process of
cultural advancement. Once again, Agassiz's belief in special creationism resulted in his employing the authority of science to support social doctrine. In this latest instance, however, he spoke out
against both physical and social equality in language stronger than
he had ever used before. The passing of time had made him more
dogmatic, a characteristic strikingly evident in his next public endeavor.

Beginning in December of 1861, Agassiz subordinated all his
other activities to the cause of bringing about an informed public
opinion through the popularization of knowledge. Once more he
took up the role of lecturer and teacher of science. The fact that
the American public was at this time exposed to the most dangerous of scientific and philosophical ideas—the concept of evolution
—made it all the more imperative for him to serve the interests of
education by teaching his version of the truth about nature. The
decision to combat the idea of evolution through popular lecturing and writing was of primary significance. It meant that Agassiz
abandoned his professional defense of classical natural history, for
he never again published an objection to evolution in a professional journal or scientific monograph.

In the light of his growing commitment to cultural democracy
in America, this was a perfectly understandable decision for Agassiz. The progress of science required an informed public opinion
capable of resisting error such as, for example, the appealing suppositions of the evolution idea. True intellectual progress could
not be built on fallacious conceptions, and conceptions, true or
false, had social consequences.

A lecture tour in Utica, New York, initiated this phase of Agassiz's attempts to popularize knowledge. It was followed by a return
to the scene of early triumphs in a series of lectures on "Methods
of Study in Natural History" delivered at the Lowell Institute
during the winter season of 1861–62. These addresses were infor-

mal interpretations, without documentation or new critical analysis. They contained a sketch of the history of natural science, designed to show the genius of Aristotle, Linnaeus, and Cuvier, men who had correctly appreciated the meaning of nature by constructing systems of classification which affirmed the ultimate coherence and plan of the universe. Like Agassiz himself, these intellects had concluded that: "Nature is the work of thought, the production of intelligence, carried out according to plan. . . . In our study of natural objects we are approaching the thought of the Creator, reading his conceptions, interpreting a system that is his not ours."[4]

Lowell Institute audiences were shown, just as they had learned in 1847, that every fact of nature supported this all-inclusive theory of being. There was the example of species themselves, ideal forms illustrating divine purpose despite the efforts of some misguided naturalists to identify them as mutable and transitory objects. Man's selective, artificial power to alter species was simply his way of effecting temporary differences and was not an index to what actually occurred in nature. There was the proof gathered from Agassiz's years of study of the coral reef formations, which showed that the coral animals of the oldest formations were just as advanced in structure and function as the animals of the youngest formations. Finally, there was the evidence from physiology, anatomy, and embryology, which demonstrated in a magnificent way how the organization and growth of each animal had been thoughtfully planned and permanently designed.

This evidence, to Agassiz, all proved a higher purpose. As for empirical data:

The facts must be collected, but their mere accumulation will never advance the sum of human knowledge. . . . It is the comparison of facts and their transformation into ideas that lead into a deeper insight into the significance of Nature. . . . Facts are the works of God, and we may heap them together endlessly, but they will teach us little or nothing till we place them in their true relations, and recognize the thought that binds them together as a consistent whole.[5]

These sentiments reached a wider audience than the Bostonians present at the Lowell lectures, for they were published as a series in the *Atlantic Monthly* beginning in January of 1862. By November of that year, people of New England and the nation had had the opportunity to read a full, serialized version of the story of creation illustrated by finely executed drawings.[6] In 1863 these articles were incorporated in the book *Methods of Study in Natural History.*

This was Agassiz's most successful volume from the standpoint of popular appeal; it went through nineteen editions between 1863 and 1884. The preface contained an unequivocal statement of its purpose. Agassiz wished to take the opportunity to "protest against the transmutation theory, revived of late with so much ability, and so generally received." He observed that "naturalists are chasing a phantom, in their search after some material gradation among created beings." Admitting that this was a "personal opinion," and that other scientists, "quite as reverential" as he, might believe in evolution as scientific truth, Agassiz maintained nevertheless that "this theory is opposed to the processes of Nature. . . ."[7]

This was but a part of Agassiz's campaign to educate the public against the new theories which were currently in circulation. After concluding his Lowell Institute lectures, he repeated the same general subject matter to appreciative audiences at the Brooklyn Academy of Music. These people, among the most influential members of the community, were so taken by the force of Agassiz's arguments that they had his lectures printed in volume form in 1865. The work was so well received that it required a second printing.[8]

Readers of the *Atlantic Monthly* had no sooner finished the "Methods of Study" when Agassiz presented a new series of twelve articles. More specialized in content than the first group, these dealt with subjects close to Agassiz's heart, geology and paleontology. They were concerned with fossil bones, proving that men lived before Adam, and spoke of animal remains which showed that the plan of creation had been in existence from the inception of time. Agassiz also discussed the history of the glacial period and how mountains originated.[9]

Even in this more specialized aspect of natural history, Agassiz lost no opportunity to delineate his interpretation of nature. The history of the earth, as exemplified in the facts of geology, proved that

a coherence . . . binds all the geological ages in one chain . . . a stability of purpose that completes in the beings born to-day an intention expressed in the first creatures that swam in the Silurian ocean . . . a steadfastness of thought . . . between the facts of Nature and . . . the . . . results of an intellectual power.[10]

These last *Atlantic Monthly* articles were also published in permanent form in 1866.[11] In the space of five years, therefore, Agassiz gave three courses of lectures and published twenty-one articles and three books, all proclaiming the truth of special creationism.

For Agassiz these days of writing and lecturing for the nation at large were happy ones and he seemed thoroughly to enjoy his chosen mission. He had excellent and devoted publicists and supporters, among them Edward P. Whipple, a dominant figure in Boston literary circles, who visited the museum often and wrote glowing reports about its progress. The same service was performed by the *Atlantic* editor, James T. Fields, whose admiration of Agassiz made this leading journal into a virtually exclusive forum for the arguments of special creationism. James Russell Lowell and Oliver Wendell Holmes voiced their praise and admiration for the man who seemed so confident and certain about the fundamental truths of natural history. Holmes could hardly contain his pleasure:

I look with ever increasing admiration on the work you are performing for our civilization. It very rarely happens that the same person can take at once the largest and deepest scientific views and come down without apparent effort to the level of popular intelligence. This is what singularly gifts you for our country. . . . You have gained the heart of our purpose; you have taken hold of our understandings by your familiar lectures and writings, and you are setting up a standard for them which will gradually lift the students of Nature among us to its own level in aspiration if not in performance. I did not think it necessary to say these . . . words, but I wanted the privilege, because I feel them sincerely.[12]

These popular denials of the evolution idea had practical consequences for American society and Agassiz's professional relationships. Theologians and publicists, emboldened by the authority of

Agassiz's reputation, adopted his arguments against Darwinism. His early heresy against revealed religion in the form of the advocacy of man's plural origin was quickly forgotten, and he became in clerical opinion "the prince of naturalists" who was waging the good fight against the evil of Darwinian evolution. Every argument Agassiz employed in his contradictions of evolution—from his jibes at Darwin's story of the bear who had learned to swim in the course of the struggle for existence to his attacks upon the materialistic philosophy of the *Origin of Species*—was incorporated into the initial popular and religious reaction against the new concept.[13]

While the public in general and the clergy in particular lauded the brilliant man who taught them why evolution was a mistaken notion, most professional naturalists reacted with attitudes ranging from indifference to amusement to downright disgust. Most offended was Asa Gray. To Gray, the clergy's "prince of naturalists" was a "prince of charlatans." The botanist gave vent to his feelings in a letter to Hooker in May, 1863. He reported the contents of an Agassiz article in the *Atlantic*, "announced as the highest science of the day." He longed to expose the fallacies it contained but was fearful of precipitating a public argument. His opinion of Agassiz was a severe condemnation: "This man, who might have been so useful to science and promised so much here has been for years a delusion, a snare, and a humbug, and is doing us far more harm than he can ever do us good."[14] Gray wrote to Darwin in similar vein: "Agassiz is writing very maundering geology and zoology, and worse botany . . . in the Atlantic Magazine. . . . He is 'joined to his idols,' and I have no expectation that he will ever be of any more direct use in nat.[ural] history. I hope better of his son, who may do something."[15]

The most valid reason Gray and other defenders of Darwin had for their annoyance with Agassiz was that in the light of his own important contributions to natural history and the challenging questions posed by the increasing evidence for evolution, it seemed clear that the man "who had once promised so much" was not interpreting the facts of natural history in a responsible fashion to the public. Gray, therefore, had a different view of Agassiz than did Holmes. "How I could tease Agassiz, if I could see him; only he is of late so cross and sore," the botanist reported to Darwin.[16] From Gray, Darwin learned of Agassiz's dogmatic pronouncements. "He seems to grow bigoted with increasing years," the Englishman observed to Gray. Gray could substantiate this opinion by citing an Agassiz analysis of glacial geology in the *Atlantic:*

It will not strain your brain. . . . It may your diaphragm to read how the hairy elephants and the bears . . . were luxuriating in a tropical climate, with their thick coats on to keep them comfortable, when, all of a sudden it changed to bitter winter—so suddenly that they could not run away, nor even rot, but were *frozen up* before there was time for either![17]

Jeffries Wyman provided a reliable gauge of scientific opinion. To Dana he wrote: "Agassiz is often inconsistent with himself in the application of his principles, and his own students cannot make them work. What justice is there in placing a butterfly higher than a bee, or a wasp, or an ant?"[18] As for Dana, he, too, realized the force of this assumed authority when he received a sharp rebuke from Agassiz for having disputed an interpretation of geological evidence made by his Cambridge colleague.

Agassiz, convinced he was performing a public service of the highest order, never allowed the growing evidence of professional disenchantment to tarnish his self-image. If he needed personal reassurance, he could always find it in the halls of the museum, whose exhibits testified to Cuvier's taxonomic genius, whose workrooms housed active student assistants, and whose storerooms continued to bulge with the accumulated treasures of world natural history. By 1864, the full amount of the original state grant had been paid, Massachusetts had made an additional appropriation of $10,000 to pay for a printed catalogue of the museum, and the annual income of the institution from all sources came to nearly $10,000. If, as was often the case, Agassiz spent money in advance of its availability, he knew he could turn to the community at large. This happened in 1864, for example, when special needs resulted in still another popular subscription that provided an additional $6,200 for Agassiz's favorite public occupation. Agassiz's belief in special providence in natural history seemed to be justified.

But, as was almost predictable in the life of this public man, success was once more tempered by circumstance. The very individuals who were in Agassiz's view the primary beneficiaries of his intellectual career, his students, began to turn against him in the early 1860's. No incident in Agassiz's life hurt so deeply as the loss of esteem on the part of his students. He could rationalize the antagonism of scientific colleagues as the result of false adherence to false views of nature. He could never understand what seemed to him sheer ingratitude on the part of students like Morse, Verrill, Putnam, and the almost complete membership of the Agassiz Zoological Club. By 1864, these once admiring and worshipful disciples had left the museum, convinced their professor was a "Great Annihilator," running a veritable monarchy of which he was the absolute scientific dictator and they his unhappy serfs.

The sources of Agassiz's unhappy relationships with his students were deeply rooted in the intellectual and social character of his American career. He felt he was giving these students an opportunity never before available, to learn the meaning of natural history in a new institution, their living costs subsidized by the wealth of Boston, their intellectual development supervised by a dedicated teacher.

From the viewpoint of the students, however, a different picture of the Agassiz influence emerged. As Wyman suggested, students and professionals alike found the practical applications of special creationism at best a difficult task. Young men like Hyatt, Morse, Verrill, and Putnam had all read Darwin, and they became weary of classifying natural history materials according to the principles laid down by Agassiz in the *Essay on Classification*. In March of 1860, for example, Morse was still able to believe Agassiz's teaching. He wrote a friend about Darwin:

Now John stop and think of it for a moment and don't you perceive that if his theory were true it would leave one without a God? Most certainly it would, for the origin of species according to his idea would be simple chance and nothing else. . . . Darwin does not support his theory by a single fact in Nature.[19]

By October, 1861, however, Morse's thinking had changed to the extent that he made the terse entry in his journal: "Attended Prof's lecture at Lowell Institute tonight. He regarded cats and dogs as belonging to the same family. Walked out."[20] Morse was the first student to leave Agassiz, departing from the museum in December of 1861. "Saw Prof and told him I was going home. He appeared blue enough and when I asked him for money, he said he had not $10.00 in his pocket or in the world." The next day Morse's decision to leave was all the stronger:

Prof came to me and said he had misunderstood me last night and was astonished and indignant that I was going to leave on such short notice. He showed himself in his true light. . . . As long as one will toady to him and be content to live on nothing . . . so long will everything go smoothly; but when one asserts his independence then is the man vexed and indignant.[21]

Alex Agassiz placed a different interpretation on this and other recent events:

Father had quite a flare up the other day with one of the students who had undertaken to tell father what he ought to do and what he should not do which resulted in Morse's leaving the Museum for which I am not sorry. . . . He was the laziest and most troublesome of the lot and things have gone on much more smoothly since. The students have been rather quarrelsome lately and have several times attempted to dictate to father. . . . It is discouraging to see in what spirit the very men for whom father has done everything, are working. There is nothing good enough for them and they impede instead of assisting in what is going on. I do not think the Museum will be on a good footing till this is rooted out completely.

Lyman learned from Alex that the only "loyal" worker at the museum was Henry James Clark. Alex predicted that if the students did not change their attitude, all would follow in Morse's footsteps.[22]

Alex was quite right. Darwin's ideas were not sufficient excuse for the students' new feeling of dissatisfaction with Agassiz. In February, 1862, Alex felt compelled to write critical letters to a student, rebuking the young men of the museum for speaking disrespectfully of his father in public and for broadcasting the charge that Agassiz treated them in a dictatorial manner. At this same time, Agassiz discovered the existence of a secret group among his students, called the "Society for the Protection of American Students from Foreign Professors." This was a striking departure from the spirit which had fostered the Agassiz Zoological Club in 1859, and Agassiz was sick at heart at his loss of contact with his students. By 1864, the mental anguish and disappointment Agassiz suffered actually caused physical illness, a situation Alex attributed directly to his father's unhappiness over his students.

The bases of disagreement between master and disciples were different in particular instances; some students felt they should be paid more for their efforts at the museum, others that Agassiz was a tyrant in his management of the institution, still others that he retarded their desire for intellectual independence. Packard, Verrill, Putnam, Morse, Shaler, Albert S. Bickmore, and Samuel H. Scudder were all bright young men. All had come to Cambridge and to Agassiz certain that they were to receive an education that would make them in time as distinguished as their teacher. Agassiz gave of himself without reserve to these confident young men. He personally arranged for Nathaniel Thayer to pay salaries for some museum assistants, paid the costs of furnishing Zoological Hall out of his own pocket, and gave all of them free tuition at the institution. In the happy, exciting, and rewarding days of the museum's origin and establishment, all went well, and Agassiz seemed a spiritual sun, giving off rays of inspiration, encouragement, and knowledge without reserve.

Within a few years, however, this encouragement made even the master himself appear much less noble than he had seemed at an earlier day. Agassiz's insistence on intellectual independence in his students resulted in his becoming the first victim of their newly won self-confidence. Late in 1863, a disagreement arose between Agassiz and Frederick Ward Putnam concerning the amount of money due the student for labors performed and Putnam's desire to publish the results of his investigations. Verrill, too, felt confined by the yoke of Agassiz's authority. On his part, Agassiz felt that his students were not yet ready to publish their research and refused to grant them permission to do so. Since their work was based on materials owned by the museum, he felt within his rights in this refusal. He wrote a friend of the situation:

I have spared no effort or shrunk from no sacrifice to educate assistants. . . . I am extremely disappointed with the spirit manifested by those for whom I have thus far done everything I could. . . . It is not my fault if these young men are so conceited that one of my constant efforts must be to show them that they should not make themselves ridiculous in the eyes of those who really understand science and thus injure the Museum; it is not my fault that they undertake to meddle with what is none of their business and constantly deserve to be rebuked. . . . They know so little yet . . . are under the delusion . . . that I could not do without them. . . . If they do not soon behave themselves I shall have to remove them all.[23]

In November, 1863, Agassiz announced new regulations for the government of the museum, and it was these rules that were the immediate cause of the student rebellion. Some of them were only restatements of existing procedures; others were a clear reflection of the failure of an administrative policy of which Agassiz was the sole interpreter. Now, the regulations clearly stated that all museum appointments were to be subject to the full approval and control of the faculty, financial compensation to be set by this body as well. As before, no one could own a private collection of specimens. However, no museum employee was allowed to work on personal research during museum hours, and any work done at the institution was to be considered its intellectual property. Most important of all, no employee was to be allowed to publish or communicate research results "without the previous consent" of Agassiz, who was to examine all such efforts prior to their presentation.[24]

When Putnam and Verrill asked Agassiz to approve of work they had done, he informed them it was not yet ready for professional publication. They in turn accused their teacher of thwarting their intellectual progress and of making it impossible for them to gain employment as naturalists in their own right. Agassiz insisted they were not ready for such responsibilities. More argument ensued, with Agassiz particularly furious over the fact that Verrill, Putnam, Bickmore, and others had tried to arrange for positions in Salem, Massachusetts, and New York City without his knowledge. Bickmore went so far as to solicit a popular subscription for a private research project of his own among the wealthy people of Boston, which would seem to indicate that he had learned much from association with Agassiz.

By December, 1863, true to his warning, Agassiz had had enough of such behavior and, as was his right, refused to recommend to the faculty the permanent appointment of his students as museum assistants. He felt they had violated the regulations of the institution in an intolerable fashion, talking behind his back, trying to arrange independent positions for themselves, insisting on publishing their researches, and using the time of their employment at the museum to engage in personal investigations. Thus, Putnam, Scudder, Hyatt, Verrill, Bickmore, and Packard all left the museum, both Agassiz and his students by this time eager to break off the association. It was a sorry turn of events in the life of the Cambridge institution, but if students were disappointed in their master, Agassiz was no less distraught by their behavior. From this time forward, he resolved to place less confidence in young students, and the work of the museum was carried forward by established naturalists like John Gould Anthony, Phillips R. Uhler, Joel Asaph Allen, and William H. Niles, aided by the always loyal Alex.

Soon, a new group of workers began arriving in Cambridge, Europeans like Marcou, Hermann Hagen, Fritz Steindachner, and G. A. Maack, men who Agassiz felt were understanding of the need for authority and who were not concerned with their personal advancement at the expense of the museum's progress. Those students who did work as assistants never gave Agassiz the difficulty that the first members of the Zoological Club and of the foreign professor protective society gave him. But, despite the more congenial relations of the years after 1864, some vital spark had been extinguished at the museum with the departure of that first devoted band. No more beer parties or attacks on the dignity of visiting scholars, no more lively discussions of the virtues of the *Origin of Species* versus the *Essay on Classification*. Progress and maturity brought stability and a new order to the Divinity Avenue building, but it was an atmosphere that lacked the excitement of pioneering days. Agassiz's students, of course, would have left the museum eventually, for, despite his protestations, they were well trained and perfectly capable of responsible professional activity. Yet the manner of their leaving was still another reflection of Agassiz's inability to achieve lasting personal relationships with subordinates.

Aside from the natural independence of spirit that made for unhappy relationships between these students and Agassiz, the essential grounds for disagreement centered on the definition of what Agassiz termed "intellectual property." He contended that his students had conducted investigations under his advice and guidance, using materials owned by the museum, their discoveries being the result of ideas he gave them and suggestions he made. Consequently, he could not allow them to publish what he felt were actual thefts of intellectual property. Such behavior was to Agassiz utterly indefensible.

The fact is that I have always given so freely to all my pupils what could be of any use to them from my own scientific capital that they have accustomed themselves to draw from me what they wanted without ever thinking that they even owed an acknowledgement for it. To all that there must be a limit and when I undertook to draw the line I met with the most determined manifestation of ill will and egotism. . . .[25]

In this way Agassiz expressed his own sense of injustice, freely admitting that he had made serious errors throughout his life in trusting too much to the good will and loyalty of others.

The scholars who went from Cambridge into the larger world of science enjoyed considerable success. On the basis of having been educated under Agassiz's direction—an experience which they came to appreciate more as time mellowed their estimation of their teacher—they all secured high positions in the profession. Verrill went to Yale as a professor of natural history and helped reorganize the study of zoology at the university. Morse, Hyatt, Packard, Putnam, and Caleb Cooke took positions at the newly established Peabody Institute in Salem.

Albert S. Bickmore decided to found a museum of his own,

which would be run on "democratic principles" in distinction to the "dictatorial methods" employed at Cambridge. He informed Dana of his scheme, and the New Haven naturalist encouraged his efforts, agreeing that Agassiz had wronged his students. Bickmore's defiant declaration proved to be no idle boast, for he was able to interest a group of prominent New Yorkers in his idea and thereby provided the stimulus for what eventually became the American Museum of Natural History. Even negatively, it appears, Agassiz's influence upon science in America in the 1860's was apt to verge on the spectacular.[26] Agassiz himself, as if in testimony to his still unchallenged dominance in public activity, visited Chicago in 1864. He told wealthy men of the importance of natural history. The result was the establishment of Chicago's natural history museum soon thereafter.

In addition to the clashes of personality at the museum during the years 1862 and 1863, some students and assistants left the place rather abruptly to join the army, much to Agassiz's further dismay. He felt that he had instructed them in good will and that they were taking the first opportunity to show their ingratitude for his generosity toward them. Morse described his impression of Agassiz's reaction to the war as it affected the museum: "Prof looks rather cross today; he does not appear as if he liked Bertie's [Ordway] going away. If they knew in Boston how greedy Prof is of our time and how actually tyrannical he is at times and actually unpatriotic, it would kill the museum."[27] It was one thing to express patriotic sentiments to friends in high places and another to witness the actual influence of the war on the museum.

A much more serious incident, related to Agassiz's concept of intellectual property, occurred at this same time. Alex Agassiz had felt Henry James Clark to be "loyal" to his father. This proved a mistaken estimate. In the preface of the first volume of the *Contributions,* Agassiz acknowledged his debt to Clark for work performed in embryological investigations. He stated:

Mr. H. James Clark has . . . drawn, with untiring patience and unsurpassed accuracy, most of the microscopic illustrations which adorn my work. I owe it to Mr. Clark to say that he has identified himself so thoroughly with my studies since he took his degree in the Lawrence Scientific School, that it would be difficult for me to say when I ceased to guide him in his work. But this I know very well,—that he is now a most trustworthy observer, fully capable of tracing . . . the minutest microscopic investigation, and the accuracy of his illustrations challenges comparison.[28]

The high regard which Agassiz felt for Clark was reflected in his efforts to do all in his power to help the young man in his career. In June, 1859, before Agassiz left for Europe and at a time when he feared he might not live to see the museum building completed, Agassiz had written President Walker that the one man capable of filling his position as director of the museum and professor of zoology was Clark. When he returned, Agassiz was able to persuade the Corporation that Clark's services were so valuable that he should be given faculty rank.

Hence in June, 1860, Clark became an assistant professor of zoology in the Lawrence Scientific School, and Agassiz took every opportunity to praise his work publicly. Clark was given complete charge of the embryological and microscopical investigations undertaken in the museum, and he lectured to students. While the university willingly agreed to Clark's appointment, it was unable to assume the responsibility for Clark's financial compensation. Therefore, Agassiz continued to pay Clark, as he had done since 1855, out of his own funds and museum resources. Working under this unusual arrangement, Clark received a total of $1,700 a year, $1,200 of which came from Agassiz's personal funds, money derived from the *Contributions* publication subscriptions.

When the third volume of the *Contributions* appeared in 1860, Agassiz once again acknowledged the "valuable assistance" he had received from Clark and felt he had paid sufficient tribute to the young naturalist for services performed. While Clark said nothing at the time, it was apparent that as soon as Volume IV was published in 1862, its subject matter once again dealing with invertebrate zoology, he began to feel he had been dealt with unfairly by Agassiz. The volumes, in Clark's estimation, had been more of a common effort than the work of Agassiz alone, and he felt his reputation as a scientist had suffered because associates had not been given proper notice of the extent or value of his labors. In June, 1862, Clark resolved to give Agassiz no further help in this undertaking and to establish his prestige as an independent scholar by publishing a series of his own embryological investigations.[29]

By March, 1863, Clark's sense of having been treated unjustly had increased, and he decided to ask Agassiz for $475 due him as salary for the past few months. It was not unusual for Agassiz to be completely destitute of funds for short periods, and evidently such was the case now, since he informed Clark he could not pay him. The two men then exchanged heated words. Clark berated Agassiz for not living up to his financial agreement, while Agassiz argued with his assistant on the grounds that Clark had spent nearly a year on personal investigations while working at the museum in Agassiz's employ.[30]

After the first argument in March, 1863, Agassiz wrote to Clark informing him that he was to return the keys of the museum to him, as well as all books, microscopes, lenses, and apparatus that were the property of the museum. Clark interpreted Agassiz's request for the keys as a demand for his resignation from the university, an act which Agassiz had no right to perform. Clark had been appointed by the Corporation and only the Corporation had the power to remove him, as Clark informed that body.[31]

The Harvard Corporation was under the administration of President Thomas Hill, who had succeeded to the office after Felton's death in 1862. Like Felton, the Reverend Hill was very close to Agassiz, eager to aid the museum director's work as much as possible. Hill, in fact, owed his office to the influence of Agassiz and Peirce, who had worked for his appointment in alliance with John Amory Lowell, Agassiz's dependable ally on the Harvard Corporation. Naturally, he wished the entire affair to be settled amicably, with Agassiz's desires granted if possible. Hill, however, had no clear basis for making a decision, since the specific authority of the director of the museum over workers at the establishment was then vague. Agassiz quite plainly chose to ignore Clark's Corporation appointment, considering him a mere museum assistant. If Agassiz had the right to dispense with the services of a faculty member under his supervision, this would negate the power of the Corporation. If, on the other hand, Agassiz were not permitted complete freedom in administering the museum, he might leave the university, an eventuality that Harvard could ill afford.

Since both men were of an unyielding nature, the matter dragged on. Clark made the basis for his position public by publishing a three-page pamphlet in July, 1863, entitled *A Claim for Scientific Property,* which he distributed to scientists in the United States and Europe. This tract stated that Clark had written most of the second volume of the *Contributions,* most of third, and so much of the fourth that he was unable to determine what had actually been written by Agassiz. Moreover, he asserted that the most valuable parts of the *Contributions,* those representing original discoveries in embryology, had been written by himself, the result of his own independent investigations. Before he had made his charges public, Clark asked that a tribunal of impartial judges be convened to determine the validity of his contentions. When Agassiz refused to agree to such a procedure, Clark published his pamphlet and even instituted a lawsuit against Agassiz, but this never reached the courts. Thereafter, Agassiz would have nothing to do with his former assistant and turned to measures comparable in their extremity to those employed by Clark. In September he proposed a motion that Clark be excluded from attending faculty meetings of the scientific school. Agassiz's colleagues, Gray among them, refused to pass the proposal and instead voted to permit Clark to continue attending meetings. Recognizing the importance of their action, the group sent notice of it to the Corporation for its approval.[32]

Agassiz, infuriated, now even disavowed the previous statements of praise which Clark had found inadequate. The only basis upon which Agassiz would agree to allow Clark the use of the museum was if he would publish a complete retraction of his charges of plagiarism. Clark, of course, refused to do so.[33]

At this point, Judge Ebenezer R. Hoar, a member of the Corporation and a companion of Agassiz at the Saturday Club, attempted to ease President Hill's dilemma by securing a promise from Clark that he would no longer attend faculty meetings, on the condition that he be allowed to use the museum. The compromise failed, since on careful thought Clark decided the agreement would indicate an admission of guilt on his part. Agassiz, in turn, refused to allow Clark the use of the museum unless he apologized. The Corporation convened a special meeting to consider the merits of each man's complaints. Agassiz was indignant at being called upon to explain his actions in this way and expressed his anger in no uncertain terms. It is likely that he also threatened to resign from his university position unless Clark were removed from the faculty.[34]

When Harvard was faced with a decision between the value of Agassiz and the rights of an assistant professor, the ultimate choice, while not expressed in clear-cut terms, was plain. Judge Hoar and President Hill, as a committee of the Corporation, recommended in November, 1863, that Clark was not obligated to attend meetings of the scientific school faculty. Effectively, this judgment, approved by the Corporation, negated the vote of the faculty, and upheld Agassiz's position. Clark asserted his right to faculty status by continuing in the professorship he held until his five-year appointment expired in 1865, although he did not teach or work in the museum after the Corporation handed down its decision. He then resigned to join the faculty of Pennsylvania State College. In 1864–65 Clark gave a series of lectures on embryology at the Lowell Institute, which were published. In the volume he called attention to his claims to "scientific property," again citing his investigations and alleging they were appropriated by Agassiz.[35]

From the evidence it appears that Clark was at least partially justified in his claims for greater recognition. Clark was a competent scientist who merited more consideration and dignity than Agassiz accorded him. Clark surely deserved better treatment from the man who had once respected him so highly, and who had even considered him worthy to be his ultimate successor as museum director. But by the time of the disagreement with Clark, Agassiz was obviously unable to evaluate his relationships with an assistant with objectivity. Undoubtedly, the difficulties with Clark seemed to Agassiz a culmination of a long and unpleasant series of personal relationships in science that stretched back to the difficulties with Schimper and the men of the Neuchâtel days. Clark thus reaped the fruits of Agassiz's frustrations and sense of ill-treatment regarding collaboration with subordinates.

One result of the Clark affair was the establishment of the new regulations for the museum's operations, since a basic Agassiz grievance against his assistant was recognized by the provision that no museum employee could work on personal projects within the establishment. Yet rules such as this also provided the spark that finally touched off the student rebellion, so that Agassiz was once more faced with the stark fact that his directorship of museum affairs had resulted in unpleasant personal relationships. Moreover, the entire series of difficulties had become well known to colleagues like Wyman, Gray, and Dana. Wyman and Dana thought Agassiz too harsh in dealing with his students, while Gray felt Clark had been treated unfairly by Agassiz.

Agassiz was not content to have the museum serve as the sole reflection of his authority. A man capable of dismissing the helper he had once so admired, and of forcing other students and workers to leave the museum, experienced no difficulty in planning the reform of science and education. The year 1862 witnessed the beginning of Agassiz's new effort to test his power at Harvard and in the nation. Fundamentally, Agassiz imagined this ambition to be a public service of the first order, and he needed the help of men of similar motivation and conviction.

The scientific Lazzaroni had first flexed their muscles in the 1850's, when, inspired in part by Agassiz's absorption in the politics of American science, they had moved to organize themselves into a co-operative body, to take over the American Association for the Advancement of Science, to control the University of Albany and the Dudley Observatory, to propose the establishment of a federal scientific organization, and to formulate alliances with key public men and administrators. The core of Lazzaroni strength by the 1860's was to be found in Agassiz, Peirce, Bache, and Wolcott Gibbs. James Dwight Dana never seemed really enthusiastic about Lazzaroni activity, and his admiration of Agassiz was by this time diminished because of the Marcou incident and his dissatisfaction with Agassiz's weak opposition to evolution. Joseph Henry, while grateful for Lazzaroni support during times when an uninformed public opinion threatened his freedom to direct Smithsonian Institution affairs as he thought best, was also temperamentally unsuited to the tough game of Lazzaroni politics. He came to have less of a role in policy-making, although his friends knew the value of his name and position for their schemes. Significantly, opposition to the Lazzaroni stemmed from the same men who were Agassiz's opponents in Cambridge and Boston, Asa Gray and William Barton Rogers. Like Gray, Rogers was disenchanted with Agassiz as a research scientist. He had fought Agassiz in the recent evolution debate, and now had good reason to challenge Agassiz's dominance in Boston society because he had just been chosen president of the recently chartered Massachusetts Institute of Technology and needed financial support from the very groups who had for so long backed Agassiz's projects.

But the Lazzaroni still had strong resources. Their numbers benefited from the addition of the aggressive astronomer Benjamin Apthorp Gould, now a Cambridge resident, a good in-fighter when the chips were down, and a man very highly regarded by Bache. The Cambridge Lazzaroni could also count on support from the newly powerful Republican politicos of Massachusetts, men like Governor John A. Andrew and Henry Wilson. Such influence could be readily extended to Civil War Washington, where Bache was already powerful. In these days of national crisis the Lazzaroni stood four square for the Union, for progress, for nationalism. If the occasion demanded, they could wave the bloody shirt and accuse those who stood in their way of "old fogeyism" in cultural matters, a charge equivalent to treason in their eyes. Equally important, the Lazzaroni, in times of change, were ready with vast programs. They had ideas and innovations to propose; they looked toward the future; they symbolized a new approach to the organization of science in universities and the nation. Their proposals would inevitably excite the imagination of men of power and influence; even Joseph Henry, moderate though he was, could not help being fascinated with the talk of Agassiz and Bache as they gathered at his Washington home during the war years and discussed the future. In the exciting days of the Civil War, the leaders of the Lazzaroni were fully prepared to capitalize on the spirit of the times and to realize their dreams of the 1850's. Unlike their opposition, these men had a program for the future. None of them was fundamentally a research scholar; they had ample energy, then, for the public involvements they were constitutionally suited to enjoy.

Asa Gray was the man most affected by the new Lazzaroni impulse, because Agassiz was the guiding genius behind Cambridge Lazzaroni activity and a central figure in their national efforts. It was not only that Gray found it impossible to accept Lazzaroni assumptions regarding science at Harvard and its national role. Men of good will would always disagree about policy and procedure when the tasks were large and the culture encouraged social and intellectual innovation. The central fact that emerged from the complex of the social relations of science in New England and the nation at this time was that Gray, by the 1860's, had clearly found Agassiz's public authority intolerable. Strengthened by the prestige of his role in the evolution debates, Gray determined to oppose Agassiz at Harvard and in the councils of national science.

His decision was based both on intellectual conviction and personal motivation. The botanist had never been a powerful teacher in the Agassiz manner. He lacked Agassiz's gift for popularizing science. Agassiz's magical ability to attract large sums of money was without precedent, and Gray, who found it difficult to acquire funds to repair the fence around his botanic garden, found himself relegated to an Agassiz-inspired obscurity in the vital realms of social and material support for science. Gray was compelled to challenge Agassiz's dominance. He could not hope to match Agassiz as a social figure, yet with the reflected glory that came from his involvement with Darwinism, he could now hope to oppose Agassiz in matters of public policy. Since Gray was as committed as Agassiz to Harvard and to professionalism, the confrontation was bound to be a test of strength.

The first clash between Agassiz and Gray was a relatively mild one, precipitated by Agassiz's effort to test and to extend his Harvard influence. But it was a preview of more to come at Harvard and in the nation, and Agassiz and Gray probably realized this. As Agassiz became convinced that the crisis of the Civil War demanded dramatic cultural and intellectual accomplishment, he determined that Harvard should meet the challenge of the times as he chose to define it.

Starting with the presidency of Felton and continuing through the administration of President Hill, natural and physical science assumed an increasingly important role in the curriculum of Harvard. With Agassiz a national figure, Gray making fundamental contributions in botany, Wyman and Holmes both distinguished anatomists, Peirce a respected authority in mathematics, and young Charles William Eliot a talented teacher of chemistry, the faculty of the Lawrence Scientific School was an impressive one. This faculty quite naturally began to consider ways to improve the instruction offered in the scientific school both to undergraduates and to special students.

In October, 1861, a faculty committee appointed to recommend a new curriculum for the school made its report. Delivered by Agassiz, the committee report asked authority to plan a two-year program for elementary studies before special education began. Gray moved that this recommendation be accepted, and the Agassiz committee was given authority to formulate such a program of study.[36] On November 4, Eliot, now acting dean of the school, presented the full recommendation of the committee for implementing the "preparatory course" of instruction. This plan, one that called for two years of instruction in the natural and physical sciences as well as courses in rhetoric, French, drawing, and German, was of Eliot's inspiration.[37] On November 18, however, Agassiz objected to certain features of the plan his own committee had recommended. A faculty argument followed, with Gray, Wyman, Joseph Lovering, and Eliot supporting the innovation, and Agassiz and Peirce in opposition. No decision was reached. Early in December, the faculty met again, and this time it was Gray who submitted a series of proposals that seemed to back up the Eliot report. Agassiz was not present at the meeting, but the faithful Peirce urged that "Agassiz had prepared a plan for the better organization of the School." President Hill, who attended the meeting, was asked by the faculty to consider the Eliot and Gray proposals and suggest some "scheme which would be unanimously approved by the Faculty." Hill, who could not help sympathizing with whatever Agassiz wished, "said that he should consult Prof. Agassiz and consider any plan he might suggest," in reference to the proposals before the faculty.[38] Agassiz had nothing to propose, and the scientific school reorganization plan was abandoned, since Hill sided with Agassiz and Peirce. Just what the academic position of these two Lazzaroni was, is somewhat of a mystery, although Hill justified his reasons for siding with Agassiz and Peirce on the grounds that the Eliot and Gray proposals diminished the professional character of the school and lowered standards.[39] Agassiz, with Peirce's help, had blocked reform of the Lawrence School because he understood that such change might be identified with Eliot and Gray and not with him and the Lazzaroni. If Agassiz's actions seemed mere negativism, they were designed to promote a program for science and education at Harvard under Lazzaroni sponsorship. With Gray and Eliot defeated on the Lawrence School front by late 1862, Agassiz could move forward in the accomplishment of really vital goals.

By this time Agassiz had given much thought to the entire question of educational policy at the university and had formulated a program that he felt would raise the intellectual level of Harvard from what he then considered to be its low state. "Professionalism" was one of the primary aspects of his argument and his appeal; but certainly there was nothing in the proposals of Eliot, Gray, and other scientific school faculty to lower this standard. Yet "professionalism" and new schemes for Harvard had, of necessity, to be the product of Agassiz's intellect.

In contrast to some other members of the faculty, Agassiz was by experience and mental outlook unconcerned with undergraduate education. From the beginning of his association with Harvard, he envisioned the Lawrence School as a graduate institution, and when progress toward this end seemed slow, he could identify the museum as a primary agency for advanced education. Loyal to his education and to concepts of German academic life, he held that the only true pedagogy was centered in a university, where an enlightened and professional spirit of research could prevail.

Agassiz felt that Harvard College was little more than a "high school," the professional schools existing as "excrescences of the College." This was a situation he found unworthy of a state or nation where education was such an important social and economic factor. He thus envisioned Harvard as a great university with strong graduate departments of law, medicine, theology, humanities, and science, which would be far superior to any in the country. This was the plan of "the great universities of the old world," and Agassiz believed that "it is high time that we should begin to compete with them, if we ever expect to raise our intellectual culture to the level of that of Europe."[40] It was with this end in mind that Agassiz proposed to Governor John A. Andrew late in 1862 that the state take full control of the university. Then, the men of the State House could raise a magnificent "private subscription" which, along with state appropriations, could be used to appoint notable professors to the faculty. University lecture courses could be instituted, each branch of higher education would have ample funds at its disposal, the library could be increased by an appropriation of ten thousand dollars, and a true university would be established at Harvard. As for the college, this should remain a private institution, and Agassiz had little hope for any improvement in this domain.[41]

Agassiz well knew that such a proposal was not only visionary but opposed to the entire educational spirit of the university. This did not trouble him in the least. These were times of momentous change, and Harvard University under Agassiz's leadership could join the march of cultural progress. He even went so far in his idealization of the new education to propose to Governor Andrew that the federal government be the agency to supply the great sums of money needed for the reorganization. He urged that the recent Morrill Act providing subsidies for academic institutions teaching agricultural and "mechanical" subjects and offering military instruction be exploited, Harvard establishing departments in these fields and thereby gaining for Massachusetts a rightful share of government funds.[42] As the man who had always idealized the University of Berlin, Agassiz thus put the matter to the governor of Massachusetts:

. . . there does not yet exist a real University in the United States . . . an institution . . . offering information of the highest character upon every topic of human knowledge and so superior to all other institutions that students from every part of the country would flock to it. . . . The students of the University of Berlin are not Prussians. . . . They are the youth of Germany in all. . . . Now is the time, and Massachusetts the State to lead this movement. . . . Be not afraid to offend the friends of Harvard. Those who love this university will feel the truth . . . and those who dislike it join in to sneer. . . .[43]

This was the compulsion that made a reform of the scientific school small by comparison and that made it necessary for any change to be Agassiz-inspired. The "friends of Harvard" might well have been shocked and distressed had they known of these proposals, the entire scope of which was never made public, but once again Agassiz was not to be deterred by the traditions of small-minded men. He thought Gray one of those who would "join in to sneer" at such progressive schemes, and began to evaluate his colleague as too much identified with undergraduate education and the time-worn recitation method. As for Gray, he could not have looked with favor on this latest Agassiz effort, coming as it did just when Agassiz was opposing his and Eliot's proposals for the scientific school. Harvard had *"lectures* on Zoology and Geology—of no use to a young beginner and very little to an older hand,"* Gray confided to Engelmann in estimating Agassiz's worth. But Agassiz was hardly deterred by such opinion. The establishment of "university lectures" delivered by outstanding scholars of Harvard and other institutions seemed to him a fine first step toward real graduate education, even though his proposals to Governor Andrew went unrealized. Encouraged by slight success, Agassiz now worked to bring to Harvard an ally who shared his views, a man who would be dean of the scientific school. That this man happened to be a fellow member of the Lazzaroni, Wolcott Gibbs of New York, reflected the power Agassiz and Peirce knew they enjoyed at Harvard. Since their fellow Florentine, Gibbs, was also a chemist of outstanding reputation, a synthesis between Lazzaroni aims and Harvard education could be achieved.

When Eben N. Horsford resigned his position as Rumford professor early in 1863, this event raised the question of his successor. Charles William Eliot, who had been performing some of Horsford's duties and who was serving as acting dean of the scientific school, expected that his services would be recognized by the transference to him of the Rumford professorship. The Corporation, moreover, was faced with the problem of deciding on a permanent dean for the school, and since Horsford had held this post until recently, Eliot might have expected to inherit this position as well. President Hill, however, felt that the best he could do in this situation was to offer Eliot a temporary professorial appointment, with scarcely sufficient guarantees of his salary. In June, 1863, the Corporation chose Wolcott Gibbs as Rumford professor, and, by September, Gibbs had been elected dean of the scientific school faculty. Eliot, who could not accept the arrangements proposed for him, was thus forced by circumstance to leave Harvard.[44]

Agassiz believed Gibbs to be by far the "superior man" for the positions of dean and Rumford professor, and, in collaboration with Peirce, exerted all his influence with President Hill and Judge Hoar to gain the appointments for his friend. To Agassiz, such an accomplishment was a vital element in the campaign to make Harvard a truly major university. Gibbs, he felt, was the first chemist of the nation, and his appointment would give Harvard the character of a "real university." He was willing for the young and to him still immature Eliot to remain at the university, but not in a position that would give him a permanent role in formulating educational policy. President Hill accepted these arguments, and with the support of Agassiz and Peirce made Gibbs's appointment possible, an event Agassiz viewed as the "keystone" for his future Harvard plans.[45]

With Eliot gone, with Agassiz triumphant in the Lawrence School and in the matter of Gibbs's appointment as dean and Rumford professor, it must have seemed to Gray that Harvard science was under the entire domination of Agassiz and his allies. No wonder, then, that Gray looked upon Agassiz's treatment of Clark as very shabby; this was an excellent and objective demonstration that Agassiz was not to be trusted. When the Lawrence School faculty voted not to exclude Clark from its meetings, Gray probably led the opposition to Agassiz's request for this action. It was one thing to resent Agassiz's actions and behavior at Harvard; it was quite another to have to stand idly by, in ignorance of what was happening, as the Lazzaroni marched on to another conquest.

At the very time when Agassiz, with Hill's support and Peirce's help, was blocking Eliot and outmaneuvering Gray in Cambridge, the Florentines were busy on another front where the stakes were even higher.

Harvard in 1863 was the home of leading Lazzaroni: Agassiz, Peirce, Gibbs, and B. A. Gould. If these men could use their influence successfully in Cambridge, there was no reason why they should not extend their power to the councils of national government. The Lazzaroni had never lost sight of their plans of the 1850's for the founding of a truly "American" university and a national scientific establishment. If Harvard seemed a congenial home for a "national" university, Washington was the natural site for a central agency of science. In 1860 and 1861, Agassiz had been instrumental in gaining membership for the "Chief" of the Lazzaroni—Bache—in both the Royal Society of London and the French Academy of Sciences. What made more sense than the establishment on American soil of a native counterpart of the French Academy, staffed by Lazzaroni and their allies? Bache had been urging such a scheme since 1851, and his dreams as well as Agassiz's now seemed to stand a good chance of fulfilment. In Agassiz's view, the Academy would at once be a prime example of the professionalism required of science in America, and place scholarship in this field on a truly national basis:

I am so deeply impressed with the importance of such an organization . . . as an evidence to go before the civilized world at large of the intellectual activity displayed even in these days of our public troubles, that I am satisfied it would tell strongly in our favor, wherever learning is duly appreciated.

Thus Agassiz informed Senator Henry Wilson of Massachusetts, a man who needed little convincing, of the national needs of science as defined by the Lazzaroni.[46]

As superintendent of the United States Coast Survey, Bache found an important ally for the Lazzaroni scheme in Charles Henry Davis, now a rear admiral of the United States Navy and a man who had achieved a distinguished combat record during the first year of the war. Davis was also a professional astronomer, a member in good standing of the "Cambridge clique" of the Lazzaroni. When Davis was recalled to Washington to head the Bureau of Navigation in November of 1862, he and Bache discussed measures whereby a larger role for science in government affairs might be achieved. The first accomplishment of the two promoters was the establishment by Congress on February 11, 1863, of a "Permanent Commission" composed of Bache, Davis, and Joseph Henry. Patterned after the "Select Commission" of the British Navy, this body was charged with the responsibility of reporting to the Navy on scientific research which might be useful in the war effort.[47]

The "Permanent Commission" was but the beginning of efforts to secure government aid and encouragement for science. In February of 1863, Senator Henry Wilson nominated Agassiz to a vacancy on the Board of Regents of the Smithsonian Institution. His appointment approved by Congress, the Harvard professor came to Washington on February 19, 1863, ostensibly to visit Secretary Henry and discuss his new duties. Agassiz did not see Henry for three days. His time was spent at Bache's home, where he was closeted with Wilson, Davis, Benjamin Peirce, and B. A. Gould. In the course of the evening of February 19, the potentialities represented by the founding of the "Permanent Commission" were discussed by important Cambridge and Washington Lazzaroni.[48] The idea of a "National Academy of Sciences" was then brought up, and all present agreed that it was an excellent suggestion.

Wilson was a key man in these deliberations, because he could make dreams of government recognition come true. He had been well primed by Agassiz, who as early as February 5 had written him of the Academy idea: ". . . if you think favorably of this suggestion you have in Bache, to whom the scientific men of the country look as upon their leader, a man who can draft in twenty four hours a complete plan for you, such as we have again and again discussed for several years past."[49] Significantly, when Agassiz wrote Joseph Henry on this same date, he said nothing about these plans, probably aware of Henry's feeling that the scheme would be "opposed as something at variance with our democratic institutions."[50] Aware of such dangers, Agassiz had counseled Wilson that the times were favorable and prejudices of this sort could be silenced when "men of the people have taken hold and given their hearty support to those institutions which unfortunately thus far have been looked upon as aristocratic."[51] Years later, when the Academy had weathered a storm of early opposition, Agassiz candidly confessed to Henry that "a better acquaintance with American ways has satisfied me that we started on a wrong track."[52] But in the flush of optimism that emboldened the Lazzaroni in 1863, these men of the people made certain that their aims and ambitions were achieved.

On the evening of February 19, it was agreed that the Academy would be composed of fifty "incorporators" chosen secretly by the men gathered at Bache's home, supposedly representing the leading scientists of the nation. These scholars would gather once a year to discuss matters of import to their profession and to decide how their united efforts might help the government. The government would in turn designate these men as the officially recognized "National Academy," would hopefully pay their expenses when they traveled to meetings, and through the Permanent Commission would suggest problems in science and technology that might be solved in the interests of national policy.

Henry Wilson introduced a bill in the Senate on February 21, 1863, incorporating the National Academy of Sciences according to the arrangements discussed in Bache's home. The bill was passed by both houses of Congress and signed into law by President Lincoln on March 3. With official sanction, the American version of the French Academy came into being.[53] The fathers of the Academy were Bache, Davis, Agassiz, and Henry Wilson. Joseph Henry did not discover until March 5 that the Academy bill had become law. Henry, however, was pictured in a painting depicting the signing of the Academy bill by Lincoln. He was clearly out of place. Agassiz, however, smiled as he stood at Lincoln's side. Frederick William III would have been proud of his former spiritual subject.

The membership of the Academy was limited by law to fifty scientists. In a gesture clearly designed to secure governmental support, eleven of the "incorporators" were Army or Navy men who could hardly be classified as "scientists," while the other names represented the Lazzaroni choice of the outstanding scientists of the nation. It was impossible, of course, to fail to include men of the caliber of Gray, Rogers, and others not of the inner Lazzaroni circle. There were, however, some notable omissions. Spencer F. Baird was not a member, nor were such savants as George P. Bond, Elias Loomis, or John William Draper. Yet every member of the Lazzaroni was included. Bache was president, Dana was vice-president, and Agassiz was foreign secretary. The director of the Museum of Comparative Zoology could now write his colleagues in Europe and invite them to become "corresponding members" of the National Academy of Sciences.[54]

A few of the scientists who had been appointed to membership in the Academy were not happy about the honor bestowed upon them without their knowledge or participation in the affair. William Barton Rogers was understandably unsympathetic and attacked the undemocratic plan of giving the president life tenure. Joseph Henry agreed reluctantly to accept membership, but more out of a desire to please Bache and Agassiz than from any fundamental conviction of the necessity for an academy. Joseph Leidy and others were highly indignant over the fact that the council of the Academy, namely Bache, Dana, Agassiz, Gibbs, and Fairman Rogers, decided to make an "oath of loyalty" to the United States a condition of membership. A vigorous floor fight over this provision of the constitution ensued during the first meeting.[55]

There were other scientists who simply disapproved of the Academy as a whole. George Engelmann refused to take part in its meetings because he considered this but another instance of Agassiz's quest for power. Joseph Leidy eventually dropped his membership, as he could not see that the Academy was serving any useful purpose. Jeffries Wyman held similar views. Gray saw the whole business as a quest for power on the part of the "Coast Survey and Agassiz Clique" and disapproved of the secret way that the Academy had been planned. Gray was also opposed to the plan to have expenses entailed in attending meetings paid by the government. While Gray could not be enthusiastic about the Academy, he had no real choice but to do nothing for the time being. The

Academy might turn out to be a good thing, and Gray was too honest a person to let his feelings color his judgment on a question of this sort. So he decided to wait things out; perhaps, with the help of W. B. Rogers, Henry, Dana, and others, he could use his own influence to counter that of Agassiz and his friends in the organization.[56]

From Agassiz's point of view, he was taking the opportunity to serve his country in a time of national crisis by helping to found the National Academy of Sciences and by acting as a regent of the Smithsonian Institution. Undaunted by the mixed professional reaction to the Academy, he moved next to widen the scope of American cultural organization by presenting Charles Sumner, senior senator of Massachusetts, with a plan to organize two other institutions under government sponsorship, an Academy of Letters and an Academy of Moral and Social Sciences. It was only fair that Sumner have his chance to found an Agassiz academy, since Wilson had already served the cause. He freely advised Sumner who should be members of these bodies; in the Academy of Letters, more than half of the proposed names were Saturday Club luminaries, while men of Boston who could not be included in this literary category were given Agassiz-inspired membership in the other academy. Emerson, Holmes, Lowell, Whipple, Motley, Ticknor, Charles Francis Adams, Jared Sparks, and Samuel Gridley Howe were all designated by their naturalist friend as worthy of such honor and distinction. To Agassiz, "all the eminent men of the country" should be grouped together in societies of this sort. That most were New Englanders was not surprising, since this region contained the Saturday Club and the leader of the Cambridge Lazzaroni.[57]

From a public standpoint, it seemed that Agassiz expanded in stature with the passing of years. His hair was graying, his once stalwart frame showed the signs of increasing stoutness, his eyes continued to trouble him, but his spiritual vision was clear, his smile as infectious as ever, his mental outlook continually optimistic. The National Academy and the Smithsonian represented his cosmopolitan interests; the condition of Harvard and the progress of the museum his everyday concerns. He refused to leave

Cambridge for any reason except the most pressing business—such as the establishment of the Academy—and continued to identify the progress of the museum with the intellectual stature of the nation. On a day in the spring of 1864 two national heroes met; General Grant came to see the American center of natural history study, and Agassiz guided him through the museum. Sumner sent Agassiz and Lizzie rare wines and cases of books; Longfellow continued to write poems in his honor, one literary effort of Christmas 1864 being attached to six bottles of fine French wine, each instructed by the poet in verse to find their way to the happy "Chez Agassiz" on Quincy Street. The museum director addressed a letter of rebuke to Emerson for some public remarks the Concord sage had made which to his friend seemed critical of science, and Emerson, under the spell of one whom he ranked next to Thomas Carlyle in his list of "my men," wrote a quick apology for having unintentionally offended his dear friend.

In England and on the Continent staunch Agassiz admirers were kept fully informed of the progress of the Union. Richard Owen, for example, learned that there were now university lectures at Harvard, the state legislature had given the museum ten thousand dollars for scholarly publication, a National Academy of Sciences had been organized, and all these events had one meaning. "I trust when such facts become known in Europe it will be evident . . . that we are alive to the higher interests of humanity and that the North is not only fighting for power."[58]

Lest the world forget Agassiz, the year 1863 witnessed another large project stemming from his fertile imagination. As firmly convinced as in 1859 that Darwin was mistaken about the interpretation of nature, Agassiz decided to amass further evidence by using the resources of the museum to gather a complete sampling of fishes native to every river system in the world. A decade before, Agassiz had been content to ask only for materials from North America, but as his influence grew, so did his demands and conceptions. The knowledge to be gained from this new enterprise would, he assured potential collectors, be very likely "to deal a crushing blow to Darwinism and similar isms in Zoology." George

Ticknor was urged to exploit his Spanish friends for this purpose, because "Spain alone might give us the materials to solve the question of transmutation *versus* creation." Once again, men with world contacts were pressed into Agassiz's service; Admiral Davis, Bache, Henry, and all agencies of the federal government that could help were asked to do so.[59] Two years later, the Department of State instructed American consulates all over the world to aid a similar Agassiz endeavor, the collection of types of domesticated animals from the four corners of the globe. In these ways, the resources of the Cambridge museum continued to grow by leaps and bounds, and if these materials did not prove Darwin wrong, their recipient knew he was gathering data of fundamental importance to the progress of national science.

By 1865, the income of the museum had grown to over twelve thousand dollars a year, another evidence of Agassiz's stature in community, state, and nation. No matter how the pecuniary condition of the museum advanced, Agassiz was more than equal to the opportunity presented. Every appropriation and special public subscription was usually spent before it was received; European scientists sold their collections to Agassiz and then had to wait long months for payment. Trustees tried in vain to introduce some kind of budgetary planning into Agassiz's scheme of operation, but they were no more successful than Alex Agassiz. As the son wrote his best friend Lyman about his father:

Of course you cannot stop a steam engine going down an inclined plane any more than I can stop father and yet I have to. . . . I can think of no one else who can talk to father and to who father will be willing to say what he is doing. . . . He has so strongly the belief that I do not approve of all he has done and would like to retrench still more that it has materially altered his confidence in me. I wish for the sake of my own peace I could see things as he does but I cannot.[60]

The steam engine that was Agassiz was no more capable of stopping than he was of budgeting money. As he candidly confessed to the trustees of the museum:

It may perhaps be said that I ought to have gone on more slowly, and saved something out of the appropriations. . . . To this I have only one answer, that the appropriations were at no time equal to the plans adopted for the Museum . . . for no man in his senses could have imagined that a Board like this, composed of the first officers of the State and some of the most prominent citizens . . . would be organized by the legislature merely to supervise a school or college museum.[61]

Other scientists might have thought this, but never Agassiz. The great museum of the age needed to be constantly replenished with both money and specimens to satisfy his insatiable desire to master the world's natural history.

It did not matter who or what a man was, if he was in a position to help the museum he sooner or later heard from Agassiz. Cyrus Field, working off the coast of Newfoundland in his attempt to lay the Atlantic cable, stopped long enough to pack a shipment of bird skins and send them to Agassiz. Dom Pedro II, emperor of Brazil, ordered his subjects to gather rare and unusual fishes for the American professor. Secretary of War Edwin M. Stanton, who in January of 1865 had more pressing problems than the demands of Agassiz, heard from the naturalist as follows:

Now that the temperature is low enough . . . permit me to recall to your memory your promise to let me have the bodies of some Indians; if any should die at this time. . . . All that would be necessary . . . would be to forward the body express in a box. . . . In case the weather was not very cold . . . direct the surgeon in charge to inject through the carotids a solution of Arsenate of soda. I should like one or two handsome fellows entire and the heads of two or three more.[62]

For Agassiz, every action was related to science. A fire in a Cambridge stable brought the professor running from his museum to the scene of the blaze, where he helped firemen in their tasks, comforted the stablemen on the loss of their beautiful horses, and arranged to have the carcasses shipped to the museum for study.

Meanwhile, Asa Gray was one Cambridge resident who did not lionize Agassiz. Although the two scientists did not break off relations at any time during the years after 1857 when they found themselves at opposite poles on almost every issue, their once close friendship had been seriously strained. Gray's indictment of Agassiz was based on feelings and ideas that symbolized the essential qualities of science and scientists in America during years of profound intellectual and institutional change. Gray was the quiet, competent researcher, his efforts increasing the quality of national scholarship. Agassiz was the symbol of natural history in the popular mind, a man who inspired grand projects by force of personality and will.

Given this contrast of personality and scholarship, there were three aspects of Agassiz's authoritarianism that were probably more distressing to Gray than any quarrel of the moment. First, Gray must have felt that Agassiz was not using to best advantage the tremendous monetary and scientific resources he had gained. Gray would have felt much better about Agassiz if he could believe that the Agassiz performance matched the Agassiz promise and that the financial wealth of Boston and the zoological materials acquired by the museum through Agassiz's organizing talent were actually being used for the advance of knowledge. Instead, it must have seemed to Gray, as it seemed to some others, that Agassiz was merely "carnivorous" regarding money and specimens; he demanded more and more, with little evidence that what had already been given was actually ever used to real advantage. Second, Gray was plainly disgusted with Agassiz's irresponsible public attacks on the evolution idea; every aspect of the botanist's personality demanded that intellectual disputation be carried on within a professional framework. But, instead, no sooner had Agassiz been challenged by Gray in intellectual debate than he shrank from further interchange and took refuge in the security of an admiring public. Third, it was apparent that Gray resented the prominence Agassiz enjoyed and the power he wielded through the Lazzaroni. The Lazzaroni had never invited Gray to become part of the inner circle, or even acknowledged his existence when they planned grand scientific projects for America. It was quite understandable that Gray did not share the "Lazzaroni spirit" and was unprepared to think that a national academy was necessary for the progress of American science. Content to enjoy the familiar society of the American Academy of Arts and Sciences or at best the American Association for the Advancement of Science as the only necessary institutional connections for communicating ideas and improving the profession, Gray very likely believed any wider activity premature.

An ironic change had occurred in the social roles of Agassiz and

Gray. In 1846 it was Agassiz who symbolized the international, cosmopolitan character of science for many Americans, and Gray was as impressed as anyone with this quality. As the years passed, and Agassiz became an American with a New England wife, his children Americans by adoption, his closest friends the elite of Boston and Cambridge society, he became an American nationalist in cultural terms. *Contributions to the Natural History of the United States* was a perfect title to reflect Agassiz's cultural iden-' tity from the 1850's onward. It was the improvement of national science, of American institutions, of popular knowledge, that Agassiz worked for, wanting to transform them into native models of what he remembered of Paris and Berlin. Significantly, Agassiz's correspondence with Europeans diminished as the years passed, and the great bulk of it had to do with gaining European specimens for the Harvard museum. It is also noteworthy that in an age of steadily improving communication with Europe, Agassiz only returned once to the land of his birth.

Gray, on the other hand, became more and more of an American representative of the most modern internationalism in science. He brought Darwin to the attention of Americans and educated his colleagues to show an open-minded respect for the character and potentialities of the new biology; he was perhaps more honored in England by men like Darwin, Hooker, and Lyell than he was in his own country. Gray had his Lazzaroni too, but this society could not sit down for many oyster dinners; its existence was centered on grand ventures, not for the improvement of America's scientific institutions, but for winning intellectual victories for the ideas of Darwin.

It was within this framework that Agassiz and Gray finally had one more dispute that was sufficient, in the light of all that had gone before, to end their friendship. Never one to hide his feelings, it was Agassiz who precipitated the break, but Gray had provoked him enough, he felt, to justify his course. In November, 1864, Agassiz wrote his close friend Harvard treasurer Nathan Silsbee of the current state of affairs at the university, giving vent to his anger with Gray in a very rare instance of placing personal, emotional reactions on paper:

If you knew the internal relations of the Professors here, you would
be aware that there is a set who would have everything stationary . . .
with recitations as the only mode of teaching and who pursue with
their interpretations everything which is doing to improve the state of
things. . . . I perceive now that a part of the scheme is to *make* the
President unpopular and to *raise* difficulties between him and the students and unless this is stopped you will see very soon that the lazy
ones among the Professors will succeed in making you believe that the
men who are active about the University are a dangerous set of re-
formers. . . . That you should at once be on the right track I will add
that I am of all the members of the University the most obnoxious to
the retardative set. The progress of the Museum is a thorn in their side.
The prime mover against every action which would disturb the status
quo is Dr. Asa Gray. I leave it to you to find out what are his motives.
I am not called upon to criminate any of my colleagues. But I warn
you that what is going on . . . is very serious and may have disastrous
results for the Undergraduate department and the Scientific School.
I am too old to care what is said of me; but I think it my duty . . . to
. . . put you on the track of the state of feelings among the Professors.
It is still a covered fire which may . . . any day blaze away.[63]

For Agassiz and Gray, the fire had blazed away some months before. In the summer of 1864 Agassiz went to the National Academy of Sciences meeting at New Haven, refreshed by the success
of a grand lecture tour through the Middle West and a great reception in Chicago. He arrived in New Haven eager to exercise his
power of unchallenged dominance. He left the university town an
unhappy, frustrated, and indignant man.

The issue was, on the surface, hardly as important as the result.
Agassiz had never really respected Spencer F. Baird, even though
he had been partly responsible for that naturalist's Smithsonian
appointment. With the passing of time, he came to think of Baird
as retarding progress in natural history, an attitude that was intensified as Agassiz demanded more and more government specimens,
and Baird became increasingly unwilling to part with them. Baird
usually co-operated in the long run, but only reluctantly. Agassiz's
attitude toward Baird was complicated by old hatreds. One of Agassiz's complaints against Desor was that his secretary had been
very rude to Charles Girard, the Swiss student who had followed
his teacher to America. When Desor was dismissed, Girard, for

reasons incomprehensible to Agassiz, left his employ. Agassiz interpreted this as patent ingratitude. Annoyance turned to anger when Girard turned up as Baird's assistant in herpetology at the Smithsonian. Girard had found the position on his own, Baird had not consulted Agassiz about the appointment, and from this time on Agassiz would have nothing to do with Girard. When Baird and Girard published any joint paper or book, as they often did, it was bound to receive a slashing review from Agassiz.

The New Haven meeting opened with Baird proposed for membership in the Academy. Agassiz began to rally his forces for the showdown vote, thinking he could easily prevent Baird's election. But when the vote was in, Baird had become an Academy member. Agassiz was too shrewd an analyst of men and had too much experience in affairs of this sort, not to suspect an alliance against him. He knew, from the facts of the case, that Baird had been supported by Henry, Gray, and Dana. He could understand Gray's role. What he could not understand was how Gray had got the support of the majority of those scientists voting on the Baird membership.

He interpreted the event as an effort by Gray, with important help from Dana and Henry, to influence the Academy membership against him, since a vote for Baird was, he understood, a vote against Agassiz. Even though Agassiz had continued to be friendly with Dana, he suspected the New Haven naturalist of disloyalty. The Marcou affair still rankled. On top of this, Dana's growing disenchantment with Agassiz's philosophical defense of special creationism could not have gone undetected. Still worse, Agassiz knew that Dana had encouraged students like Verrill and Bickmore in their display of independence in the museum revolt, and Dana had hired Verrill at Yale. Since Dana voted the wrong way on the Baird issue, it was plain to Agassiz that he had joined forces with Gray. Agassiz confronted Dana with this charge after the hectic meeting was over, and insisted that Dana remove his name as associate editor of the *American Journal of Science*. Dana charged this request to the heat of the moment and ignored it. Two years later, however, the *Journal* carried a review by Verrill of Henry James Clark's volume *Mind in Nature,* wherein Verrill referred to the *Contributions* controversy, and praised the book highly. Agassiz was incensed that Dana would allow this to be printed, knowing the feelings of one of his editors, Agassiz, regarding Clark. He now demanded removal of his name as editor, writing Dana, "Our habits of friendly and open discussion of all scientific matters . . . have been interrupted of later years." His request was granted.[64]

Agassiz could understand how Dana had been motivated to help Gray, but he could not understand Henry's position. His basic argument against Baird's election was that the Smithsonian assistant secretary had made no original contributions to knowledge, and he did not see how Henry was capable of evaluating a man not in his field. Agassiz was very bitter against Henry, more so because he felt Henry had allowed Gray to use him for his own ends.[65]

Henry could be dealt with, but Agassiz could not forgive Gray. He undoubtedly was mortified to realize that a man entirely outside the Lazzaroni network could play the Lazzaroni game with such skill and could invade Agassiz's own domain, the Academy, and capture its membership in one sally. Hurt, angry, feeling the power of years slipping from his grasp, Agassiz could not understand that another man could have powerful allies too. What he did not realize was that the years of splendid authority had left him almost entirely unaware of Gray's influence with many naturalists. For the moment, Agassiz elected to close his mind to the real meaning of Gray's victory—that this was not a personal triumph for Gray but a measure of unrest over his own dominance—and interpreted the botanist's action as a purely personal, vindictive attack. Gray's role in the Baird controversy was very likely the cause of a bitter argument between the two men. Gray reported to Engelmann that he had to break off all personal relationships with Agassiz in the summer of 1864, the result of Agassiz's having "insulted me so foolishly and grossly." Thus two Harvard scientists, each a dominant figure, came to an unhappy parting.[66]

As for Henry, he took Agassiz's criticism of his behavior in stride, writing his friend a long letter recalling how he had not been informed when the Academy had been planned, that he did not approve of the secret methods employed, that the vote in favor of Baird was in fact a vote to prevent "the few who organized the academy" from governing it by themselves, and that had Baird lost, a majority of the naturalists would have resigned their memberships. Disavowing any role in an intrigue against Agassiz, Henry urged him to be more charitable toward others and to stop taking upon himself the burden of overseeing the entire structure of science in America. He offered words of sincere advice, the kind Humboldt might have proffered:

You have already done good service by your presence in this country,— by your immediate instruction and by the enthusiasm and sympathy, which you never fail to awaken. You are formed to lead men by the silken words of love rather than to urge them on by the rough method of coercion. Let me beg of you . . . to first take care of your health and secondly to devote yourself for the remainder of your life to those investigations which have given you so wide and permanent a reputation and in which . . . you can elevate yourself in your own self esteem as well as in the admiration of the world. It is lamentable . . . how much time, mental activity, and bodily strength have been expended among us during the last ten years, in personal altercations, which might have been devoted to the discovery of new truths. . . .[67]

Agassiz replied to this well-meaning and excellent advice by writing: ". . . difference of opinion never has had . . . an influence upon my estimation of my friends' character or my intercourse with them, as long as such differences are pressed with the proprieties and urbanity which we must all observe toward one another, if we would live in peace."[68] Apparently, these requirements were not met by Clark, Gray, or Dana.

Agassiz was paying both a physical and psychological price for carrying the world of American natural history, the condition of national science, and the organization of professional activity on his shoulders. By the winter of 1864–65, his health, as Henry had predicted, began to fail. He grew increasingly tired and short-tempered and seemed weary of the world. The world of Civil War days, with the student rebellion, the Clark affair, the Gray difficulty, and similar disturbances had taken its toll. It was plain to one who knew him as well as Henry that he was heading for a breakdown if he did not slacken his pace. Once again, at just the right time, fortune intervened in the life of this perennially fortunate man.

During the winter of 1864–65, Agassiz found eager audiences for another series of lectures at the Lowell Institute. This time he used glacial action in South America to prove the fallacy of organic evolution. The continental ice sheet had destroyed all life in its path, and hence there could be no connection between the species of the past and present. At the close of his last lecture, Agassiz expressed the belief that it would be a contribution to science of the greatest value if a naturalist were to explore Brazil and the Andes Mountains in order to gather direct evidence of glacial action. Such an expedition could also provide much fascinating data on the natural history of the region.

Nathaniel Thayer, one of the wealthiest of Boston's businessmen and an active trustee of the museum, realized that in such a locale Agassiz would be able to engage in important studies and regain his health. To make this possible, Thayer offered to defray the costs of an entire exploring party, to be composed of four assistant naturalists besides Agassiz. Thayer did not have to persuade his friend to accept this proposition. As with all of Agassiz's projects, a small undertaking grew enormous almost overnight. A journey to restore Agassiz's health suddenly developed into the Thayer Expedition to Brazil.[69] The author of *Brazilian Fishes* would travel to a familiar land.

Ever since his student days he had longed to command an exploring party that would travel to an exotic region for a lengthy sojourn. The United States in 1846 had seemed promising territory, but civilized Boston and an admiring public had been the primary beneficiaries of the Agassiz wanderlust. The brief trip to Lake Superior had only stimulated this desire. Von Martius, Spix, Humboldt, Alfred Russel Wallace, Henry W. Bates, and other naturalists had visited the Amazon Valley with its impressive mountains and endless system of rivers. Darwin and many other illustrious Europeans had gone on journeys of discovery and received the plaudits of the scientific world for their findings. Now that he was director of a museum, foreign secretary of the National Academy, professor of zoology and geology, and author of many works in natural history, the only remaining area in which Agassiz had not achieved real distinction was that of explorer.

The planned scientific enterprise soon took on aspects of an educational and social mission. Agassiz was determined that his wife accompany him on the journey and also decided to take along, in addition to the assistant naturalists, six students of Harvard, including William James and Stephen V. R. Thayer, a son of the expedition's patron. They would hear lectures from Agassiz on all branches of natural history during the course of the trip and would also aid the trained scientists in gathering specimens. The group of assistant naturalists included nearly the entire professional staff of the museum, with the exception of Alexander Agassiz, who was left behind in Cambridge to superintend the museum. The paid helpers were Joel Asaph Allen, ornithologist; John Gould Anthony, conchologist; Orestes St. John, paleontologist; and Charles Frederick Hartt, geologist.

In addition to Thayer's initial generosity, the expedition benefited from the aid and co-operation of many people anxious to do all they could to assist Agassiz. Secretary of the Navy Gideon Welles issued an order instructing all naval officers to be alert to the needs of the Harvard professor should they meet him in his travels. Secretary of State William Seward brought Agassiz's mission to the attention of the American minister to Brazil, James Watson Webb, who was also requested to render whatever assistance he could. The officials of the Pacific Mail Steamship Company, learning of the expedition, placed the steamship "Colorado" at the complete disposal of Agassiz and his fellow voyagers, providing them with free passage to Rio de Janeiro.[70]

This grand journey in the cause of science prompted Agassiz to write his mother: "I feel like the spoiled child of the country, and I hope God will give me strength to repay in devotion to her institutions and to her scientific and intellectual development all that her citizens have done for me."[71]

The Thayer Expedition left New York on April 1, 1865, and returned on August 6, 1866. Agassiz and his assistants collected over eighty thousand items of natural history, certainly one of the largest assemblages gathered in a single exploration. Fishes from all the tributaries and branches of the vast Amazon River system were piled waist-high on the decks of the steamers supplied by the Brazilian government and the Amazonian Steamship Company. Agassiz divided his party into a number of small groups, each responsible for surveying a particular area. The expedition concentrated its activities for three months in the environs of Rio de Janeiro and spent the remainder of the sojourn in the lush tropical jungles and slow-moving streams of the Amazon Valley.

A highly original phase of the expedition was Agassiz's lectures to students and assistants on board the "Colorado" and in camp along the route. In these talks, the professor taught his young disciples how to collect specimens, identify, label, and sort materials, and pointed out the significance of particular groups of natural objects. William James wrote of his impressions of Agassiz as a man and as a teacher in a letter to his father:

I have profited a great deal by hearing Agassiz talk, not so much by what he says, for never did a man utter a greater amount of humbug, but by learning the way of feeling of such a vast practical engine as he is. No one sees farther into a generalization than his knowledge of details extends, and you have a greater feeling of weight and solidity about the movement of Agassiz's mind, owing to the continual presence of this great background of special facts, than about the mind of any other man I know. . . . I delight to be with him. I only saw his defects at first, but now his wonderful qualities throw them quite in the background. . . . I never saw a man work so hard.[72]

James's observations provide, in fact, an excellent portrayal of the happy and exciting Brazil days:

The Professor has just been expatiating over the mass of South America, and making projects as if he had Sherman's Army at his disposal. . . . The Prof now sits opposite me with his face all aglow holding forth to the Captain's wife about the imperfect education of the American people. . . . *Offering* your services to Agassiz is as absurd as it would [be] for a S. Carolinian to *invite* Gen. Sherman's soldiers to partake of some refreshment when they called at his house.[73]

Of more personal contacts between young man and older master, James had this to say: "I am getting a pretty valuable training from the Prof. who pitches into me right and left and makes me up to a great many of my imperfections. This morning he said I was 'totally uneducated.' "[74]

Lizzie Agassiz impressed James tremendously. He thought she was "one of the best women I have ever met—her good temper never changes and she is so curious and wide awake and interested in all that we see . . . that she is like an angel. . . ."[75] The Agassiz urge to encompass all of Brazil's natural history had its lighter and more pleasant moments according to James's private account in his diary:

I went to the photographic establishment we have set up on board ship. . . . On entering the room found Prof engaged in cajoling three Mocas whom he called pure Indians but whom I thought afterwards . . . had white blood. They were very nicely dressed. . . . Apparently refined, at all events not sluttish, they consented to the utmost liberties being taken with them and two without much trouble were induced to strip and pose naked. While we were there Senhor Tavares Bastos came in and asked me mockingly if I was attached to the Bureau of Anthropologie.[76]

Agassiz's enthusiasm made an impression beyond the members of his party. Dom Pedro II of Brazil was delighted with Agassiz's company and could not do enough for his comfort. The Thayer Expedition personnel were not allowed to pay for any transportation, meals, or incidental expenses. Government steamers for interior travel were provided, an army major was appointed to act as a constant guide, and even the emperor embarked on a personal collecting expedition for some rare fishes Agassiz wanted to study. In the course of these kindnesses Agassiz and his wife told the emperor of life in the United States and the great good will the government entertained for the Brazilian nation. It is likely, as Agassiz reported to Representative James A. Garfield, that these discussions induced the emperor to open the Amazon River to international commerce in 1867.[77]

The cultured people of Rio de Janeiro, as the prominent folk of Boston before them, heard the former Swiss naturalist extol the "plan of creation" in the animal kingdom in a series of highly successful lectures. These addresses, moreover, signaled the emancipation of women in Brazil, for at the insistence of both Louis and Elizabeth Agassiz, the emperor for the first time permitted women to attend public gatherings at which males were present. Agassiz even gave Dom Pedro personal instructions in the science of natural history. The two men, both grand figures in their way, thus commenced a friendship that lasted for years afterward.

The most remarkable of all these experiences for Agassiz was his exploration of the Amazon Valley. The naturalist found himself in a new world of luxuriant tropical foliage, strange and exciting plants and animals, fascinating geological phenomena, and Indians who never ceased to arouse his curiosity. Relaxing on board the steamer "Ibuchy" after a hard day's collecting, Agassiz wrote a letter to his benefactor, Nathaniel Thayer, telling of his present state of mind. He was, he wrote, happy in what he was doing, but he was also tired. This was his last great enterprise. He had wasted too much valuable time and energy on social and public activity during the past two decades. Now, when he returned to the United States, he would resign his directorship of the museum, turn the institution over to the care of his son, and devote the rest of his days to solitary intellectual activity. He would even resign his professorship, if he could secure a small sum to supply his simple wants and those of his wife. His debt to America and its people had been repaid by his exertions of the past years, and now he wanted nothing more than to retire to a quiet, undisturbed existence so that he might fulfil his ambitions in the scholarly contemplation of nature.[78]

With thoughts like this in mind, Agassiz returned to America in August of 1866. Mountains of work awaited him in Cambridge. Alex Agassiz and Theodore Lyman had kept the museum operating only by effecting the most rigorous economies. It was plain to both men that with the golden touch of "papa" removed, the museum was a mundane establishment with pressing bills to be paid, and debts contracted over long periods that had somehow never been brought to light before. Alex, in direct charge of the museum, was happy to see his father return.

As for Agassiz, he might well have followed through with the intentions expressed to Thayer, established the management of the museum under different authority, and returned to intellectual labor. Fellow scientists would have liked nothing better. Even Gray was prepared to believe that Agassiz had seen the folly of his ways, and the fact that Agassiz, in October, apologized for former behavior meant that cordial relationships were once more restored between the two men.[79]

With the nation once more at peace, it remained to be seen whether Agassiz could still his gregarious urge to do more and more for science in America as he had defined the needs of the age. If Agassiz could find the internal peace that seemed so inviting in the Brazilian jungles, this would be a happy occasion indeed. He had, in an important sense, dedicated his life in the Civil War years to the advancement of national culture. By some process stimulated by time for introspection in Brazil, perhaps both America and Agassiz's spirit would allow him the calm reflection of mature years.

9

THE PAST AND THE PRESENT

1866-1873

AGASSIZ'S RESOLUTIONS made in the solitude of the Brazilian jungles faded into a half-remembered dream upon his return to the United States. Without any sense of interruption, he resumed his public life, pursuing the elusive goal of shaping the American intellect according to his conception of an enlightened and progressive civilization. There were always new goals and grander missions and therefore more need to acquire material support. His was a genuine if compulsive search for an ideal becoming ever larger and more challenging with each new accomplishment. Ends became new means by which further ends might be reached. The Museum of Comparative Zoology, the publication of the *Contributions to the Natural History of the United States,* the National Academy of Sciences, the Thayer Expedition to Brazil, all were but partial accomplishments which had to be reinforced by constant attempts at improvement. Agassiz's aspirations and motivations were too deeply intrenched. To cease striving was to cease to live.

To be sure, there were times when Agassiz was torn by the twin drives that had given impetus and direction to his life. "I am falling behind in my influence among scientific men," he told Lyman in a rare moment of self-appraisal. "I do not write enough, and especially not enough of that stuff which is neither attractive nor instructive to the many, but which might check some of the weak productions with which we are flooded." What could a man with Agassiz's public responsibilities do to still the gnawing realization of intellectual obligation? The museum was like a weight upon him. He would like to throw it off, but it was his child, and he was convinced, with a conviction rooted in years of authority, that if he abandoned the product of his devotion to American science and culture, it would die. More money than ever before was required, and only he could supply the lifeblood to nourish his beloved institution.

Such an effort would mean, he knew, a continual drain on his physical and intellectual energy. Lizzie, who with Lyman and Alex understood Agassiz perfectly, hoped fervently but vainly that her husband could cease his public activity. But lectures were always a source of funds, and the fiscal condition of the museum, strained by Agassiz's absence and the expenses of caring for the vast Brazilian collections, demanded such efforts of him. He gave of himself so entirely in these public performances that he was left for days in an exhausted state. Yet life must go on, and without the $200,000 Agassiz thought he needed to keep the museum functioning, current expenses had to be met by personal effort and sacrifice. Besides, another Agassiz compulsion told him he had an obligation to inform his public of the marvels of science found in South America.[1]

Even during his absence Agassiz had been unable to refrain from continuing his usual role. He and his wife had written a series of three articles for the *Atlantic Monthly,* explaining the purpose of the Thayer Expedition and its accomplishments. Agassiz's public was thus informed of the delights of an "Amazonian Picnic," which included meeting a charming emperor, collecting strange and wonderful specimens of natural history, making exciting voyages along the Amazon, and holding informal instruction sessions on zoology and geology. There were also accounts of Agassiz's important discoveries: the finding of a great variety of fishes in the rivers of Brazil, some observations on the physical and cultural characteristics of the native population, and the important revelation that glacial remains had been found in the mountains and valleys of South America.[2]

On August 12, 1866, less than a week after returning to Boston, Agassiz visited Washington to read a paper on "Traces of Glaciers under the Tropics" to the annual meeting of the National Academy of Sciences. This remarkable address advanced hypotheses which were as challenging as those of his glacial theory in 1837. Glacial action, he reported, had covered great areas of Brazil. He noted that he had been previously convinced by his geological explorations that glaciers had existed in North America; his experiences among Brazilian land formations and mountain chains now yielded similar data for South America. Agassiz concluded that the flora and fauna of America had been created anew after the glacial ice had receded and that no genetic relationship was possible between animals and plants which had lived prior to and after the glacial epoch. Here again was undeniable evidence of the mistaken conceptions of evolutionists.[3]

The address to the Academy was followed by a series of lectures before the Lowell Institute in September and October devoted in general terms to the evidences of glaciation in South America. Continuing his dual efforts for museum money and public instruction, Agassiz also spoke to an appreciative assembly of New Yorkers on the same subject.

Scientists in England and the United States were astounded to learn of Agassiz's new pronouncement. Their reaction stemmed from the fact that Henry W. Bates and Alfred Russel Wallace had spent many years carefully investigating the natural history of the Amazon Valley and had discerned no signs of glaciation whatever. Nor had other naturalists who had visited the region. Agassiz, after spending less than a year in the area, contradicted all existing knowledge on the subject. But Agassiz furnished no proof for his claims. Sir Charles Lyell epitomized scientific opinion concerning Agassiz's glacial assertions:

Agassiz . . . has gone wild about glaciers. . . . The whole of the great [Amazon] valley, down to its mouth was filled by ice. . . . He does not pretend to have met with a single glaciated pebble or polished rock . . . and only two or three far-transported blocks, and those not glaciated. As to the annihilation during the cold of all tropical . . . plants and animals, that would give no trouble at all to one who can create without scruple not only any number of species at once, but all the separate individuals of a species capable of being supported at one time in their allotted geographical province.[4]

Words like these, coming from a scientist who had not yet indicated his agreement with Darwin, were devastating. Gray interpreted Agassiz's current excesses as further indication that the naturalist was willing to oppose Darwinian evolution even without data. Since Agassiz had announced before leaving for Brazil that he expected to discover glacial remains, his reports led Darwin to write that his "predetermined wish partly explains what he fancies he observed."[5]

In January, 1868, scientists and the general public had yet another opportunity to read of Agassiz's Amazonian discoveries in *A Journey to Brazil,* an attractively bound book written by "Professor and Mrs. Louis Agassiz." This narrative included a diverting account by Elizabeth Agassiz of the adventures which befell the Thayer Expedition, an essay which had been published in part in the *Atlantic Monthly.* Agassiz told of the many new discoveries he had made, especially the phenomena of glaciation in Brazil. "I am prepared to find," he wrote,

that the statement of this new phase of the glacial period will awaken among my colleagues an opposition even more violent than that by which the first announcement [in 1837] of my views on this subject was met. I am . . . sure that, as the theory of . . . glaciers in Europe has gradually come to be accepted by geologists, so will the existence of like phenomena . . . be recognized sooner or later. . . .[6]

Agassiz failed to see the dissimilarity between his work of 1837 and that of 1866. As a young scientist he had spent nine summers exploring the Alpine glaciers and foothills, assisted by a corps of trained assistants. He had announced his results in scientific journals and before professional societies, buttressing his theories with ample documentation. He had published two books on European glaciation, and had visited England and Scotland to discover further evidence. His latest conception was presented in forty-two pages of a semipopular volume, two lecture series, an address to a group of scientists of whom only one member was a geologist, and two popular articles which appeared in a literary magazine.

Agassiz's concept of glaciation in Brazil was significant in that it provides a revealing insight into his intellectual efforts at this stage of his life. He was compelled to approach natural history, in both its broad and special aspects, within the restrictive framework of special creationism. This attitude of mind led him to an actual disregard for that empiricism that had marked his work three decades earlier and encouraged generalizations derived from the scantiest of data.

It is not likely that Agassiz found reliable evidence for glaciation in Brazil. He admitted that he had not seen the unmistakable evidences of ice he had discovered in former investigations, namely, the furrows, scratched rocks, and moraines that had marked a glacier's path. He did claim, however, that his field studies in the great valley of the Amazon and the region around Rio de Janeiro had yielded "evidence" also demonstrating glacial action. He interpreted as glacial drift material such as loose rocks and pebbles contained in unstratified clay deposits, asserting this was in fact morainic debris and therefore evidence of glaciation. He also claimed that he had seen a few erratic boulders and so-called *roches moutonnées,* or polished rocks, which to him were suggestive of glacial action. He admitted the difficulty of not finding what he had himself identified as necessary signs of ice, but claimed that the tropical climate with its heat and great rainfall had destroyed any such evidence.

The kind of evidence Agassiz submitted in support of his latest extension of the glacial theory was explained by geologists like Lyell and other contemporaries as due to causes other than former glacial action. If these men were right, then the polished rocks Agassiz saw were merely the weathered surface of exfoliating gneiss, and the morainic debris and boulders he discerned were to be explained as the result of the weathering of rocks in the sites where they were originally deposited.

On the other hand, Agassiz could very well have observed evidences of glaciation in Brazil, but if he did, he entirely misinterpreted his discovery. He alleged that his findings confirmed the fact of an Ice Age in Brazil, wherein ice had covered this region as it had central Europe, Great Britain, and areas of the United States. Thus Agassiz spoke about his finding as if it were part of the same process he had identified with Pleistocene history in other parts of the world. His motive was plain: the more ice he could discover all over the world as marking the earth's history at the close of the Tertiary and beginning of the Quaternary periods, the greater the barrier he could construct between the animals and plants that had immediately preceded the present creation and the more authoritative could be his affirmation that no genetic affiliation existed between life in the present and in the past. Pleistocene glaciation, then, was the great catastrophe inspired by supernatural fiat that disproved the transmutation of species.

Subsequent findings have shown that glaciation did in fact occur in parts of Brazil. But this glaciation took place in the Permocarboniferous period, an age separated by about two hundred and thirty million years from the Ice Age of the Pleistocene that Agassiz had done so much to identify. It was this Ice Age (or more properly the "Ice Ages") that provided men like Darwin and Lyell with such a suggestive hypothesis regarding the geographical distribution of contemporary animals and plants. However, if Agassiz actually saw traces of former glaciation in Brazil, these were signs of the most ancient Ice Age of Permocarboniferous times. Had he understood this presumed discovery correctly, it would not have served his ideological purpose, because his sole effort was directed at covering the whole world with ice before the introduction of modern plants and animals so as to construct the obstacle he imagined between past and present.

It is ironic that explorers of South America before Agassiz had noticed the same deposits he had, but had not identified them with recent geological times. Humboldt, for example, had seen some of the same deposits that Agassiz inspected. Humboldt had not, of course, interpreted these as signs of any Ice Age, but had dated them as belonging to the history of the earth in the Devonian period, about three hundred and fifty million years before the time of Pleistocene glaciation. Agassiz was fully aware of Humboldt's identification, but prided himself on being able to correct his mentor's work by greatly advancing the era of Brazilian earth history and also by demonstrating that the deposits Humboldt had examined were signs of ice action. Had Agassiz been less compelled to cover the world with ice, he might have had greater respect for his master's insight, since Humboldt's identification, while it was too early for the facts of the case, was nevertheless much more accurate than Agassiz's "modern" interpretation.

The Brazil volume received little attention from professional scientists, except when they registered amazement over its geology. As a romantic adventure the book was well received by people eager to read of faraway places and strange customs. The Agassizes sent advance copies as Christmas presents to a number of their close friends and received expressions of sincere appreciation in return. Longfellow wrote of the expedition account: "There has been nothing like it since Hipparchus sent his fifty-oared galley to bring Anacreon to Athens."[7] Anna Eliot Ticknor told her friend Lizzie Agassiz: "Please tell Mr. Agassiz there is not a word too much of science to my perception, though of course *I* do not understand it. What a beautiful book it is, so handsomely and accurately printed."[8] Holmes expressed a common opinion: "So exquisitely are your labors blended, that as with the Mermaidens of ancient poets, it is hard to say where the woman leaves off and the fish begins. The delicate observations from the picturesque side relieve the grave scientific observations."[9] Emerson was equally appreciative: "A very cheerful book to read: I am glad the expedition was sent, and in such dark times . . . glad that it attained such adequate results . . . glad that the social side should have been so honored . . . glad . . . that Mrs. Agassiz . . . was to be the angel of woman in Brazil."[10] George Bancroft, William James, George Ticknor, and many others added similar words of praise.

Despite these signs of continued public favor, Agassiz could not resist the feeling that he was becoming more and more isolated from the mainstream of research and study in natural history. By 1868, he had virtually ceased his dogmatic and popular attacks on evolution—the Brazil volume representing a final effort in this direction—and was trying to discover the sources of the Darwinian conviction by more objective analysis. He was also concious that the years of peace had encouraged a revitalized effort at new investigation and research in his discipline, an effort conducted without his superintendence. This was especially true in the realm of paleontology, where government surveys, geological field parties, and individual field workers were making notable discoveries. The arena for all this activity was the trans-Mississippi West, where the trail blazed by the builders of the Union Pacific was followed by energetic workers like Fielding B. Meek, Edward Drinker Cope, Ferdinand V. Hayden, and Othniel Charles Marsh, men who had gained their knowledge without the Agassiz imprint but were nevertheless uncovering materials of first importance for geology and paleontology. Agassiz's former students, too, such as Verrill, Hyatt, Morse, and Shaler, were all active researchers, convinced as they worked that the evolution idea was the most reasonable explanation of the data of natural history.

The day of the universal naturalist was drawing to a close. Agassiz's own students, and men like Marsh and Cope, with their specialized research interests heralded the end of the era when a man like Agassiz could confidently generalize on the entire sweep of creation. The work of these men, in distinction to the efforts of Silliman, Hitchcock, and Agassiz in an earlier day, was confined to precise observations, without reference to metaphysical speculations regarding the role of a first cause in nature. Aware of this trend, Agassiz felt swept along on a tide of change he did not fully understand or quite comprehend. He determined, however, that the new school of natural history would not be without the Agassiz impact.

Agassiz longed to involve himself once more in the intellectual realm of original investigation and was especially interested in the new paleontological discoveries being made in the West. He had

planned to visit the western territories in 1847 but his popularity
in the East had not allowed time for such an expedition. In 1853
he had visited St. Louis, but only briefly on a lecture tour. Early
in 1864 he spent some time in Burlington, Iowa, and was excited
by the fossil remains he discovered in this region, but his time for
investigation was limited by the requirements of the lecture tour
he was then undertaking to raise funds for the museum. Now that
he had served America in so many public ways, he hoped to undertake a significant journey to the trans-Mississippi West.

Consequently, when Agassiz's old friend, financial supporter,
and museum trustee Samuel Hooper, now a Massachusetts con-
gressman, offered to pay the costs of a surveying party which would
observe the progress of the Union Pacific and explore the Rocky
Mountain region, Agassiz at once accepted leadership of the enterprise. The trip began in July of 1868 at Chicago, where a group of
"tired members of Congress and business men" gathered to journey through the West with the learned professor and looked
forward to promised lectures on glacial geology in the course of
the trip. General William Tecumseh Sherman gave them a personal cavalry escort across the prairies, to the end of the Union
Pacific tracks at Green River Station, Wyoming. Here Agassiz saw
limestone beds rich with fossils of every description. Judge Amos
Carter of Fort Bridger in the Wyoming Territory urged Agassiz to
inspect some even more interesting deposits in the Grizzly Bear
Buttes, where an old trapper had come upon some fossilized bones
which had excited great curiosity. Agassiz, however, was not equal
to the physical exertion, having suffered a mild heart attack in
March of 1868. He declined to visit the area and thus missed finding what later proved to be an important discovery in reconstructing the evolution of vertebrates, a fossil of the genus *Palaeosyps*
Leidy.

Agassiz, however, did find significant evidence establishing the
extent of glacial action in the Rocky Mountains, although he never
published these findings. In the course of his travels he met Ferdinand V. Hayden, and the younger field geologist wrote Joseph
Leidy that "Prof Agassiz has just left in high glee, he has . . . obtained a fine jaw and teeth of a species of *Mylodon* and several

species of fish with fossils. He says he now understands my work as he never understood it before."[11]

Having completed his journey, Agassiz could feel that he had become more informed about recent discoveries in an area of science in which he had made his reputation. He did not return directly to New England. Instead, he spent two months in Ithaca, New York, where he helped in the founding of Cornell University. Cornell was very close to his heart, because he saw in the institution and the progressive spirit of President Andrew D. White the potential realization of educational concepts that had long been primary among his aspirations for national culture. As Agassiz had urged in regard to Harvard in 1863, he advised President White to work for a "true university," in accord with the intellectual progress of the century and the demands of the new age. Such a great university would embody "democratic principles" and yet rival the institutions of the Old World. Once again, a typical Lazzaroni program for accomplishment was advanced by Agassiz, and he urged White to bring the greatest minds of the day to Cornell to make the dream possible. It was not even necessary to have such men in permanent residence; the outstanding intellects of the country could come to Ithaca for brief periods each year and serve as visiting professors. This view, one shared entirely by White, was realized when Agassiz himself, James Russell Lowell, and other distinguished men accepted non-resident lectureships in the new institution.

The Agassiz imprint on Cornell was thus permanently established. The institution always seemed to him one bright sign of educational progress. As for Cornell, its faculty, and its first students, there always remained—and continues to remain—a deep and lasting appreciation of the Agassiz spirit in Ithaca. Their admiration was heightened by the fact that two of Agassiz's best students—men entirely devoted to their master—Burt G. Wilder and Charles F. Hartt, were appointed on Agassiz's recommendation as the first teachers of natural history at Cornell. The Agassiz spirit at Cornell was also exemplified in the stirring address he delivered at the opening of the university in October, 1868, wherein he repeated the by now well-known philosophy of higher education he and fellow Lazzaroni had done so much to advance and inspire. Agassiz's few months in Ithaca were, like his earlier ventures in Albany and Charleston, characteristic enterprises with characteristic results. A deity of science and culture descended for a time to charm his subjects, and, perhaps because such visitations were temporary, the folk memory of the grand inspirer was always worshipful.[12]

The western trip and the Ithaca sojourn restored Agassiz's confidence. Previous doubts about a life centered on public involvement at the expense of intellectual activity faded away. Returning to Cambridge in November, 1868, Agassiz was willing to believe that he could pursue each of these aims with the authority and dominance of old. He was pleased, moreover, at the prospect of important assistance from Alex, who had returned to the museum after spending the past year superintending a fabulously successful investment enterprise in Michigan copper mines. Theodore Lyman, too, was a valued worker in the common cause, having taken the post of treasurer of the museum. With Alex and Lyman working for the Cambridge establishment, the progress of knowledge as Agassiz defined it seemed well in hand.

Before leaving for the West in the summer of 1868, Agassiz had been in low spirits, with Alex away and museum finances never equal to his ambitions for the institution. He had even written a letter to President Hill, tendering his resignation from all his Harvard positions. The letter reflected his feeling that he was sacrificing personal intellectual drives for institutional involvements. Hill resigned the presidency of Harvard in September, 1868, and there is no indication that Agassiz's letter, if it was ever sent, reached the Corporation.[13]

Back in Cambridge in the fall of 1868, Agassiz, having forgotten both the contents of the letter and the conditions inspiring its words, threw himself once again into the world of decision-making with a vigor that signaled his resolution of previous doubts. There were, moreover, important decisions to be made at Harvard, and Agassiz could not rest without taking a hand in the affairs of the day. The presidency of Harvard had owed much to the Agassiz influence since the 1850's, and now that a new chief officer was to be chosen, Agassiz had to give evidence of his authority.

The only problem was that Agassiz had no candidate he could support wholeheartedly. The Corporation nominated Charles William Eliot to the presidency in March, 1869. It was not until May, however, that the Board of Overseers approved the nomination, having once returned it to the Corporation, an action close to rejection. Prior to the decision of the Corporation in Eliot's favor, and during the time the nomination was before the Overseers, Agassiz and Peirce fought to prevent the appointment. Having blocked the chemist from the Rumford professorship and scientific school deanship in 1863, they could hardly tolerate the thought that Eliot might be president of the university they had dominated for so long. But in 1863 these staunch allies had a candidate who was Eliot's superior in reputation, and now their opposition was merely negative. At best, Agassiz and Peirce, unable to come up with such a sympathetic supporter as Felton or Hill, would have preferred a man who could be certain not to interfere with their activities, someone like the historian Ephraim Gurney or the theologian and university preacher Andrew P. Peabody.

In this situation, all Agassiz and Peirce could do was complain that Eliot, now a professor at the Massachusetts Institute of Technology, where W. B. Rogers had offered him refuge after the defeat of 1863, would be a symbol of "practical" science. Actually, Agassiz and Peirce were very afraid that Eliot might reform science education at Harvard. With no thoughts of the presidency, Eliot had written two articles for the *Atlantic Monthly* early in 1869, where he repeated his earlier ideas for a reform of the Lawrence School, emphasized the need for a new outlook in science instruction, a broadening of the curriculum, the desirability of an elective system for undergraduate courses of study, and greater stress on graduate work. Having singled out the deficiencies of the scientific school, Eliot had also pointed to the virtual isolation of the Museum of Comparative Zoology from any co-ordinated scheme of science education. This was an obvious slap at Agassiz's personal domination of advanced zoology instruction. It was understandable that Dean Wolcott Gibbs, whose administration had been attacked and whose naturalist friend had been criticized, answered Eliot's charges by a passionate public defense of existing practices. But by the time Gibbs's statement appeared in the *Atlantic*, Eliot had been nominated for the presidency, although not yet confirmed by the Overseers.[14]

With so much at stake it was understandable that Agassiz employed the full force of his influence with the Overseers to defeat the Eliot nomination in this body. After all, men such as these had a long and honorable record of moral and material support for Agassiz causes. William Gray was Agassiz's agent in this contest, and quite probably managed to delay the inevitable by working for the return of the nomination to the Corporation, but his efforts came to nothing, and the nomination was finally approved by the Overseers.[15]

This was a bitter defeat for Agassiz and the Lazzaroni. It represented the waning of their once undisputed power at Harvard. The Lazzaroni were now without significant influence at Harvard or in the councils of major national scientific institutions. New men of authority like Gray, Rogers, and Eliot had replaced them. Bache's death in 1867 had symbolized the end of an era, and now Gibbs, Peirce, and Agassiz were men of an old order, faced with the prospect of living in an age where they were the conservatives, and their former enemies the successful and welcomed innovators. The Lazzaroni could gain some comfort from their continued authority in certain institutions that had provided the focus of their national power. Peirce had succeeded Bache as head of the Coast Survey, and Agassiz had his museum.

With Eliot now president of Harvard, Agassiz knew he could no longer have a direct voice in university affairs, and so he turned to his devoted friend Lyman, who was also very close to Eliot, for service as an intermediary whenever he needed approval from the administration for plans or projects. As for Eliot, he wisely avoided any direct conflict with Agassiz and shrewdly worked through his good friends Lyman and Alex Agassiz when he thought it necessary to make changes in natural history instruction. And so an uneasy truce prevailed between president and museum director.

Taking no part in university affairs after 1869, Agassiz was now isolated at Harvard, a situation that was in sharp contrast to the power he had enjoyed in earlier days. He would now and then write to Ticknor and others protesting against an Eliot scheme for reorganization and complaining that the new professors chosen by the new president were inferior men, hardly worthy of the honor of Harvard appointments. If Harvard was to progress it had to appoint "superior men" and live up to the "university ideal." Yet Agassiz's great men were vaguely remembered shadows of the past; the ideal university that had seemed such a grand concept at Albany in the 1850's was now being realized in Cambridge, but other men were making possible the transition to modern times. It was revealing that Agassiz had spoken in glowing terms of the present need for bold new ideas in his Cornell address of 1868, identifying the Ithaca spirit as truly progressive and Harvard ways as antiquated. Sadly, Agassiz could not become associated with the new reform ideal now prevailing at Harvard.

Agassiz's only course was to take refuge in the reality of the museum, an institution he could identify with completely. In fact, he had thought of little else since returning from Brazil, and personal and public drives prevented him from abandoning what seemed to him his very reason for being. By late 1865, while Agassiz was still in South America, the museum had begun to show a deficit in its operating funds. Economies had to be practiced, and President Hill showed his understanding of the situation in September, 1865, when he wrote an employee who had been dismissed from the museum:

The sudden crippling of the resources of the Museum by Mr. Agassiz's departure makes us sensible of the entire identification of the institution with its founder. . . . I am confident that on Professor Agassiz's return he will make diligent endeavors to regain his Aladdin's lamp, and . . . he will attempt to draw you to his side again.[16]

Hill's prediction of Agassiz's future course of action proved most accurate.

Any thought Agassiz might have had of leaving his post vanished as he became convinced of the need for his talents. He proceeded in what was by this time his customary fashion. Early in 1867 he made a report to the trustees stating that the museum was "rather a store-house than a well-arranged collection," which was quite true. Trustees, realizing who had made the museum what it was, might have paused to consider its future in the light of this statement. But Agassiz anticipated their objections:

Such a result may seem to show bad management, and might be fairly criticized, had our primary object been that of forming a museum for public exhibition. . . . But the *tacit* understanding . . . has been to work at the building up of a scientific institution, which should rival the most extensive establishments of that kind in Europe.[17]

In this regard, the museum now possessed some of the best collections to be seen anywhere in the world. He went on to say that it had not been his original intention to make such extensive collections in Brazil, but the generosity and kindness of the United States government, the emperor, and many private Brazilian and American citizens had placed such marvelous resources at his disposal that he could not avoid accumulating a great store of materials. What was needed now was a new addition to the northern wing to be built onto the present building, so that this vast array of natural history could be housed, sorted, and displayed effectively. All that was required was $100,000 to pay for the cost of construction and an additional $100,000 to provide for twice the number of staff and assistants now serving the museum.

Agassiz admitted that this might seem a large and unreasonable demand, but he felt that since the prospect of restored national unity was apparent, it would not be long before ". . . the thoughtful and far-sighted may still find the means to do for the arts of peace, and the culture of the people, something commensurate with the increase of our material prosperity, wealth and power." He closed the statement with an observation which was as revealing of Agassiz's own nature as it was descriptive of the subject at hand: "If some are inclined to criticize the costliness of such establishments, I can only answer that it is with museums as with all living things; what has vitality must grow."[18]

Agassiz was now forced to work harder than ever to meet new financial problems. He approached his tasks with what seemed greater energy than ever before, working over the sorting and

identification of specimens until late in the evening, coming home
exhausted, and returning to the museum early the next morning.
His assistants were driven relentlessly in the effort to bring some
order to the mass of materials brought back from Brazil. He had
an annual income of about $12,000 a year for this purpose, to-
gether with his salary, which was now $2,500, and a special annual
appropriation of $2,000 from the university. At this time he ac-
tually thought in terms of stringent economy, insisted on bal-
ancing the budget of the museum, wiped out old debts, and put
accounts in order before embarking on plans for expansion. By
early 1868, Agassiz felt that times were propitious to renew his
quest for additional funds.

Addressing himself once more to the trustees, Agassiz expressed
his gratitude for a supplemental state appropriation of $10,000
made in the spring of 1867. But he needed much more than this
if the museum were to continue its functions. He wished to double
the size of the exhibit spaces, enlarge all workrooms, erect an addition to the present building, increase staff salaries, and improve
the library and publications of the museum. In 1867 he had anticipated the cost of all this at $200,000; now he asked the trustees for
$500,000. He justified this request by reference to what had been
accomplished with the limited means available to date: the sorting
and classification of many marine animals, mollusks, and vertebrates in the institution. In 1859, Agassiz affirmed, no museum in
the nation could boast of anything but a local collection of curious
objects. European scientists never thought of the United States
when they considered where they might locate special materials
needed for their research. Just nine years later, the Museum of
Comparative Zoology had furnished data for monographs of the
highest distinction written by leading naturalists of the Old
World. The museum was no longer a mere state or college institution. It enjoyed national and international importance and deserved support commensurate with its role in the world.[19]

Agassiz's most recent effort to build up the museum's resources
was aided by an 1866 gift from the philanthropist George Peabody
to Harvard University of a $150,000 trust fund to found both a
museum and a professorship of "American Archaeology and Eth-

nology." There now existed an additional impetus for Agassiz to seek an enlarged museum, for he could envision an entire cluster of museum buildings on the Divinity Avenue grounds. In the spring and early summer of 1868, friends and trustees began an active campaign for funds among Bostonians and men of the legislature. When Agassiz returned from Cornell in October, he learned that the legislature had appropriated $75,000 for his use. This money was to be paid over a three-year period, on the condition that an equal sum be donated by private individuals. By January, 1869, $25,000 had been given from each source. It had taken just over two years for Agassiz to assure the construction of the new addition to the present building. Agassiz had asked first for $200,000, then for $500,000, and had received $150,000. By asking for the impossible, he had, as usual, acquired funds that might not have come his way had he made realistic appeals based on realistic expectations.

Agassiz was sufficiently impressed by this latest example of confidence in his purposes to write in his annual report:

Scientific investigation in our day should be inspired by a purpose as animating to the general sympathy as was the religious zeal which built the cathedral of Cologne or the basilica of St. Peter's. . . . It is my hope to see . . . a structure rise among us which may be a temple of the revelations written in the material universe. If this can be so, our buildings for such an object can never be too comprehensive, for they are to embrace the infinite work of Infinite Wisdom.[20]

Words like these appealed deeply to statesmen, trustees, and philanthropists. They were lost on fellow scientists, who could only wonder how the money of Boston and Massachusetts was spent in the Agassiz museum. The Brazilian fishes were not even unpacked as late as 1869. Alexander Agassiz, always understanding of his father's compulsive penchant for collecting, was depressed when he viewed the barrels and cans piled helter-skelter in the basement. Jeffries Wyman, now director of the Peabody Museum, recognized that Agassiz could accomplish little of lasting scientific value with such an enormous amount of material to take care of. Writing to Joseph Leidy, in reply to a request for some fossil teeth, he complained that they were lost, and reported further: "My impression is that I loaned them to Prof. Agassiz several years since, but that they have never come back. It were a hopeless task to find them in his collections in the state in which they now are. . . ."[21]

Agassiz never could delegate authority to others, a shortcoming which placed impossible responsibilities on his own shoulders. Although his unhappy experience of 1863 might have influenced him against allowing too much freedom to his staff, he now had a group of loyal and competent European and American naturalists working for him, but he persisted in keeping all work under his personal control.

Yet Agassiz's exploits fired the imagination of all who knew him, whether they approved of his actions and beliefs or not. At this stage of his career, Agassiz was clearly the popular symbol of natural history, just as Humboldt had been earlier. It was thus entirely fitting that he delivered a two-hour oration on September 14, 1869, in commemoration of the centennial anniversary of Humboldt's birth. Bostonians heard him tell of Humboldt's life with the reverence and praise due one of the intellectual giants of the nineteenth century. Agassiz's talents as a biographer and general orator, hidden until now, seemed nearly equal to his capabilities as a popularizer of scientific knowledge.[22]

Immediately after returning home from this noteworthy occasion, Agassiz suffered a cerebral hemorrhage which left him completely paralyzed and unable to speak. His body finally exacted a penalty for his many years of tireless activity. Fortunately, he was attended by one of the world's leading physicians, Dr. Charles Edward Brown-Séquard, a French medical man who had come to the United States at Agassiz's urging and had a received a professorship at the Harvard School of Medicine through his friend's influence. Brown-Séquard nursed Agassiz with untiring care, consulting with his colleagues Holmes and Morrill Wyman, both outstanding neurosurgeons. In time, the paralysis disappeared, but the patient was forbidden to stir from his bed. His physicians realized that he had complained of pains in his eyes and forehead for many years, and apparently diagnosed the condition as severe hypertension that resulted in cerebral clots. Brown-Séquard forbade Agassiz one of his greatest pleasures—cigar-smoking—warning that the introduction of any such irritants into the blood vessels might prove fatal. A long period of convalescence was prescribed as an absolute necessity.

Agassiz remained bedridden for many months, finding the inactivity almost unbearable. He worried about his collections at the museum, the work of his assistants, the progress of his drive for funds. He wrote almost pathetic letters to Lyman, pleading for still more money to be raised, expressing his deep fear that all would be lost without his presence, and giving advice about details large and small. The museum, he feared, would be permanently damaged by any small-minded policies. Money had to be spent, not invested; the new wing needed the treasures of European natural history in its cases, and Lyman and his fellow trustees were begged to keep pace with the goals Agassiz held out to them. He called museum workers to his bedside for frequent consultations, giving them detailed instructions on what should be done. It was difficult for him to accept the fact that his establishment could function without him. He dictated plaintive notes to friends, begging their forgiveness for his inability to discuss subjects of mutual interest in detail. He wrote regretfully to the administrators of Cornell University, resigning the visiting professorship he had recently accepted and hoping that natural history at the university would not suffer as a consequence.

His mood and appearance were aptly described by Marcou:

His speech was affected . . . and physicians forbade every exertion. . . . This last privation was the most painful. . . . Lying on a lounge in his sick-room, he looked like a lion loaded with chains and encaged in an iron box. His splendid and strongly built body was no longer at the command of his will; his inquisitive, brilliant, and intelligent eyes followed closely every visitor, as if to inquire what they thought of his sickness.[23]

By November, 1870, Agassiz was able to resume his normal occupations. Constant nursing by his wife, a convalescent period in Deerfield, Massachusetts, and good care by physicians had produced excellent results. Agassiz felt strong and robust, eager to take up his life where he had stopped fourteen months before.

Agassiz's refreshed spirit was reflected in his determination to see the museum placed on a financial basis that would insure its world supremacy. He wrote a moving appeal to Nathaniel Thayer: "It has become plain to me that I have not many years before me and that the work I had hoped to accomplish for the benefit of science in this country will be left undone, unless unusual—I may say extraordinary measures are taken to furnish me with the means for doing it." He went on to say that, his desires granted, the museum could, in a short time, secure a position of the highest rank in the world for Harvard University. In Europe, France and Germany were wasting their resources in barbarous warfare, and America should take the lead in promoting the progressive arts of peace. This could best be accomplished by taking from the Old World those natural history treasures which it did not properly appreciate.

Basing his opinion on information provided by Alexander, who had just returned from Europe on a scientific mission, Agassiz informed Thayer that for a sum of $250,000 he could ". . . by a bold stroke of policy, silently carried out, without ostentation so as not to awake competition, place our Museum . . . in a position which the civilized world would look upon either with admiration or envy." In addition to their inherent value, these European collections had the added attraction of representing "the intellectual work of about one half the number of the most eminent intellects of our age, which we could thus import and make a part of our progress and civilization."[24]

Agassiz did not rely solely on Thayer's generosity. He wrote Governor William Claflin to inform him that Massachusetts should appropriate at least $50,000 for the museum in 1871. The $75,000 state grant made in 1868 for three years would soon expire, and would not provide enough to finish the new wing now under construction. The private subscription of a similar amount, begun in 1868, was pushed to completion by the publication of an attractive pamphlet written by Tom Cary in 1871, detailing the history of the museum since 1859 and pointing out how public and private aid had resulted in donations of $473,000 to Agassiz and the museum since Harvard had first purchased his collections in 1853. The meaning of this history was clear: nearly half a million dollars (an accounting that was actually too low by about $100,000) was money well spent considering the results obtained for American science. The pamphlet had been suggested and outlined by Agassiz, just as he had included in his appeal to the governor a beautifully worded statement on the museum's needs which he suggested be incorporated into the annual message to the legislature. The result of this activity signaling the return of Agassiz to public life was the appropriation of $50,000 by the state and the successful conclusion of the popular subscription begun in 1868. Now, Agassiz could look forward to the completion of the new building, an event that occurred in 1871, and the means, if only partial, of filling its halls with materials purchased in Europe and gathered from remote corners of the world.[25]

These accomplishments prompted a letter from Gérard Paul Deshayes, who held the post of director of the Museum of Natural History in Paris that Agassiz had once been offered:

How happy you are, and how enviable has been your scientific career since you have had your home in free America! . . . You have the means of carrying out whatever undertaking commends itself to you as useful. Men and things, are drawn to your side. You desire, and you see your desires carried out. You are the sovereign leader of the scientific movement around you, of which you yourself have been the first promoter. What would our old Museum not have gained having at its head a man like you! We should not be lying stagnant in a space so insufficient that our buildings, by the mere force of circumstances, are transformed into store-houses. . . . It depressed me to read your letter.[26]

Life in Cambridge, however, could not entirely hold Agassiz's interest, particularly in view of a renewed determination to engage in fundamental research. Shortly after his recovery he received a proposal from Benjamin Peirce, who, as head of the Coast Survey, offered to place a new steamship at Agassiz's disposal for the scientific study of marine life. Peirce suggested that Agassiz travel aboard the "Hassler" down the eastern seaboard and Atlantic shore of South America, through the Strait of Magellan, north along the Pacific Coast to Central America, and make a final landfall at San Francisco. In 1869 Agassiz had experimented with new dredging equipment developed by Coast Survey technicians in an exploration of the Florida Keys; now Peirce informed him that this machinery had been more fully perfected and could be relied on to dredge to greater ocean depths than man had ever probed before. "Your proposition leaves me no rest," Agassiz wrote his friend; "I do not think anything more likely to have a lasting influence upon the progress of science was ever devised."[27]

The scientific world was also interested in the prospect of this voyage, hoping that Agassiz would once again turn his unquestioned talents to productive research. Darwin wrote his young friend Alexander Agassiz: "Pray give my most sincere respects to your father. What a wonderful man he is to think of going through the Strait of Magellan."[28] With the route agreed upon, apparatus available, all expenses paid by the government, and his wife to accompany him, Agassiz organized a staff of assistants to help his labors. His former European student, Count François Pourtalès, headed the group of scientists. As might be expected, a popular subscription of $20,000 was raised to pay for the preserving of specimens, and Nathaniel Thayer provided additional funds.

The cruise of the "Hassler" was delayed until December 4, 1871, when the Agassiz party left New York harbor. Following the precedents of the Thayer Expedition, Agassiz again gave regular lectures to his staff on zoology and ichthyology, and directed the gathering of specimens, this time from the ocean bottom. He was delighted at the opportunity to obtain materials previously unknown to any other naturalist. He determined to make an even greater reputation for himself as an ocean explorer and to discover finally whether the idea of organic evolution was true or false.

In the years since 1866 when he had announced his incredible concept of Brazilian glaciation, Agassiz had undergone a decided change of attitude toward the question of evolution. His knowledge of the fossils discovered by Marsh and Hayden, together with his own experiences in the Far West, seemed to have opened his mind to a realization that there might be more to the Darwinian hypothesis than he had ever been willing to concede. For these reasons he had canvassed the museums of Europe and the United States, attempting to build up a great collection of vertebrate paleontology. Agassiz had also ceased his constant reiteration of metaphysical causes which he had previously relied upon so heavily as the primary reasons for rejecting Darwinism. He began to emphasize factual deficiencies in the theory. The exchange of ideas between Agassiz, Gray, Darwin, and other advocates of evolution took on a more cordial tone. Agassiz was far from converted to the doctrine of evolution, but he had learned to play the role of devil's advocate with a challenging skill that could contribute real insights and encourage deeper analyses by those who accepted Darwin's formulations.

As the "Hassler" skirted the coast of Nicaragua in the early months of 1872, Alexander Agassiz back in Cambridge was writing to Darwin of his own near-acceptance of evolution.[29] Thereafter, aboard ship, Agassiz wrote his thoughts on the subject in a long letter to his German colleague, Carl Gegenbaur. "My main occupations are of course sea animals," he began,

but I had a special purpose in this journey. I wanted to study the whole Darwinian theory free from all external influences and former prejudices. It was on a similar voyage that Darwin himself came to formulate his theories! I took along few books . . . mainly Darwin's principal works. I have lived through the epoch of Naturphilosophie, the influence of Cuvier's work, and the period of the evolution concept, and I have pondered over these changes. I do not argue with the fiery enthusiasm I would have exhibited years ago, but I now examine questions step by step, and I must admit that up to now I have not made any great progress in my conversion to the growing doctrine of evolution.

Agassiz went on to admit that the Darwinian idea might very well be true, if evidence could be gathered from all parts of the world which would show a genetic relationship between species he had previously considered distinct and separately created. "That is why I am on this voyage; and I have taken particular pains to get together such collections as will contain sufficient material to allow thorough comparisons between related living types of the northern and southern hemispheres." But Agassiz could not entirely shake off a cosmic approach to the universe, and added, "P.S.: Don't you find a meaning in nature? . . . Is not this world full of the most wonderful combinations;—just think of man himself. . . . From where comes intelligence, and how come that we ourselves can stand in a connection with this world which we can understand?"[30]

These phrases, written to Gegenbaur in strict confidence, indicated the way in which Agassiz was sincerely trying to keep pace with the beliefs of his day. All he wanted was factual proof so that he could be convinced by the perceptions of his own mind. This was the most Gray or Darwin had ever asked of him. They realized there were few scientists better equipped through intelligence, training, and the opportunity presented by the museum collections, to decide the question of the origin of species. His final words of wonderment at the seeming design and purpose of the universe in the face of the possible power of secondary forces would have found a completely sympathetic reader in Asa Gray, who had been writing in this way since 1860.

That Agassiz was never able to gather evidence of the sort he required on this journey, or to study the materials he collected in an intensive fashion, was tragic. The sea could offer fundamental information concerning evolution, as Alexander Agassiz would demonstrate time and time again in his future oceanic explorations. As Agassiz realized, the deepest waters of the ocean represented the only natural environment where the scientist might discover what life was like millions of years ago. Here were duplicated the conditions of existence that first prevailed in the shallow seas which covered the world in earliest geological time. If sea creatures from the lowest reaches could be compared with those of the upper levels of the ocean, it would be possible to discover whether any genetic relationships existed between these different species, as well as the manner in which various types had adapted to different environmental conditions.[31]

But technology failed to come to the aid of biology at this crucial stage of Agassiz's thinking. The "Hassler," an experimental steamship of deep-draft and double-hulled construction, was constantly in need of repair, and much valuable time was spent waiting for parts to fix unforeseen damage. Even worse, the new dredging equipment would not work properly, and for long stretches of the voyage would not function at all. Even when it did operate smoothly, it was unable to reach the depths anticipated by Peirce. Consequently, the flora and fauna of the deepest water eluded Agassiz. Furthermore, the naturalist was in poor health during the trip. The mere physical exertion of lecturing and directing scientific operations left him constantly weak and fatigued. Thus there was no fundamental change in Agassiz's evaluation of the evolution idea, but the motivation that prompted the journey and the letter to Gegenbaur were signs of a more open-minded attitude toward views that differed from his own.

The voyage did result in certain significant experiences. Agassiz had expected to find evidence of glacial action in the Strait of Magellan and the Pacific coastal regions of South America, convinced that the same forces had operated here as he had identified in Brazil. Such a discovery would confirm his conviction that the entire South American land mass had been covered with a vast ice sheet. In distinction to Agassiz's questionable Brazil findings, he found unmistakable evidence of Pleistocene glaciation in this exploration. In the bays of the Strait of Magellan, he investigated the evidence of moraines, scratched rocks, and erratic boulders. He was delighted at the chance to study the great glaciers of the region, primary evidence of the Ice Age. He was able to observe how a wide region of the Pacific coastal area south of the thirty-seventh latitude evidenced signs of former ice action. North of this region he found that glaciers had not reached the sea but had occupied the valleys of the Andes Mountains, and that scattered traces of glacial action could be found almost as far inland as Santiago. He identified, therefore, a continuous former glacial chain extending from south to north.

Agassiz did not publish these findings. His wife offered cultured Bostonians a partial, impressionistic account taken from letters Agassiz had written to Benjamin Peirce and some observations he dictated to her. Here was a clear opportunity for a monograph comparable to *Études sur les Glaciers,* but the only report published was Mrs. Agassiz's article that appeared, appropriately enough, in the *Atlantic Monthly.* While colleagues were denied exact knowledge of this significant aspect of Pleistocene glaciation, readers of the *Atlantic* did learn how Agassiz's findings substantiated his Brazil "discoveries" and how the entire continent had therefore been covered by ice in former times.

Darwin was pleased at Agassiz's exploration effort because he might have hoped that the opportunity to visit the Galapagos Islands would provide Agassiz with the kind of insight he had received from his visit there in 1835. But Agassiz had no such experience. He admitted that the species of the Galapagos Archipelago were different from those of the South American mainland. This had been Darwin's observation, too. But Darwin had urged that these animals and plants were nevertheless related to those of the mainland, having reached the islands because of either a former continuous land formation or the agency of ocean transport. Consequently, these species had originated from a common center, instead of having been independently created where they presently existed. The flora and fauna of the islands then diverged from their parent types because of the modifications resulting from geographic separation and different conditions of life. Agassiz was willing to admit the possibility that such species had originated in South America. But, he argued, if this was the case, how had transmutation taken place? The Galapagos Islands were of recent origin, contemporaneous with the modern epoch. Given this fact, and granted that their species were related to those of the mainland, then transmutation did not require "such unspeakably long periods" as its advocates claimed necessary. Given the knowledge of the day, this was an incisive criticism. Agassiz, of course, could have no knowledge of the power of mutations to affect rapid change, or that dominant traits could change populations in a relatively short period of time. But Darwin did not know of these evolutionary mechanisms either, and given the outmoded genetics he had advocated, the Agassiz position was a perfectly plausible one. This was but another example of the way in which Agassiz's scientific criticisms of evolution often illustrated weaknesses in the concept as it was championed in his day.[32]

When the "Hassler" docked at San Francisco, Agassiz performed a characteristic service in the cause of science. Here the Agassiz party was met by a group of admiring naturalists who had begun to establish their discipline in a professional way on the West Coast. The Agassizes were feted at dinners and banquets tendered by the California Academy of Sciences, and the naturalist responded with speeches relating the adventures of the voyage and predicting great things for the progress of science in California.

The overland journey to Massachusetts was another strain on Agassiz's already overtaxed constitution. Bostonians, who had been informed of the activities of the expedition by Mrs. Agassiz's article in the *Atlantic Monthly*, welcomed their weary professor, happy to have him back in the comfortable and secure locale of Harvard Square. By October of 1872, Agassiz was again ready to take up his work for the museum, for science, and for American culture. His first thought was for his museum.

Others, however, had also been giving much thought to the future of the institution. Once more in charge of the museum during his father's absence, Alex reflected on the future of the institution built and sustained by his parent. Better than anyone else, except perhaps Lyman, Alex knew that the museum was a constant drain on the health and well-being of Agassiz. He knew, too, that his father would resume his old occupations of fund-raising, the purchase of collections, and still more finance drives when he returned. As much as Agassiz trusted Alex and Lyman with the business affairs of the museum, he still considered it his personal responsibility. And so Alex wrote to Lyman, who was spending the summer of 1872 in Agassiz's native Switzerland, asking if they could not conceive of some plan whereby Agassiz would give up his administrative control of the museum in the interests of his health and future welfare. Lyman replied in a lengthy letter. The document provides the most perceptive assessment of Louis Agassiz ever penned by a contemporary:

... the Museum ... is a large pill, and yet can perhaps be dealt with. Here is the way it is. Your papa ... is a great genius; in spite of that he is aggravating to us. As to money and plans—he is a *very great* genius. When you come to consider, 1. That he is a remarkable naturalist, having been the creator of the glacial theory, of the arrangement of fossil fishes, and besides great contributions to Embryology

. . . and an important contributor in many other branches. 2. That he has unusual powers as a teacher, and can lecture in three languages. 3. That he is an organizer, capable of building up a great series of collections (even if he can't finish them) . 4. That he is a man of most unusual social power, capable of going into refined society, and of fascinating people right and left. When . . . you consider these four points, you will see that there is not a single man in Europe today who could do *all* that he has done. You might find his equal as a naturalist; but then you also would find that he wore a dirty shirt, swallowed his knife, and could only talk German. You might find his equal in social force, but then it would be a man who didn't know the difference between an earth worm and a sparrow. Then, Papa has done more to excite and push forward Natural History in the U.S. than any man who was ever there. He has brought to his department at Harvard $600,000, and more; besides a prestige which is now considerable on the Continent. When you think of this, all you will see is that it is not in reason to hope for a *Successor* to such a steam-engine. What we find to be the case is this: there is a Museum containing really remarkable collections. These are not only not arranged, but are piled up faster than the arrangement goes on. Papa has a really grand and philosophic plan for its arrangement, but he has not the strength, nor the workmen, nor the money, nor the constancy to put it through! As long as Papa lives— (and he has lots of vitality) he can have always strength enough to indicate what he wants, and give, at least, a sketch of the work to be done; moreover he will, by hook or by crook, get more or less money, in addition to our annual income. . . .

Lyman closed by suggesting that the museum administration be divided between Agassiz and Pourtalès, that he and Alex continue as advisers, and that teaching chairs be established for Burt G. Wilder and Edward S. Morse.[33]

Lyman's analyses of Agassiz and the museum he built were both accurate commentaries on the past and fair prophecies for the future. Agassiz could never give up complete authority over the museum while he lived, although in November, 1872, he wrote a trustee expressing his wish that Pourtalès succeed him as director "in a few years." He wanted this to be accomplished quietly, so as not to excite notice or opposition, and if the European naturalist could be appointed as a professor in the Lawrence School and paid by funds contributed by the public, this would be a beginning step.[34]

Despite the pleading of Alex and Lyman, and the earlier warning from Joseph Henry, Agassiz could not relax for a moment his eternal vigil over the progress of science in New England and the nation. He wrote an angry letter to Henry when he discovered that the Smithsonian Institution was sending many original specimens and rare duplicates to Europe. Agassiz objected both as a regent of the Smithsonian and as the director of the Museum of Comparative Zoology; he felt that America was being robbed of treasures rightfully hers.[35] He wrote pleading letters to Othniel Charles Marsh, begging the paleontologist to join the staff of the museum. Marsh would be given any salary he asked, would have complete freedom of research, and, incidentally, should remember to bring his interesting fossil vertebrates with him. If Marsh would not come to Cambridge, Agassiz offered the alternative that he send his specimens to the museum, where plaster casts would be made of them. Marsh could have any number of the casts he might wish, if only the Agassiz museum could have possession of a complete series of them.[36] Having at least set these projects in motion, Agassiz again turned to the recurrent matter of finances.

As Lyman had predicted, "Papa" was equal to the needs of the moment. Early in 1873, another appeal to the legislature assured a $25,000 appropriation for the museum, with an equal amount to be supplied by donations from private citizens in keeping with earlier arrangements that had proved so effective. These people were a special group: all graduates of the Agassiz School for Girls, they thus repaid the instruction of former years. But their donations were dwarfed by an outright gift of $100,000 for the museum from Agassiz's son-in-law, Quincy Adams Shaw, who, with Alexander Agassiz, had made a fortune from an interest in the Calumet Mining Company in Michigan.[37]

Exhilarated by his success in insuring the future of the museum, Agassiz launched still another great venture in his never-ending attempt to enhance the appreciation of Americans for their natural environment. In December of 1872, Nathaniel Southgate Shaler, now a professor at Harvard, suggested to his former instructor that a summer school for teachers of natural history be established in some seaside locale. Agassiz immediately approved of the idea and drew up a program for carrying it out. An announcement of the school was printed and distributed, listing its resident faculty as Agassiz, Shaler, Burt G. Wilder, Alexander Agassiz, Pourtalès, Alpheus Spring Packard, Frederick Ward Putnam, Joel Asaph Allen, and Theodore Lyman. The fact that all of Agassiz's faculty were his former students, now recognized as professional naturalists, attested to his influence upon the development of science in post–Civil War America. In the case of some of these men, their new association with Agassiz showed how the quarrels of the 1860's had been forgotten.

Agassiz was not troubled in the least that there were absolutely no funds available to implement a scheme so fully formulated and advertised. He merely took the occasion of a visit of Massachusetts lawmakers to the museum to make an impassioned speech for funds to support his latest interest. Before the legislators had time to consider the appeal, John Anderson, a prosperous New York merchant and landowner, read a report of Agassiz's speech in the *New York Times.* He telegraphed Agassiz, whom he had never met, to do nothing further about securing support for the summer school until he heard from him. A few days later, Anderson's emissary appeared in Cambridge and offered Agassiz the gift of Penikese Island, part of the Elizabeth group in Buzzard's Bay, opposite the town of New Bedford in southern Massachusetts.[38] Agassiz was very happy to accept Anderson's philanthropy, and even more pleased when Anderson wrote on March 19 that he would give fifty thousand dollars to found the school. Agassiz replied, "I am overwhelmed by your generosity. Such a gift . . . opens visions before me such as I had never dared to indulge in connection with this plan."[39] As Deshayes had observed from his vantage point in Paris, Agassiz needed only to desire, and desire became reality.

From May to July of 1873 the once deserted island was a scene of intense activity as wooden buildings were constructed to provide dormitories and laboratories for the expected students. On July 6, a Sunday, the buildings were not yet completed, and the school was scheduled to open the following Tuesday. Agassiz made a speech to the carpenters concerning the obligations of working men to labor on the Sabbath in the cause of education, and the buildings were finished when the students arrived. However, it was discovered at the last minute that no partitions had been built between the male and female quarters of the dormitories. Nature study was not to be taken this seriously. More hasty construction went on while the assembled group of fifty schoolteachers from all over the United States stood on the dock to hear Agassiz's speech of welcome. Further ceremonies opened the school and were marked by another address by Agassiz extolling the grandeur of creation and the need to study nature itself, not textbooks, for true knowledge.

The months of July and August passed pleasantly. Excursions were made to surrounding islands, a small dredge was operated offshore, teachers were kept busy in the laboratories, and lectures were given in the open air. The *New York Tribune* reported Agassiz's talks on the "plan of creation," marine animals, glaciers, fossil fish, and other topics in zoology and geology at regular intervals, later publishing these accounts in book form. John Greenleaf Whittier wrote a poem based on the first summer of the Anderson School.[40]

When Agassiz returned to the museum in 1873, he decided to give a series of lectures to advanced students on the evolution question. He was bothered by reports of ill-feeling on his part toward Darwin, and he wanted to make clear the essential points of his intellectual disagreement. Darwin, whose view of Agassiz had been shaped by Gray's accounts of his disputes with his Cambridge colleague, had followed the early evolution debates with great interest, amused and perplexed at Agassiz's argumentation. He had quite naturally applauded Gray's efforts in combating Agassiz and had been as disturbed as Gray by Agassiz's popular attacks on evolution in the early 1860's. In all this time there is no evidence of direct correspondence between Agassiz and Darwin, except for the note that accompanied the gift of the *Origin of Species* in 1859. When Agassiz apologized to Gray in 1866 for the quarrel of 1864, and the two men resumed friendly relations, Gray thought it quite proper to ask Agassiz to answer a question Darwin had put to him regarding sexual selection in fishes. Agassiz took this occasion to write Darwin. He informed the English naturalist that he knew no better way to contradict reports of ill-feeling than by providing the information he asked for directly. The question answered, Agassiz went on to describe his personal views of Darwin's work:

It is true that I am and have been from the beginning an uncompromising opponent of your views. . . . It is equally true that I hold these views as mischievous because they lead to a looseness of argumentation which it has been the aim of science to avoid. . . . There is nothing in these feelings against yourself, as you have done original researches . . . and as to allowing my feelings to be the master of my judgement, I hope I shall never be guilty of such a mistake.[41]

Darwin replied that he thought Agassiz "had formed so low an opinion" of his work that it would have been improper to ask information of him directly, so that this indication of regard pleased him greatly. "I have never for a moment doubted your kindness and generosity," Darwin wrote, "and I hope you will not think it presumptuous in me to say that when we met many years ago I felt for you the warmest admiration."[42]

In this new spirit of toleration, Agassiz had taken the "Hassler" cruise. Just before this, in the spring of 1871, he had lectured on evolution to students at the museum. Agassiz told his students that he felt the greatest admiration for Darwin "on account of his perfect honesty and love of truth and his willingness to face obloquy or misrepresentation when convinced his belief is right." It was as if the Agassiz of the present was praising Darwin for his courage in meeting the attacks of the Agassiz of the past. With remarkable candor, he admitted that his views, and those of all scientists, were influenced by personal experiences, historical ideas, religious convictions, mental habits, and intimate feelings. He tried to free himself from such subjectivity by observing "like a sensitive statue," allowing the facts of nature to "take whatever place they would in my mind and rest there as long as they chose and by and by, by some unconscious method they would work out." Darwin had worked this way too, before 1859, but after that date his efforts had been marked by presupposition and argumentation, a failure understandable because "the English possessed this propensity for fanciful theories more than any other nation." Even worse, ". . .

evolution appeals to a general disposition on the part of the public, that is the facility in which it can be indulged, for it demands no thought, no labor. Any man with wit and plausibility can argue and present it so that it will be popular." Darwin, then, was condemned for having preconceived notions and popularizing them.[43]

In his lectures of 1873 Agassiz determined to set forth a comprehensive assessment of Darwin and the evolution idea and to contradict still persistent reports of personal animosity. He had published no direct analysis of the evolution concept since 1860, although he had taken every occasion to dispute transmutation in his lectures and popular articles on natural history and geology of the Civil War years. Agassiz determined also to publish these latest addresses as a public announcement of his considered position. Characteristically, he chose the *Atlantic Monthly* as his medium. As he wrote the first of these analyses, he was conscious of its importance for his professional standing. He wrote to editor William Dean Howells, in words familiar to authors the world over:

I have already had this MS copied three times and I would remodel it again did I not fear to delay you by retaining it longer. I send it therefore as it is, with the express condition that I shall receive *three* proofs . . . as soon as [it] is set in types. . . . I have too much at stake in this Article to be willing to allow it to appear without the severest criticism and keeping it for many days before me in its revised form.[44]

This article, "Evolution and Permanence of Type," remains Agassiz's ultimate evaluation of the transmutation concept. The passage of time had given Agassiz a respect for Darwin that he had not shown in previous published assessments. He seemed to strive desperately, however, to erase this new appreciation from his mind. Thus the carefully chosen words of his analysis represented all facets of Agassiz's scientific personality. As an observer of nature he could not help admiring the "startling array of facts" Darwin had presented over the years. He was forced, at last and reluctantly, to grant Darwin credit for originality and a status equal to that he had reserved for himself and a few others. "Darwin has placed the subject on a different basis from that of all his predecessors, and has brought to the discussion a vast amount of well arranged information, a convincing cogency of argument. . . ."[45]

Yet for Agassiz there was a disturbing parallel between the method and attitudes of Oken and the generalizations of Darwin that seemed to be an oppressive weight preventing any fresh appreciation of evolution. Both men had "claimed that the key had been found to the whole system of organic life." Darwin, he admitted, had been unique in maintaining such an all-embracing view "not upon metaphysical ground but upon observation."

These twistings and turnings in Agassiz's almost agonizing reappraisal of evolution demonstrated the unhappy effect upon his mature intellect of a combination of idealism, dedication to the sacredness of "physical fact," and isolation from professional intercourse. Darwin had offered a "key" to explain the phenomena of natural history; but encompassing views were supposed to be "metaphysical" according to Agassiz's German education. How was it possible to grant intellectual status to a theory grounded solely on observation of facts? It was gratifying to Agassiz that Darwin's interests finally led him to study expressions of emotion in man and animals. "This last phase of the doctrine is its identification with metaphysics. . . . I can only rejoice that the discussion has taken this turn."[46] But even this reassuring turn did not obviate the impressive manner in which Darwin had dealt with "facts." A physical fact, after all, was "as sacred as a moral principle." Empiricism, however, learned faithfully from Cuvier, was supposed to furnish the ammunition that would destroy fanciful theories of "development." Now, Darwin piled fact upon fact to demonstrate by "cogency of argument" his key to understanding all natural history.

Agassiz was quite plainly the victim of both his European education and his authoritative role in American science. When Darwin and his followers studied the facts of nature and arrived at objective conclusions that were at once synthetic and universal, Agassiz now felt alone, misunderstood, and one of the few defenders of the only true interpretation of nature. He could thus write in all honesty, ". . . the science of Zoology . . . has furnished, since the beginning of the nineteenth century, an amount of startling and exciting information in which men have lost sight of the old landmarks."[47] The only possible attitude for him, as a devoted student of Cuvier, was to rest upon the authority of "the old landmarks" and deny the transmutation of species, partly on idealistic grounds but much more intensively on grounds of "evidence," his observation of the "facts." Hence Agassiz's latest statement on evolution was more in the spirit of objective inquiry than anything else he had ever written on the subject. But since he chose the grounds of disputation, he inevitably predetermined the outcome of his analysis. Admitting that "it is not my intention to take up categorically all the different points on which the modern theory of transmutation is based," he persisted in his own definition of what "evolution" meant. Demonstrating the indelible influence of his education, he identified evolution with "metamorphosis" and the ontogenetic transformations demonstrated by embryology. It was a foregone conclusion that such an identification "proved" that "all embryonic growth is a normal process of development, moving in regular cycles, returning always to the same starting-point, and leading always to the same end." It was an established fact, therefore, that "metamorphoses have all the constancy and invariability of other modes of embryonic growth, and have never been known to lead to any transition of one species into another."[48]

Agassiz did try, in half-hearted fashion, to assess the fundamental and primary mechanisms that distinguished the evolution idea. Here too, the result was the same, since, as Hooker had once pointed out to Gray, facts took on meanings which differed according to the assumptions by which they were evaluated. Agassiz affirmed that all his experience with domestic varieties proved that such variations returned to their original type in a wild state, Darwin to the contrary. Agassiz observed that Darwinians placed great emphasis on the "prophetic types," which he himself had distinguished, as primary evidence supporting the concept of transmutation. He admitted that this might well be true, if it could be proved that these types were actually related by analogous internal organization to those which succeeded them in geological time. Proof of this relationship had not been established to his satisfaction. He was convinced that intermediate forms possessed anatomical and physiological characteristics which distinguished them from their supposed ancestors or descendants.

He documented his analysis with an examination of the earliest and most primitive forms of life. Some of these ancient types were duplicated in contemporary flora and fauna. They had thus remained, in terms of plan and structure, essentially unchanged. Therefore, how could evolutionists talk of "development" from simple to complex forms if "primitive types" were still in existence? This objection stemmed from Agassiz's particular definition of "lower," "higher," and "primitive" forms. While his standards were subjective, his objection was an understandable one in the light of the knowledge of his own day.

Another aspect of Agassiz's criticism of evolution was that he refused to concede that variations were permanently transmitted from parent to offspring, because he did not feel Darwin had done anything to explain the processes of heredity. Once again, Agassiz had pointed to serious gaps in the chain of Darwinian reasoning. In subsequent generations, sentimental former students would assert that Agassiz's work in geology, paleontology, and embryology had anticipated the mechanisms of evolution. Such claims, while highlighting the way Darwin used all fundamental contributions to support his conceptual scheme, entirely obscured Agassiz's role as an often incisive critic of evolution.

In a summary appraisal, Agassiz illustrated both an understanding and a misinterpretation of the idea that had been such a dark shadow over his later life:

The world has arisen in some way or another. How it originated is the great question, and Darwin's theory, like all other attempts to explain the origin of life, is thus far merely conjectural. I believe he has not even made the best conjecture possible in the present state of our knowledge.[49]

The evolution idea, he understood, was an all-embracing concept of natural processes. He persisted, however, in ranking this generalization with "all other attempts to explain the origin of life," demonstrating his inability to understand the revolutionary character of Darwin's work, the phenomena it actually explained, and the way evolution was based on conceptualizations that were foreign to ultimate, metaphysical assessments of nature's meaning.

It is impossible to ignore the tone of finality and the consciousness of ultimate and permanent evaluation that was reflected in Agassiz's words in this article. It was as if Agassiz were submitting his scientific will and testament to the world. The people of America, for whom these assessments were particularly designed, read his analysis with sadness in their hearts. On December 14, 1873, just one month before the *Atlantic Monthly* essay appeared in print, Louis Agassiz died of a cerebral hemorrhage in his Quincy Street home.

Agassiz had given no sign of failing health in the last months of his life. If anything, he was more active than ever. He had returned from Penikese bursting with new plans for the museum. The fall of 1873 found him engaged in characteristic labors, planning typical enterprises. He determined to find the money to buy the marvelous collections of James Hall, specimens that had fascinated him since he first inspected them in the glorious days following his arrival in America. He planned to make a permanent exhibit of a giant whale sent to him by an admirer from St. John's, Newfoundland. He would institute a new course of annual lectures on special subjects for the secondary-school teachers of Massachusetts, whose predecessors had listened to him tell about God's "leetle jokes." He envisioned an important place for natural history and science in general in the centennial exposition of 1876 and began planning how he could aid in this demonstration. He agreed to journey to Yale and lecture to young students of natural history.

On December 2, Agassiz delivered a lecture in Fitchburg, Massachusetts, to an appreciative assemblage of farmers, animal breeders, and state legislators, hoping thereby to gain more funds for the museum by impressing rural folk with the significant information to be gleaned from the study of domesticated animals.[50]

Returning from Fitchburg, he was as always urged on by a grand new idea. He would apply to the federal government "for a million acres of land more or less," to be granted to the museum under the Morrill Act. The museum would fulfil its part of the bargain by offering instruction in agricultural subjects. "We shall soon have a land scrip department to peddle out the acres as fast as alcohol is needed," Alex wrote Lyman half-jokingly. The fact that his father had recently been responsible for Congress' granting scientific institutions exemption from the federal tax on alcohol should have convinced Alex that an Agassiz plan was always to be taken seriously.[51]

December 5 was a happy day for Agassiz. It was Lizzie's birthday, and all the family gathered to celebrate at the Quincy Street house. The Cary relatives were there and the married Agassiz children—Alex and his wife, Anna, who were living with the older Agassizes at the time; Pauline and her husband, Quincy A. Shaw; and Ida, who had married Henry Lee Higginson in 1863, and her husband. Much good food, wine, and music were enjoyed, and Agassiz, contrary to medical instructions, smoked a cigar in honor of the occasion.

The next morning he complained of feeling "strangely asleep" and very weary. He managed to get to the museum but returned almost immediately. Resting in his room, he fell asleep and never regained his faculties. He recovered consciousness from time to time, but he never spoke again. His old friend and physician Dr. Brown-Séquard attended him, but there was little he could do except prescribe food through injections; Agassiz's throat was paralyzed. He did not suffer pain during this last illness, and on the evening of the fourteenth he died.

Two years before, a young naturalist had written him: "Without your name how will our society appear? Like a fish without scales or a glacier without its moraine. The new era in America dates precisely from the day of your arrival in Boston."[52] This was a prophetic statement of the sense of loss to be felt by countless Americans on December 14, 1873, when Louis Agassiz succumbed, at the age of sixty-six, to the constant physical and intellectual demands he had placed on himself.

Elizabeth Agassiz expressed her sorrow in words that aptly characterized the sentiments of the few people who knew Louis Agassiz well. She had lived for twenty-three years "a life that grew daily fuller and richer in happiness," through the companionship of her husband.

I can never tell anyone how delightful it was to live by the side of a mind so fresh and original, so prodigal of its intellectual capital. There was never a day when I did not feel this powerful stimulus and it seems to me now that I am in danger of mental starvation.[53]

Lyman and Alex Agassiz, writing affectionately about "papa," usually called him a "steam engine"; others thought of him as a spiritual father, still others as simply a teacher. The *Contributions* had represented one of his visions for America; they were not completed when he died. But, in a true sense, Agassiz's dreams for America, the land he loved so deeply, could never be completed, could never be realized. His was a romantic spirit, and like his constant opposition to Darwin, this spirit was consistent and never satisfied.

The day of the funeral, December 18, the entire university community ceased its activity to honor its professor. The services at Appleton Chapel were conducted by Dr. Andrew Preston Peabody, who tried to convey the meaning of Agassiz to Harvard and the United States. Vice-President of the United States Henry Wilson and president of Harvard Charles William Eliot represented the nation and the university that Agassiz had served. Scientists, teachers, students, and ordinary people all came to pay homage to this American by adoption. The Boston newspapers reporting Agassiz's death and funeral were edged in black. He was put to rest in Mount Auburn Cemetery, and a few months later an Alpine boulder weighing 2,500 pounds arrived from Switzerland to serve as a lasting monument marking his life in nature.

The Quincy Street home he willed to Elizabeth, and he gave Alex his library. One year later, $300,000 was contributed to the Museum of Comparative Zoology in the form of the Agassiz Memorial Fund. Alex Agassiz and Quincy A. Shaw gave $180,000 of this amount, and the state of Massachusetts contributed $50,000. But the $70,000 that made up the popular subscription fund would have meant the most to Louis Agassiz.[54]

John Amory Lowell had welcomed Agassiz to the United States in October of 1846; his cousin James Russell Lowell paid poetic tribute to his memory. Lowell, being in Italy, did not learn of Agassiz's death until two months later.

> . . . with vague, mechanic eyes,
> I scanned the festering news we half despise . . .
> When suddenly,

> As happens if the brain, from overweight
>   Of blood, infect the eye,
> Three tiny words grew lurid as I read,
> And reeled commingling: *Agassiz is dead!*

Lowell wrote on and on, to fill eleven pages of the *Atlantic Monthly* with tribute to his friend. Agassiz was now in a better world, the poet consoled himself, for

> . . . the wise of old
> Welcome and own him of their peaceful fold . . .
> And Cuvier clasps once more his long lost son.

> We have not lost him all; he is not gone
> To the dumb herd of them that wholly die;
> The beauty of his better self lives on
> In minds he touched with fire, in many an eye. . . .
> He was a Teacher: why be grieved for him
> Whose living word still stimulates the air?
> In endless file shall loving scholars come
> The glow of his transmitted touch to share,
> And trace his features with an eye less dim
> Than ours whose sense familiar wont makes numb.[55]

EPILOGUE TO THE
NEW EDITION

1988

WHEN I BEGAN RESEARCH into the career of Louis Agassiz, I had to pierce the web of mythology and the mystique of romance that surrounded him in his lifetime and afterwards. Agassiz was a folk hero to the general and the literate public, a "man of destiny," who brought the wisdom of Europe to nurture the young American environment. Affirmations of ability and noble purpose were fashioned by Agassiz himself and transmitted to future generations by students and admirers who wove a web of cultural invincibility about his memory. To assess properly what W. H. Auden, in his observations about biography, called "the body half undressed," it was necessary to sweep away these cobwebs of pietistic worship so as to appraise the ideas he represented and the times in which he worked.

Biography is a reconstruction of the text of an individual's ideas and the context in which they were expressed. Each must be connected and defined, especially in the realm of science, where so much is hidden from view by the superimposition of myth upon the text and personality. The intermixture of text and context enables

391

the biographer to highlight creativity and provide a definition of culture. This process was wisely described by Edmund Gosse when he wrote, "There has always been a singular interest in observing how personal character acts on the work performed. . . . To analyze the honey is one thing, and to dissect the bee another; but I find a special pleasure in watching him, myself unobserved, in the act of building up and filling the cells." That special pleasure is what I felt in reconstructing Agassiz's life, conscious that it represented a mental snapshot taken at a point in time where the life of the critic and the knowledge of his subject intersected at the moment of creative recreation. Such pictures, when exposed to the light of new insight and interpretation, and with the passage of time, change hue and shape; such is the relativity of knowledge. But if the camera was correctly designed to capture essences, these remain pristine, and that is how I see my picture of Agassiz. In this spirit, I can reassess aspects of my portrait; and I will present at the end of this epilogue some important publications about the subject written since 1960.

The relation between text and context is illustrated by Agassiz's role in the evolution debates. Agassiz came to these controversies ill prepared. He was simply not the same scientist who came to America in 1846, vibrant with achievements in natural history. Asa Gray, James Dwight Dana, and William Barton Rogers recognized this condition and lashed out at the man they had once almost deified. Thus they highlighted their intellectual growth as well as the process of turning against an individual whose personality and ideas they had once venerated.

The actual evolution argument in America has been viewed as an epic battle, a kind of western movie, with the white-hatted proto-Darwinians riding off into the sunset, while Agassiz lay wounded and dying, stripped of his ill-gotten social and intellectual status. In fact, the American arguments served to set forth views that were already established and well known to a relatively small circle of naturalists; but their ideas did not enjoy wide communication soon after the debates, either among their scientific peers or to the general public. The "battle" provided amusement for Darwin and his circle but resulted only in emotional uplift for them and to assure a fair hearing for Darwin, which was never in Agassiz's power to deny. Actually, the years of the evolution controversy and subsequent decades witnessed the continued sway of the very idealism that was supposed to have ended with Agassiz's fall from grace.

The case of James Dwight Dana is instructive. Dana was the American naturalist best suited, by virtue of wide-ranging knowledge, to evaluate and contribute to the evolution idea. His review of Agassiz's *Essay on Classification* had expressed disappointment with the static concepts of special creationism that it represented, and he called for more study of variation in species. But Dana never extended his own research in marine biology or geology either to affirm or to dispute the evolution synthesis. Despite his friendship with Darwin, Dana could not bring himself to read the *Origin* upon its publication, pleading the fatigue that, as with Agassiz, was the result of his recent efforts to advance science institutionally. By 1863, Dana had still to read the *Origin of Species*, but he nevertheless took the occasion of the first edition of his *Manual of Geology* to challenge its concepts obliquely and to defend theism in natural history. This was curious behavior from an individual who, five years before, had castigated Agassiz for defending Marcou without having read him.

Such behavior also illustrates the tension and inner turmoil the evolution idea elicited in Dana. The Yale naturalist never confronted evolution empirically, affirming instead the idealistic concept of "cephalization" as an all-embracing shield against the challenge of transmutation. This notion ranked species by the degree to which their organs of locomotion were in "service" of the brain (as with man's arms and hands), a taxonomic scheme that placed humans highest by virtue of brain development and intelligence. Much to the perplexity of Darwinians, Dana doggedly elaborated this idea, which for him became a dogma. Ultimately, in the last (1895) edition of the *Manual,* Dana accepted evolution as a viable explanation; but he also embraced the vitalistic explanation of neo-Lamarckianism as an alternative hypothesis, and he still voiced his overarching theistic view of creation.

Asa Gray, unlike Dana, contributed empirical evidence to advance the Darwinian hypothesis but still, in the 1880s continued to embrace the argument from design and the teleological significance of evolution. Theism, design, teleology, and neo-Lamarckianism were neither necessary nor logical concomitants of Darwinian evolution. Like special creation, all such concepts represented fixed, final, and closed systems. Had he lived longer, Agassiz would have been dolefully amused to witness the continuance of idealism within empirical frameworks while his dualism of idealism and natural knowledge was scorned by American colleagues.

It makes more sense to see the interplay between Agassiz and Darwin's defenders in the light of the politics of science. The years from the mid-1850s to the mid-1860s saw the beginning and maturation of a successful anti-Agassiz alliance that replaced one establishment with another. The forces of Agassiz, Alexander Dallas Bache, and Benjamin Peirce were countered by a coterie led by Dana, Gray, and Rogers, with occasional assistance from Spencer F. Baird, Addison E. Verrill, and Charles William Eliot. Each group saw in Agassiz a pivotal leader, and each member of the anti-Agassiz clique had personal reasons for his vendetta. Rogers always believed he had more right than Agassiz to a Harvard professorship, and he spent these years establishing the Massachusetts Institute of Technology, appealing to some of the same public and private sources of money that Agassiz employed. Gray chafed under Agassiz's domination of science at Harvard. He counseled Agassiz's disgruntled students, delighted in the opportunity to report Agassiz's foibles to the Darwin circle, applauded Dana's efforts to bring advanced science education to Yale, and helped engineer the defeat of Agassiz in the National Academy in 1864. Dana never forgave Agassiz for the Marcou affair, and he encouraged such Agassiz defectors as Albert S. Bickmore and Verrill to establish independent careers. He, too, delighted in the National Academy revolution and thus signaled his ultimate shift from the fringe of the Scientific Lazzaroni to a central role in the new power group. Dana hired Verrill to teach at Yale and to work at the Peabody Museum. He and his student Marsh were delighted to learn through this conduit the inner workings of the Harvard museum and Agassiz's future plans for it. These were the cultural conditions and human connections that flowed through the evolution debates and their aftermath. They combined to make the last decade of Agassiz's life a time marked by professional isolation, the sad and ironic result of the transformations wrought by personality and culture upon personal intellectual histories.

Alexander Agassiz witnessed the years of scorn and intellectual isolation that his father knew, and his memory of them prefigured and defined his own career as museum administrator and naturalist. Loyal to the ideal of his father, Alex devoted the millions he made from Michigan copper mining to completing the physical building of his father's museum. In so doing, Alex was compelled to face up to the reality that his father's conception of a teaching, research, and public museum was unworkable. The intellectual enthusiasm of Louis Agassiz created a legacy of countless specimens that became an impossible logistic, financial, and administrative burden for Alex as museum director. Beginning in the 1880s, Alex reluctantly turned against his father's vision and instituted a museum policy of consolidation, contraction, and the rational allocation of limited resources.

This conservatism was made the more necessary because of Louis Agassiz's imprint on the creation of museums in the United States. The American Museum of Natural History, the United States National Museum, and the Peabody Museum at Yale were all institutions founded and directed either by Agassiz students or by persons who determined to outdo him in the effort to institutionalize natural history. Alex Agassiz realized it was impossible for the Harvard museum to compete with the money and other resources of these newer institutions and, for this reason as well, had to abjure an encompassing dream of his father. In this spirit, Alex Agassiz bitterly attacked the aggrandized use of federal funds by Yale's Othniel Charles Marsh and John Wesley Powell for the study of vertebrate paleontology, terming such "spread-eagle bosh" dangerous to the ethos of wisely allocated research resources. When Powell accused Alex of contradicting his father's vision for American science, the truth of this observation touched the despair Alex felt at trying to reconcile his father's dreams to the practicalities of reality. In response to this despair, Alex retreated to solitary engagements with marine biology and oceanography and the singular absorption in nature those activities afforded, effectively giving up his superintendence of the Harvard museum. Toward the close of his life he was content to visit Japan, treasure beautiful jade carvings, and identify with Clarence King, Oliver La Farge, and Henry Adams, all self-styled recluses from an expansive materialism they saw as representative of their times.

It also fell to Alex to preside over the demise of another of Louis's dreams, the seaside school of Penikese, whose original concept had been fashioned by Nathaniel Southgate Shaler. The first two Penikese summers were typical of Agassiz's vision for American science, in this case the empirical study of the marine littoral environment, pursuing the ideal of studying nature first hand to become uplifted by its transcendent virtue. Alex, as businessman and administrator, had to put an end to this grandiose scheme combining science and moral uplift. But the Penikese idea resulted in the establishment of the Marine Biological Laboratories at Woods Hole in 1888. There the researches of such as Charles Otis Whitman became the result of an Agassiz inspiration. The Penikese idea, more graphically than the Harvard museum, represented the complete synthesis of research, teaching, and idealism that Louis Agassiz symbolized. Alexander Agassiz, at once the inheritor and the reshaper of his father's life in science, represented a haunted double vision of past and present and a transitional stage in the phylogeny of modern American natural history.

NOTES

Where a published source agrees substantially with a manuscript account, the published version is cited.

NOTES, CHAPTER 1

1. Quoted in Elizabeth Cary Agassiz, *Louis Agassiz: His Life and Correspondence* (2vols.; Boston, 1885), I, 12–13.

2. For descriptions of Motier by contemporaries see Jules Marcou, *Life, Letters, and Works of Louis Agassiz* (2 vols.; New York, 1896), I, 5–7; E. C. Agassiz, *Agassiz*, I, 1–4. Unless noted, sources for the early life of Agassiz are the Marcou and Elizabeth Agassiz biographies, and G. R. Agassiz, *Letters and Recollections of Alexander Agassiz* (Boston, 1913).

3. See E. Bonjour *et al.*, *A Short History of Switzerland* (Oxford, 1952), pp. 7–8, 16–19, 190; J. Christopher Herold, *The Swiss without Halos* (New York, 1948), pp. 98–178.

4. Quoted in E. C. Agassiz, *Agassiz*, I, 144–45.

5. Quoted *ibid.*, p. 146.

6. See Wilhelm Oechsli, *History of Switzerland, 1499–1914* (Cambridge, 1922), pp. 229–31, 282–85, 316, 360, 365, 368, 372–73, 423.

7. Quoted in Bonjour *et al.*, *Short History of Switzerland*, p. 203.

8. Agassiz read the first editions of these works. Lamarck, *Histoire naturelle des animaux sans vertèbres* . . . (7 vols.; Paris, 1815–22); Cuvier, *Le Règne animal distribué d'après son organisation, pour servir de base à l'histoire naturelle des animaux, et d'introduction à l'anatomie comparée* (4 vols.; Paris, 1817). Agassiz's comments on Lamarck indicate he also read that author's *Philosophie zoologique* . . . (2 vols.; Paris, 1809) at this time.

9. Agassiz, autobiographical sketch in E. C. Agassiz, *Agassiz*, I, 146–47.

10. Quoted in Marcou, *Agassiz*, I, 15.

11. Agassiz to Rodolphe Agassiz, April 24, May 24, 1826, reporting prior com-

munications, Agassiz Papers, Houghton Library, Harvard University. For descriptions of the status of higher education in the German states of the early nineteenth century see John Theodore Merz, *A History of European Thought in the Nineteenth Century* (4 vols.; London, 1896–1914), III (1912), 116–30, and Friedrich Paulsen, *German Education Past and Present,* trans. T. Lorenz (London, 1908), pp. 180–88.

12. Agassiz to Rodolphe Agassiz, May 24, 1826, Agassiz Papers, Houghton Library; Agassiz, autobiographical sketch in E. C. Agassiz, *Agassiz,* I, 148–49; G. A. Goldfuss, *Petrefacta Germaniae* . . . (2 vols.; Düsseldorf, 1826–33).

13. Agassiz, autobiographical sketch in E. C. Agassiz, *Agassiz,* I, 148–49.

14. Agassiz to Rodolphe Agassiz, April 24, 1826, Agassiz Papers, Houghton Library.

15. Tiedemann, *Anatomie und Bildungsgeschichte des Gehirns im Fötus des Menschen, nebst einer vergleichenden Darstellung des Hirnbaues in den Thieren* (Nuremberg, 1816) and *Icones cerebri simiarum et quorundam mammalium rariorum* (Heidelberg, 1821). For Tiedemann's role in the history of the recapitulation idea see Sir Charles Lyell, *Principles of Geology* . . . (11th ed.; 2 vols.; New York, 1876), II, 275.

16. To Carl Braun, May 12, 1826, in E. C. Agassiz, *Agassiz,* I, 26–27.

17. Agassiz, autobiographical sketch, *ibid.*, p. 149.

18. Quoted *ibid.*, p. 32.

19. Agassiz to Rodolphe Agassiz, [December, 1826], *ibid.*

20. Agassiz to Alexander Braun, June 10, 1827, Agassiz Papers, Houghton Library.

21. See Agassiz to Alexander Braun, May 26, 1827, in E. C. Agassiz, *Agassiz,* I, 34; Agassiz to Braun, July 13, 1827, Agassiz Papers, Houghton Library. *Naturphilosophie* was published in three volumes (Jena, 1809–11).

22. Oken, *Elements of Physiophilosophy,* trans. Alfred Tulk (London, 1847), p. 492. For analyses of Oken's system of classification see Agassiz, *An Essay on Classification* (London, 1859), pp. 336–40; Erik Nordenskiöld, *The History of Biology,* trans. Leonard Bucknall Eyre (New York, 1928), pp. 288–89; Charles Singer, *A History of Biology* (rev. ed.; New York, 1950), pp. 217–19.

23. August 9, 1827, in E. C. Agassiz, *Agassiz,* I, 43–45.

24. Agassiz, autobiographical sketch, *ibid.*, pp. 149–50. Quotation from p. 150.

25. Rose Agassiz to Agassiz, January 8, 1828, *ibid.*, p. 61, referring to prior discussions concerning permission to go to Munich. Here, and throughout the text where European monetary values are given in dollar equivalents, they have been transvalued in terms of dollar values current at the time noted in the text.

NOTES, CHAPTER 2

1. Agassiz to Rodolphe Agassiz, February 14, 1829, in E. C. Agassiz, *Agassiz,* I, 98.

2. Joseph Dinkel, manuscript account written, in English, at the request of Elizabeth Cary Agassiz, Agassiz Papers, Houghton Library.

3. Agassiz, autobiographical sketch in E. C. Agassiz, *Agassiz,* I, 156–57. For an

account of Munich in 1839 see Asa Gray, Journal of June 12, 13, 14, 1839, in Jane Loring Gray (ed.), *Letters of Asa Gray* (2 vols.; Boston, 1893), I, 231–41.

4. Von Baer, *De ovi mammalium et hominis genesi* . . . (Leipzig, 1827). See Arthur William Meyer, *Human Generation: Conclusions of Burdach, Döllinger, and Von Baer* (Stanford, 1956), p. 31.

5. Agassiz to Cécile Agassiz, November 20, 1827, in E. C. Agassiz, *Agassiz*, I, 56.

6. Agassiz, autobiographical sketch, *ibid.*, pp. 150–51.

7. Agassiz, *Contributions to the Natural History of the United States* (4 vols.; Boston, 1857–62), I, dedication following title page. See Meyer, *Human Generation*, pp. 27–28, 126; Meyer, *The Rise of Embryology* (Stanford, 1939), p. 320; Owsei Temkin, "German Concepts of Ontogeny and History around 1800," *Bulletin of the History of Medicine*, XXIV (May–June, 1950), 233–34.

8. Henry James Clark in *Proceedings of the American Academy of Arts and Sciences*, IV (1857–60) (Boston, 1860), 137 n. See Meyer, *Human Generation*, p. 28, and *Rise of Embryology*, pp. 238–77.

9. *Methods of Study in Natural History* (Boston, 1863), pp. 296–97.

10. Agassiz to Auguste Agassiz, December 26, 1827, Agassiz Papers, Houghton Library. Agassiz, "Die süsswasser Fische Mitteleuropas, 1826," manuscript on deposit at the Houghton Library.

11. J. B. Spix, *Simiarum et Vespertilionum Brasiliensium species novae* (Munich, 1823); *Avium species novae, quas itinere per Brasiliam annis 1817–1820* . . . (2 vols.; Munich, 1824–25); J. B. Spix and C. F. P. von Martius, *Reise in Brasilien* . . . *1817–1820* (3 vols.; Munich, 1823).

12. Cuvier to Von Martius, November 7, 1827, Agassiz Papers, Houghton Library.

13. October 29, 1828, Agassiz Papers, Houghton Library.

14. To Auguste Agassiz, July 27, 1828, Agassiz Papers, Houghton Library. Cuvier's work was published as Georges Cuvier and Achille Valenciennes, *Histoire naturelle des poissons* (22 vols.; Paris, 1828–49).

15. [October, 1828], in E. C. Agassiz, *Agassiz*, I, 81.

16. "Beschreibung einer neuen Species aus dem Genus Cyprinus *Linn.*," *Isis*, X (1828), 1046; N.S., IV (1829), 414.

17. October 29, 1828, in E. C. Agassiz, *Agassiz*, I, 81.

18. April, 1829, Agassiz Papers, Houghton Library. The Brazil volume was published under the title *Selecta genera et species Piscium, quos in itinere per Brasiliam* . . . *digessit, descripsit et observationibus anatomicis illustravit Dr. L. Agassiz* . . . (Munich, 1829).

19. Cuvier to Agassiz, August 3, 1829, Agassiz Papers, Houghton Library.

20. Alexander Braun to Carl Braun, February 15, 1829, Agassiz Papers, Houghton Library.

21. May 22, 1829, in E. C. Agassiz, *Agassiz*, I, 110.

22. January 8, 1828, *ibid.*, p. 62.

23. To Rose Agassiz, February 3, 1828, Agassiz Papers, Houghton Library.

24. February 21, 1828, in E. C. Agassiz, *Agassiz*, I, 65.

25. March 3, 1828, *ibid.*, p. 68.

26. March 25, 1828, *ibid.*, pp. 69–70.

27. To Agassiz, August 25, 1828, *ibid.*, pp. 77–78.

28. February 14, 1829, *ibid.*, p. 100.

29. February 23, 1829, *ibid.*, pp. 101–2.

30. *Histoire naturelle des poissons d'eau douce de l'Europe centrale.* . . . *Prospectus* (Munich, 1830).

31. Agassiz to Rose Agassiz, February 3, 1828, in E. C. Agassiz, *Agassiz,* I, 63.

32. October 20, 1830, Agassiz Papers, Houghton Library.

33. To Rose Agassiz, April, 1830, in E. C. Agassiz, *Agassiz,* I, 126.

34. April 7, 1830, *ibid.*, p. 129.

35. *Recherches sur les ossemens fossiles* . . . (4 vols.; Paris, 1812; 5 vols.; Paris, 1822–24).

36. January 18, 1830, in E. C. Agassiz, *Agassiz,* I, 124.

37. Agassiz to Auguste Agassiz, May 29, 1830, *ibid.*, p. 133.

38. Agassiz, autobiographical sketch, *ibid.*, pp. 152–53.

39. November 18, 1828, Agassiz Papers, Houghton Library.

40. See Agassiz, *Essay on Classification,* pp. 336–40; Arthur O. Lovejoy, *The Great Chain of Being* (Cambridge, Mass., 1936; 1957 printing), pp. 315–33.

41. November 16, 1830, in E. C. Agassiz, *Agassiz,* I, 139–40.

42. December 25, 1831, Agassiz Papers, Houghton Library.

43. March, 1832, in E. C. Agassiz, *Agassiz,* I, 171.

44. Agassiz to Mathias Mayor, February 16, 1832, *ibid.*, pp. 164–65.

45. February 16, 1832, *ibid.*, pp. 165–66.

46. March 25, 1832, *ibid.*, pp. 177–78.

47. [March, 1832], *ibid.*, pp. 180–81.

48. The species was *Cyclopoma spinosum* Agassiz, and is described in Agassiz, *Recherches sur les poissons fossiles* . . . (5 vols.; Neuchâtel, 1833–43), IV: *Cténoïdes* [1833–43], 20–21.

49. Agassiz, autobiographical sketch in E. C. Agassiz, *Agassiz,* I, 152.

50. *Ibid.*, p. 153.

51. *Ibid.*, pp. 152–53.

52. See Charles Singer, *A History of Biology,* pp. 223–32, and Ernst Mayr, "Agassiz, Darwin, and Evolution," *Harvard Library Bulletin,* XIII (Spring, 1959), 171–76.

53. March, 1832, in E. C. Agassiz, *Agassiz,* I, 172.

54. April 11, 1832, Agassiz Papers, Houghton Library.

55. Agassiz, *Address Delivered on the Centennial Anniversary of the Birth of Alexander von Humboldt* . . . (Boston, 1869), pp. 45–46.

56. *Ibid.*, p. 46.

57. *Ibid.*, p. 44.

58. March 27, 1832, in E. C. Agassiz, *Agassiz,* I, 187.

59. [March, 1832], Agassiz Papers, Houghton Library.

60. March 27, 1832, in E. C. Agassiz, *Agassiz,* I, 191–92.

61. Agassiz to Humboldt, May, 1832, *ibid.*, p. 193, reporting a letter from Coulon.

62. June 4, 1832, *ibid.*, pp. 197–99.

63. June 18, 1832, *ibid.*, pp. 199–200.

64. July 25, 1832, *ibid.*, pp. 200–201.

65. Agassiz to Humboldt, May, July, 1832, *ibid.*, pp. 195, 202.

NOTES, CHAPTER 3

1. Ernest Favre, "Louis Agassiz . . . ," *Annual Report of the Smithsonian Institution, 1878* (Washington, 1879), pp. 247–48.

2. December, 1832, in E. C. Agassiz, *Agassiz,* I, 213–14.

3. December 4, 1832, *ibid.,* pp. 211–13.

4. December, 1832, *ibid.,* p. 215.

5. December, 1832, *ibid.,* p. 216.

6. Humboldt to Agassiz, December 29, 1832, in Marcou, *Agassiz,* I, 53–55.

7. Humboldt to Coulon, January 21, 1833, in E. C. Agassiz, *Agassiz,* I, 217–18.

8. Agassiz to Humboldt, May 22, 1833, Agassiz Papers, Houghton Library.

9. G. R. Agassiz, *Alexander Agassiz,* pp. 3–4.

10. Humboldt to Agassiz, May, 1835, in E. C. Agassiz, *Agassiz,* I, 255.

11. The only comparable nineteenth-century work was the singular study by Arthur Smith Woodward, *Catalog of the Fossil Fishes in the British Museum* . . . (4 vols.; London, 1889–1901). See Karl Alfred von Zittel, *History of Geology and Paleontology,* trans. Maria M. Ogilvie-Gordon (London, 1901), pp. 379, 410–11, and Albert C. L. G. Günther, *An Introduction to the Study of Fishes* (Edinburgh, 1880), p. 20.

12. Sedgwick to Lyell, September 20, 1835, in John Willis Clark and Thomas McKenny Hughes, *The Life and Letters of the Reverend Adam Sedgwick* (2 vols.; Cambridge, 1890), I, 447; Lyell to Sedgwick, October 25, 1835, in Katherine M. Lyell (ed.), *Life, Letters, and Journals of Sir Charles Lyell, Bart.* (2 vols.; London, 1881), I, 457.

13. *Recherches sur les poissons fossiles* . . . , I (1833–43), title page. A separate volume of illustrated plates accompanied each volume of text. In 1843, the various parts and chapters were bound together in five volumes, and an additional bound volume of corrections and new data was added to the set. Agassiz also made additions and corrections in the earlier parts of the text at this time, so that it is not possible to determine exactly the actual date of writing of any one part, unless it can be compared with the original sections. These have not been available.

14. See William Martin Smallwood, *Natural History and the American Mind* (New York, 1941), pp. 189–90.

15. See Günther, *Study of Fishes,* p. 20.

16. *Poissons fossiles* (1833–43), I [1842?], 171.

17. *Twelve Lectures on Comparative Embryology* (Boston, 1849), p. 27.

18. *Poisson fossiles* (1833–43), I [1842?], 91.

19. *Ibid.,* I [1833?], 171–72.

20. *Ibid.,* p. 172.

21. *Ibid.,* p. 171.

22. Agassiz and Augustus A. Gould, *Principles of Zoology* . . . (Boston, 1848; 1878 ed.), pp. 417–18.

23. To Charles Lyell, September 20, 1835, in Clark and Hughes, *Sedgwick,* I, 447.

24. October, 1835, *ibid.,* pp. 257–58.

25. *Poissons fossiles,* I [1833?], xiii.

26. *Monographie des poissons fossiles du Vieux Grès-rouge ou Système Dévonien* (3 parts; Soleure and Neuchâtel, 1844–45). See also *Rapport sur les poissons fossiles découverts en Angleterre* (Neuchâtel, 1835).

27. June, 1834, in E. C. Agassiz, *Agassiz,* I, 238.

28. *Monographies d'echinodermes vivans et fossiles . . .* (4 parts; Neuchâtel, 1838–42); *Études critiques sur les mollusques fossiles* (2 parts; Soleure and Neuchâtel, 1841–42).

29. See, for example, *James Sowerby's Mineral-Conchologie Grossbrittaniens . . .* (2 vols.; Neuchâtel, 1837–44); *Conchyliogie minérale de la Grande Bretagne, par James Sowerby . . .* (2 vols.; Neuchâtel, 1837–45); *Geologie und Mineralogie in Beziehung zur natürlichen Theologie von W. Buckland . . .* (2 vols.; Neuchâtel, 1839); *Nomenclator zoologicus . . .* (Soloduri, 1842).

30. December 2, 1837, in E. C. Agassiz, *Agassiz,* I, 267–71.

31. Agassiz, "Discours prononcé à l'ouverture des séances de la Société Helvétique des Sciences naturelles," *Actes de la Société Helvétique des Sciences naturelles,* II (July, 1837), v–xxxii, reprinted in Marcou, *Agassiz,* I, 89–108.

32. See Tilly Edinger, "Agassiz Lebt" [Part II], *Natur und Volk,* LXXXII (November, 1952), 360.

33. The German edition of *Études sur les glaciers* appeared as *Untersuchungen über die Gletscher* (Solothurn, 1841). See also Agassiz, "On Glaciers and the Evidence of Their Having Once Existed in Scotland, Ireland, and England," *Proceedings of the Geological Society of London,* III (November 4, 1840), 327–32, *La Théorie des glaciers et ses progrès les plus récens* (Geneva, 1842), and Edward Desor, *Excursions et séjours dans les glaciers . . . de M. Agassiz . . .* (Neuchâtel, 1844).

34. Agassiz, *Geological Sketches* (Boston, 1866), p. 208.

35. [1838], in E. C. Agassiz, *Agassiz,* I, 289.

36. "Discours prononcé à l'ouverture," in Marcou, *Agassiz,* I, 107.

37. February 13, 1841, Agassiz Papers, Houghton Library.

38. October 15, 1840, in E. C. Agassiz, *Agassiz,* I, 309.

39. March 1, 1841, Agassiz Papers, Houghton Library.

40. Agassiz, *Geological Sketches, Second Series* (Boston, 1876), pp. 57–58.

41. December 22, 1837, in E. C. Agassiz, *Agassiz,* I, 273.

42. *Notice sur la cause probable du transport des blocs erratiques de la Suisse* (Paris, 1835), reprinted in Marcou, *Agassiz,* I, 77–82. See George P. Merrill, *The First One Hundred Years of American Geology* (New Haven, 1924), p. 619.

43. Charpentier, *Essai sur les glaciers, et sur le terrain erratique du bassin du Rhône* (Lausanne, 1841), pp. vii–viii; Agassiz to Charpentier, June 28, 1841, in Marcou, *Agassiz,* I, 162–63; Agassiz to Charpentier, December 25, 1843, Agassiz Papers, Houghton Library.

44. Tilly Edinger, "Agassiz Lebt" [Part I], *Natur und Volk,* LXXXII (October, 1952), 322; Agassiz, "Discours prononcé à l'ouverture," in Marcou, *Agassiz,* I, 101, acknowledging the stimulus of Schimper's idea.

45. *Erwiderung auf Dr. Carl Schimper's Angriffe* ([Neuchâtel], November, 1842).

46. Agassiz, *A Reply to Mr. James D. Forbes on the Laminated Structure of Glaciers* (Neuchâtel, April 11, 1842, 10 pp.); *Système glaciaire . . .* (Paris, 1847), pp. 436–47; John Campbell Shairp, Peter Guthrie Tait, and A. Adams Reilly, *Life and Letters of James David Forbes, F.R.S.* (London, 1873), pp. 256–73, 544–61; Agassiz to Forbes, June 12, 1842, to [J. Murray & Sons], February 13, 1842, in Marcou, *Agassiz,* I, 196–97, 200–201; Agassiz to Sir Roderick Murchison, November 1, 1842, to Sir Philip Egerton, December 15, 1842, Agassiz Papers, Houghton Library; Alexander and Elizabeth Cary Agassiz, manuscript defense of Agassiz *in re* Forbes, 1874, Agassiz Papers, Houghton Library.

47. Shairp *et al., Forbes,* pp. 544–50; Agassiz to Sir Roderick Murchison, November 18, December 15, 1842, April 3, 1843, Agassiz Papers, Houghton Library.

48. November 18, 1842, Agassiz Papers, Houghton Library.

49. Sir Charles Lyell, *Principles of Geology . . .* (11th ed.), I, 367–68.

50. See James Geikie, *The Great Ice Age . . .* (3d ed.; New York, 1901), esp. pp. 28, 511; Reginald Aldworth Daly, *The Changing World of the Ice Age* (New Haven, 1934), esp. pp. 1–2, 23–30; T. C. Chamberlain, "Preliminary Paper on the Terminal Moraine of the Second Glacial Epoch," *Annual Report of the United States Geological Survey,* III (Washington, 1883); A. Penck and R. Brückner, *Die Alpen im Eizeitalter* (3 vols.; Leipzig, 1901–9).

51. *Anatomie des salmones par L. Agassiz et C. Vogt* (Neuchâtel, 1845).

52. August 15, 1840, in E. C. Agassiz, *Agassiz,* I, 313–14.

53. Journal for August 18, 1841, in Shairp *et al., Forbes,* p. 266.

54. *Embryologie des salmones, par C. Vogt* (Neuchâtel, 1842); *Anatomie des salmones par L. Agassiz et C. Vogt* (Neuchâtel, 1845). For the conflicting claims of each man see Agassiz, *Essay on Classification,* p. 122, n. 1; Karl Vogt, *Aus meinem Leben* (Berlin, 1896), pp. 196–97, 201; William Vogt, *La vie d'un homme: Karl Vogt* (Paris, 1896), pp. 34–35.

55. Vogt, *Aus meinem Leben,* p. 196.

56. Agassiz to Charles Lucien Bonaparte, [July, 1845], Agassiz Papers, Houghton Library.

57. The most reliable account of Agassiz's domestic life in this period appears in John Amory Lowell *et al.,* "Process Verbal, Agassiz versus Desor, February 15, 1849," Agassiz Papers, Houghton Library. Agassiz hints at an unhappy family life in a letter to Benjamin Silliman, May 28, 1847, Daniel Coit Gilman Papers, Johns Hopkins University Library.

58. Agassiz to Charles Lucien Bonaparte, April 3, 1843, Agassiz Papers, Houghton Library.

59. Agassiz to Charles Lucien Bonaparte, March 7, 1845, reporting a letter from Humboldt; to Sir Roderick Murchison, April 16, 1845; Murchison to Agassiz, March 11, 1845, Agassiz Papers, Houghton Library. Agassiz to Augustus A. Gould, May 6, 1845, Augustus A. Gould Papers, Houghton Library.

60. Agassiz to Silliman, January 6, 1835, in George P. Fisher, *The Life of Benjamin Silliman* (2 vols.; New York, 1866), II, 150–53; Silliman to Agassiz, April 22, 1835, Agassiz Papers, Houghton Library; Agassiz to Silliman, November 12, 1837, Benjamin Silliman Papers, Rare Book Room, Yale University Library; *American Journal of Science,* XXVIII (April, 1835), 193.

61. November 11, 1838, Edgar Fahs Smith Memorial Collection, University of Pennsylvania.

62. Charles Lucien Bonaparte to Agassiz, April 4, June 7, September 1, 1842, Agassiz to Bonaparte, April 3, 1843, November 19, December 10, 1844, Agassiz Papers, Houghton Library; George Ord to Bonaparte, September 11, 1842, George Ord Papers, Academy of Natural Sciences of Philadelphia.

63. Agassiz to Bonaparte, March 7, 1845, Agassiz Papers, Houghton Library, reporting the letter to Humboldt after the Prussian grant had been announced. Charles and Mary Lyell to Agassiz, February 22, 1845, Agassiz Papers, Houghton Library, reporting Agassiz's request.

64. Baron Werther to Agassiz, March 10, 17, 1845, Agassiz Papers, Houghton Library.

65. Lyell to Agassiz, May 1, 1845, Agassiz to Bonaparte, [July, 1845], Agassiz Papers, Houghton Library.

66. September 16, 1845, in E. C. Agassiz, *Agassiz,* I, 381.

67. February 1, 1846, Agassiz Papers, Houghton Library.

68. 2d ser., I (May, 1846), 451.

69. August 14, 1846, in Fisher, *Silliman,* II, 159–60.

70. *Geological Sketches, Second Series,* p. 77.

71. Agassiz to Gould, May 6, 1845, Gould Papers; Silliman to Samuel S. Haldeman, July 20, 1845, Samuel S. Haldeman Papers, Academy of Natural Sciences of Philadelphia; Gould to Lewis R. Gibbes, November 21, 1845, Robert W. Gibbes to Lewis R. Gibbes, June 24, 1846, Lewis R. Gibbes Papers, Manuscripts Division, Library of Congress.

72. February 1, 1846, Agassiz Papers, Houghton Library.

NOTES, CHAPTER 4

1. Agassiz to Rose Agassiz, December 2, 1846, Agassiz Papers, Houghton Library.

2. Agassiz to Rose Agassiz, December 2, 1846, to Henri Milne Edwards, May 31, 1847, Agassiz Papers, Houghton Library.

3. Agassiz to Henri Milne Edwards, May 31, 1847, Agassiz Papers, Houghton Library.

4. November 8, 1846, James Hall Papers, New York State Museum, Albany.

5. October 13, 1846, Asa Gray Papers, Gray Herbarium, Harvard University.

6. October 26, 1846, Joseph Leidy Papers, Academy of Natural Sciences of Philadelphia.

7. November 17, 1846, Haldeman Papers.

8. To G. A. Mantell, November 16, 1846, in Fisher, *Silliman,* II, 219.

9. November 12, 1846, to James Hall, Hall Papers.

10. Agassiz to Rose Agassiz, December 2, 1846, Agassiz Papers, Houghton Library.

11. Longfellow, Journal for January 9, 1847, in Samuel Longfellow (ed.), *Life of Henry Wadsworth Longfellow* (2 vols.; Boston, 1886), II, 74.

12. Quoted in Edward Waldo Emerson, *The Early Years of the Saturday Club* (Boston, 1918), p. 31.

13. *Boston Daily Evening Transcript*, November 7, 9, 15, 1846; Asa Gray to John Torrey, December 9, 1846, January 24, 1847, Gray Papers; Ferris Greenslet, *The Lowells and Their Seven Worlds* (Cambridge, Mass., 1946), p. 35.

14. Diary for January 13, 1847, in Fisher, *Silliman*, II, 81.

15. January 4, 1847, Gray Papers.

16. Agassiz, *Introduction to the Study of Natural History* (New York, 1847), p. 6. This was a full account of Agassiz's New York lectures, but they were the same as the Boston addresses, which were not reported in full.

17. To John Torrey, January 24, 1847, Gray Papers.

18. Agassiz, *Introduction*, p. 25. The author of the *Vestiges*, as is well known, was Robert Chambers.

19. Agassiz, *Introduction*, p. 25.

20. To Chancellor Favargez, December 31, 1846, in E. C. Agassiz, *Agassiz*, II, 431.

21. To Charles Henry Davis, February 21, 1847, Charles Henry Davis Papers, Henry Ford Museum, Dearborn, Mich.

22. This is reported in Agassiz to John Amory Lowell, January 18, 1849, Agassiz Papers, Houghton Library.

23. Peter Lesley, Jr., to Susanna Elizabeth Lesley, January 18, 1847, in Mary Lesley Ames, *Life and Letters of Peter and Susan Lesley* (2 vols.; New York, 1909), I, 143–44.

24. To John Torrey, February [20], 1847, Gray Papers.

25. Agassiz to [John Amory Lowell], September 12, [1847], Etting Collection, Historical Society of Pennsylvania.

26. April 10, 1847, Spencer F. Baird Papers, Smithsonian Institution Archives.

27. Agassiz to Benjamin Silliman, May 28, 1847, Daniel Coit Gilman Papers, Johns Hopkins University Library.

28. August 18, 1847, Record Group No. 23, United States National Archives and Records Service.

29. *Proceedings of the American Association for the Advancement of Science*, I (1848), 5, 8–12.

30. To Henri Milne Edwards, May 31, 1847, Agassiz Papers, Houghton Library.

31. 2d ser., IV (November, 1847), 449.

32. *United States Magazine and Democratic Review*, XIX (November, 1846), 413.

33. Records of the Corporation and Fellows of Harvard College, VIII (December 27, 1845, November 28, 1846, February 13, 1847), 290, 330, 339, Harvard University Archives (cited hereafter as "Corporation Records");

Harvard College Papers, 2d ser., XIII (February 27, 1846), XIV (February 14, 1847), 180–85, 213, Harvard University Archives (cited hereafter as "College Papers"); Edward Everett to Abbott Lawrence, August 19, 1847, Harvard College Letters, Vol. I, Harvard University Archives (cited hereafter as "College Letters"); *Addresses at the Inauguration of Hon. Edward Everett* (Boston, 1846), pp. 34–39.

34. Edward Everett to Eben Norton Horsford, February 19, 1847, to Jeffries Wyman, April 9, 1847, College Letters, Vol. I; Corporation Records, VIII (April 3, 1847), 352–54.

35. Abbott Lawrence to Samuel Atkins Eliot, June 7, 1847, Corporation Records VIII (June 7, 1847), 361–65. See Richard J. Storr, *The Beginnings of Graduate Education in America* (Chicago, 1953), pp. 46–51, for a full discussion of the origins of the Lawrence Scientific School.

36. Abbott Lawrence to Samuel Atkins Eliot, June 7, 1847, in Corporation Records, VIII (June 7, 1847), quotations from pp. 362, 360.

37. Corporation Records, VIII (February 13, 1847), 340, IX (March 11, 1848), 27; Samuel Atkins Eliot to Agassiz, September 16, 1848, Agassiz Papers, Houghton Library.

38. Agassiz to T. G. Cary, Jr., September 14, 1870, Agassiz Papers, Houghton Library.

39. Greenslet, *The Lowells*, p. 235.

40. Hall Papers. This fact was also reported in Henry D. Rogers to [William Barton Rogers], June 25, 1847, in Emma Rogers (ed.), *Life and Letters of William Barton Rogers* (2 vols.; Boston, 1896), I, 272.

41. To the President and Fellows of Harvard College, July 19, 1847, in Corporation Records, VIII (July 29, 1847), 376. Lawrence's conception of the binding character of this document is stated in Lawrence to Samuel Atkins Eliot, September 20, 1849, in Corporation Records, IX (October 3, 1849), 112.

42. Everett's offer is reported in Everett to Leonard Horner, July 29, 1847, Edward Everett Papers, Vol. LXXXII, Massachusetts Historical Society. Agassiz to Everett, October 3, 1847, College Papers, XV, 155, reporting the request to the Prussian minister.

43. College Letters, Vol. I. Everett to Abbott Lawrence, September 10, 1847, College Letters, Vol. I, refers to conversations between the two men regarding the Agassiz appointment.

44. Corporation Records, IX (September 25, 1847), 4.

45. [October, 1847], Hall Papers.

46. Agassiz to Edward Everett, October 3, 1847, College Papers, XV, 155. Agassiz to Abbott Lawrence, September 15, 1849, College Papers, XVII, 29, refers to the appointment for a three-year period.

47. Rose Agassiz to Agassiz, November, 1846, in "Process Verbal . . . February 15, 1849," Agassiz Papers, Houghton Library.

48. Agassiz to Abbott Lawrence, September 15, 1849, College Papers, XVII, 28–29, mentioning his state of mind in 1847.

49. *New York Daily Tribune*, October 3, 1847.

50. Agassiz, *Introduction to the Study of Natural History* (New York, 1847), reprinting *Tribune* articles of October, November, and December, 1847.

51. *Introduction,* pp. 4, 5.

52. John Torrey to Asa Gray, November 27, 1847, Historic Letter File, Gray Herbarium, Harvard University; John Watson *et al.,* to Agassiz, November 25, 1847, Agassiz Papers, Houghton Library; *American Journal of Science,* 2d ser., V (January, 1848), 139.

53. December 2, 1846, Agassiz Papers, Houghton Library.

54. Agassiz to Robert W. Gibbes, December 23, 1847, Academy Collection, Academy of Natural Sciences of Philadelphia; Mrs. John E. Holbrook to Agassiz, January 6, 1850, Agassiz Papers, Houghton Library. For an example of Agassiz's southern prestige see F. S. Holmes, "Notes on the Geology of Charleston, S.C.," *American Journal of Science,* 2d ser., VII (March, 1849), 187–201.

55. Journal for January 8, 1848, in Samuel Longfellow (ed.), *Longfellow,* II, 105–6.

56. *Boston Daily Evening Transcript,* January 27, 1848.

57. Agassiz to James Hall, January 27, February 15, April, 1848, Eben Norton Horsford to Hall, March 20, 1848, Hall Papers.

58. Agassiz to Redfield, January 6, 22, October 20, 1848, Joseph Henry to Redfield, October 20, 1848, William C. Redfield Letter Books, Rare Book Room, Yale University Library; Agassiz to Baird, April 26, December 26, 1848, Baird Papers.

59. Agassiz to Edward Everett, February 22, 1848, College Papers, XV, 333.

60. Everett to Eliot, March 16, 17, 19, 1848, College Letters, Vol. II.

61. April 26, 1848, Baird Papers.

62. Reported in Burt G. Wilder, "What We Owe to Agassiz as a Teacher," *Popular Science Monthly,* LXXI (July, 1907), 11.

63. To Lyell, April 5, 1848, in Anna Ticknor and George Hilliard (eds.), *Life, Letters, and Journals of George Ticknor* (2 vols.; Boston, 1876), II, 231.

64. Agassiz and Gould, *Principles of Zoology.* . . . Part I: *Comparative Physiology* (Boston, 1848).

65. In 1878 Thomas Wright published a revised new edition which continued to have a wide sale, especially in England.

66. To William Dwight Whitney, June 25, 1848, in Edwin Tenny Brewster, *Life and Letters of Josiah Dwight Whitney* (Boston and New York, 1909), pp. 95–96.

67. Agassiz, *Lake Superior . . . with a Narrative of the Tour by J. Elliot Cabot* (Boston, 1850), p. 72.

68. Agassiz to Benjamin Silliman, August 22, 1848, Gilman Papers; George P. Merrill, *The First One Hundred Years of American Geology* (New Haven, 1924), pp. 276–77.

69. *American Journal of Science,* 2d ser., IX (May, 1850), 455, X (July, 1850), 83–101, reprinting the chapter on glaciation; Baird to Agassiz, May 26, 1850, Baird Papers; C. S. Kendall to James Hall, April 11, 1850, Hall Papers.

70. June 15, [1850], in E. C. Agassiz, II, 469.

71. *Lake Superior*, p. 144.

72. *Ibid.*, p. 145.

73. *Ibid.*, pp. 153, 186–87.

74. *Ibid.*, pp. 247–48.

75. *Ibid.*, p. 377.

76. *Ibid.*, p. 142.

77. *Proceedings of the American Association for the Advancement of Science,* I (1848), 30–32, 41, 68–71, 79, 94, 99.

78. G. R. Agassiz (ed.), *Alexander Agassiz*, pp. 12–13; Marcou, *Agassiz*, II, 18–19.

79. October 2, 1848, College Papers, XVI, 174.

80. *Twelve Lectures on Comparative Embryology* (Boston, 1849).

81. Augustus A. Gould to James Hall, January 8, 1849, Hall Papers.

82. Benjamin Silliman, Diary for May 2, 1847, in Fisher, *Silliman*, II, 83; John M. Clarke, *James Hall . . .* (Albany, N. Y., 1923), p. 219; Desor to James Hall, May 23, July 29, 1847, Hall Papers; Charles Henry Davis to Alexander Dallas Bache, July 24, 1848, Record Group No. 23, United States National Archives and Records Service.

83. Rose Agassiz to Agassiz, November, 1846, in "Process Verbal . . . February 15, 1849," Agassiz Papers, Houghton Library.

84. *Catalogue raisonné des familles, des genres et des espèces de la classe des échinodermes, par L. Agassiz et E. Desor,* (Paris, 1846–48); *Proceedings of the Boston Society of Natural History*, III (1848–50), 11, 13–14, 17–18 (published in May, 1848). See Agassiz, *Essay on Classification*, pp. 145, n. 1, and 146 for his evaluation of what he termed "one of the most extraordinary cases of plagiarism I know of."

85. Agassiz to Desor, April 8, 1848, transcript in "Process Verbal . . . February 15, 1849," Agassiz Papers, Houghton Library.

86. [1848], Record Group No. 23, United States National Archives and Records Service.

87. E. Desor and L. Agassiz, Agreement on Arbitration, October 24, 1848, Agassiz Papers, Houghton Library.

88. John Amory Lowell *et al.*, "Process Verbal . . . February 15, 1849," Agassiz Papers, Houghton Library.

89. January 13, 1849, in *Trial of the Action of Edward Desor . . . versus Chas. H. Davis* (Boston, 1852), p. 60.

90. Lowell *et al.*, "Process Verbal . . . February 15, 1849," Agassiz Papers, Houghton Library. This document contains the full report of the arbitrators on the Agassiz-Desor relationship and is the primary source for the events of 1848–49.

91. Gould to James Hall, February 14, 1849, Hall Papers.

92. *Trial of the Action of Edward Desor . . . versus Chas. H. Davis,* legal proceedings on pp. 3–52; Desor to Hall, November 12, 1851, February 2, 10, 1852, Hall Papers.

93. To George Engelmann, August 28, 1848, Gray Papers.

94. Leidy to Samuel S. Haldeman, February 17, 1849, Leidy Papers.

95. *Proceedings of the American Association for the Advancement of Science*, II (1849), 68–77, 85–91, 100–101, 140–43, 157–59, 389–96, 411–23, 432–38.

96. "On the Differences between Progressive, Embryonic, and Prophetic Types . . . ," *ibid.*, pp. 432–38.

97. July 22, 1849, Hall Papers.

98. Agassiz to Abbott Lawrence, September 15, 1849, College Papers, XVII, 28–29.

99. Lawrence to Samuel Atkins Eliot, September 20, October 3, 1849, Corporation Records, IX (October 3, 1849), 112–16, 117.

100. Eliot to Agassiz, June 1, 1850, Agassiz Papers, Houghton Library.

NOTES, CHAPTER 5

1. March 8, 1850, Agassiz Papers, Houghton Library.

2. Quoted in Louise Hall Tharp, *Adventurous Alliance: The Story of the Agassiz Family of Boston* (Boston, 1959), p. 5.

3. Quoted in G. R. Agassiz (ed.), *Alexander Agassiz*, pp. 441–42.

4. October 2, 1849, Charles Sumner Papers, Houghton Library.

5. Agassiz Papers, Houghton Library.

6. [January 7, 1850], Gray Papers.

7. Lyell to Charles Bunbury, January 17, 1850, in Katherine M. Lyell (ed.), *Sir Charles Lyell*, II, 158.

8. January 6, 1850, Agassiz Papers, Houghton Library.

9. [April, 1850], Radcliffe College Archives. I am indebted to Mrs. Louise Hall Tharp for a copy of this letter.

10. [April, 1850], Radcliffe College Archives. I am indebted to Mrs. Louise Hall Tharp for a copy of this letter.

11. Quoted in Lucy Allen Paton, *Elizabeth Cary Agassiz* (Boston, 1919), p. 35.

12. To James Hall, June 20, 1850, Hall Papers.

13. Agassiz, "Report to the Committee of the Overseers . . . November, 1851," Reports to the Overseers of Harvard College, Vol. IX, Harvard University Archives (cited hereafter as "Overseers Reports").

14. Quoted in G. R. Agassiz (ed.), *Alexander Agassiz*, p. 441.

15. William Dallam Armes (ed.), *The Autobiography of Joseph Le Conte* (New York, 1903), pp. 128–29.

16. Quoted in Tharp, *Adventurous Alliance*, pp. 110–11. For the Florida journey and its results see Armes (ed.), *Joseph Le Conte*, pp. 130–35, 139; Joseph Le Conte, "On the Agency of the Gulf Stream in the Formation of the Peninsula of the Keys of Florida," *American Journal of Science*, 2d ser., XVII (January, 1857), 48–60; Agassiz, "Report on the Florida Reefs," *Memoirs of the Museum of Comparative Zoology*, VII (1880–82), 1–61.

17. Records of the Faculty of the Lawrence Scientific School, June 4, 9, July 8, 1851, Harvard University Archives (cited hereafter as "Scientific School Records"); Armes (ed.), *Joseph Le Conte*, pp. 128–29, 140–42, 156–57; Samuel Eliot Morison, *Three Centuries of Harvard, 1636–1936* (Cambridge, Mass., 1936), p. 280.

18. May 2, 1850, Leidy Papers.

19. To Asa Gray, July 29, 1851, Historic Letter File, Gray Herbarium.

20. Agassiz to Hall, November 17, 1849, February 28, March 3, 1850, Benjamin Silliman to Hall, December 5, 6, 1849, Eben Norton Horsford to Hall, November 21, 1850, James Dwight Dana to Hall, February 25, 1850, Hall Papers. The chart is reproduced in John M. Clarke, *James Hall,* pp. 206–7, 208–9.

21. Clarke, *James Hall,* pp. 206, 208, 210; Agassiz to James Dwight Dana, November 21, December 6, 1850, Dana Papers; Augustus A. Gould to James Hall, March 21, 1851, Hall Papers.

22. "Report of Professor Agassiz," in Edward Everett, *Report of the Committee of the Overseers . . . 1849* (Cambridge, Mass., 1850), pp. 13–15.

23. Bache, "Presidential Address," *Proceedings of the American Association for the Advancement of Science,* VI (1851), xli–lx. See A. Hunter Dupree, *Science in the Federal Government . . .* (Cambridge, Mass., 1957), pp. 115–19, for a full discussion of the Bache proposals and their background.

24. Agassiz to Hall, August 3, 1851, Peirce to Hall, October 27, 1851, in Clarke, *James Hall,* pp. 193–96; Agassiz to James Dwight Dana, February 9, 1852, Dana Papers. See Storr, *Graduate Education,* pp. 67–74.

25. Agassiz to T. Romeyn Beck, Simon Gratz Papers, Historical Society of Pennsylvania.

26. Agassiz to James Hall, November 3, 1853, Hall Papers; Edward Waldo Emerson, "Benjamin Peirce," in Edward Waldo Emerson, *Saturday Club,* pp. 102–3; Peirce to Alexander Dallas Bache, January 27, 1855, cited in Storr, *Graduate Education,* p. 172, n. 1.

27. Benjamin Peirce to Alexander Dallas Bache, May 8, 1855, Benjamin Peirce Papers, Harvard University Archives. I am indebted to C. Norman Guice for a copy of this letter. Alexander Dallas Bache to Benjamin Peirce, October 15, 1856, cited in Storr, *Graduate Education,* p. 179, n. 45. Cf. Storr, pp. 82–93.

28. Peirce to Alexander Dallas Bache, May 8, 1855, Benjamin Peirce Papers, Harvard University Archives.

29. Benjamin A. Gould to James Dwight Dana, November 22, 1856, Dana Papers.

30. James Dwight Dana to Bache, January 9, 1857, Agassiz to Bache, January 14, [1857], Wolcott Gibbs to Bache, June 7, December 13, 1857, Henry Q. Hawley to Bache, December 17, 1857, John F. Frazer to Bache, December 23, 1857, Alexander Dallas Bache Papers, Manuscripts Division, Library of Congress.

31. C. F. Winslow to John Warner, October 2, 24, 1857, John Warner Papers, American Philosophical Society.

32. James C. Welling to William Pepper, June 17, 1890, William Pepper Collection, University of Pennyslvania Library, reporting a conversation with Peirce and Agassiz in 1858.

33. J. H. Easterby, *A History of the College of Charleston* (Charleston, S.C., 1935), pp. 109–11; Agassiz to James Hall, August 3, 1851, in Clarke, *James Hall,* p. 193.

34. Agassiz to Spencer F. Baird, June 20, [1853], Baird Papers; to Samuel S. Haldeman, July 9, 1853, Agassiz Papers, Houghton Library; "Recent Researches of Professor Agassiz," *American Journal of Science,* 2d ser., XVI (July, 1853), 135–36.

35. Agassiz to Charles Wilkins Short (printed circular, July, 1853), Charles Wilkins Short Papers, Library of Congress.

36. Agassiz to Alexander Dallas Bache, July 28, 1853 (including printed circular), Bache Papers.

37. Copies of the Agassiz circular are to be found in nearly every manuscript collection relating to American natural history in this period.

38. Spencer F. Baird, "Report of the Assistant Secretary . . . ," in *Annual Report of the Smithsonian Institution, 1854* (Washington, 1855), pp. 79–97.

39. November 23, 1853, Baird Papers.

40. October 11, 1853, Baird Papers. On Agassiz's encouragement of museum activity see Lawrence Vail Coleman, *The Museum in America* (Washington, 1939), p. 223, and Frederick A. Lucas, *Fifty Years of Museum Work* (New York, 1933), p. 9.

41. August 25, 1853, Agassiz Papers, Houghton Library.

42. September 27, 1853, Agassiz Papers, Houghton Library.

43. June 28, 1853, Agassiz Papers, Houghton Library.

44. Agassiz to Eliot, August 15, 1853, Eliot to Agassiz, August 29, 1853, Agassiz Papers, Houghton Library; Corporation Records, IX (August 27, 1853), 272–73.

45. November 1, 1853, College Papers, XX, 234.

46. College Papers, XXI (April 26, 1854), 43; Overseers Reports, Instruction Series, Vol. I (October 31, 1855).

47. March 27, 1854, Davis Papers.

48. May 20, 1854, Agassiz Papers, Houghton Library.

49. September 28, 1854, College Papers, XXI, 320.

50. January 9, 1855, in E. C. Agassiz, *Agassiz,* II, 515, 517.

51. Armes, *Joseph Le Conte,* pp. 161–62.

52. "Extraordinary Fishes from California . . . ," "Synopsis of the Ichthyological Fauna of the Pacific Slope . . . ," *American Journal of Science,* 2d ser., XVI, XIX (November, 1853, March, May, 1855), 380–90, 71–99, 215–31, are representative papers of this sort.

53. January 30, 1855, Agassiz Papers, Houghton Library.

54. Ernst Mayr, "Agassiz, Darwin, and Evolution," *Harvard Library Bulletin,* XIII (Spring, 1959), 174, citing an analysis of Agassiz's tendencies in species designation as revealed in "Notice of a Collection of Fishes from the Southern Bend of the Tennessee River . . . ," *American Journal of Science,* 2d ser., XVII (March, May, 1854), 297–308, 353–69.

55. *American Journal of Science,* 2d ser., XIX (January, 1855), 91.

56. Agassiz to Samuel S. Haldeman, May 31, 1855, Haldeman Papers.

57. May 5, 1856, Agassiz Papers, Houghton Library.

58. Agassiz, *Contributions to the Natural History of the United States* (4 vols.; Boston, 1857–62), I, x.

59. Agassiz to Lewis R. Gibbes, "Private Circular," May 28, 1855, Gibbes Papers.

60. Agassiz to Spencer F. Baird, May 31, 1855, Baird Papers.

61. May 28, 1855, Baird Papers.

62. Francis C. Gray to Spencer F. Baird, June 6, 1855, Agassiz to Baird, June 6, 1855, Baird Papers. The "Prospectus" and "Private Circular" are to be found in almost every manuscript collection relating to American natural history in this period.

63. Some small sampling of Agassiz's enterprise and the extent of the subscription campaign is to be found in Agassiz to Baird, June 6, 8, 12, 1855, Baird Papers; Agassiz to Dana, May 28, 1855, Dana Papers; to James Hall, May 28, 1855, Hall Papers.

64. February 8, 1856, Agassiz Papers, Houghton Library.

65. [November, 1855], in E. C. Agassiz, *Agassiz,* I, 537–38.

66. *Contributions,* I, xvii–xliv.

67. Jared Sparks, "Professor Agassiz," *National Intelligencer,* May 15, 1855; Agassiz to Sparks, May 8, 1855, Jared Sparks Papers, Houghton Library.

68. [C. C. Felton], "Professor Agassiz," *Boston Advertiser,* May 22, 1855.

69. April 2, 1855, Ralph Waldo Emerson Memorial Association Papers, Houghton Library.

70. August 30, 1855, Agassiz Papers, Houghton Library.

71. Quoted in Emerson, *Saturday Club,* p. 35.

72. Journal for May 28, [1857], in Bliss Perry (ed.), *The Heart of Emerson's Journals* (Boston, 1938), p. 278.

73. "The Fiftieth Birthday of Agassiz, May 28, 1857," manuscript copy in Agassiz Papers, Houghton Library, reprinted in E. C. Agassiz, *Agassiz,* II, 544–45.

74. Volume I (January, 1858), 320–33. See James C. Austin, *Fields of the Atlantic Monthly* (San Marino, Calif., 1953), p. 26.

75. Humboldt, May 9, 1858, Ticknor to Agassiz, June 4, 1858, to Humboldt, July 8, 1858, Agassiz Papers, Houghton Library. The *Boston Courier* of June 9, 1858 reprinted Humboldt's letter.

76. Agassiz, *Contributions to the Natural History of the United States,* Part II, "North American Testudinata," I, 235–452; Part III, "Embryology of the Turtle," II, 451–622, with twenty-seven plates.

77. *Contributions,* I, 8.

78. *Ibid.,* p. 143.

79. *Ibid.,* pp. 57–63. See Ernst Mayr, "Agassiz, Darwin, and Evolution," *Harvard Library Bulletin,* XIII (Spring, 1959), 168.

80. *Contributions,* I, 133.

81. [James D. Dana], "Agassiz's Contributions to the Natural History of the United States," *American Journal of Science,* 2d ser., XXV (May, 1858), 341.

82. Dana to Agassiz, March 2, July 25, 1853, Agassiz Papers, Houghton Library; [James D. Dana, preliminary notice of the *Contributions*], *American Journal of Science,* 2d ser., XXV (January, 1858), 126–28.

83. [James D. Dana], "Agassiz's Contributions to the Natural History of

the United States," *American Journal of Science,* 2d ser., XXV (March, May, 1858), 202–16, 321–41; cf. 328–29, 332–41.

84. Dana to Asa Gray, November 13, December 4, 1857, March 9, 1858, Historic Letter File, Gray Herbarium; Gray to Dana, November 7, 1857, Gray Papers; Arnold Guyot to Dana, March 17, 1858, Dana Papers.

85. March 31, 1855, Gray Papers.

86. Quoted in Emerson, *Saturday Club,* p. 31.

87. To Charles Wilkins Short, July 5, 1855, Short Family Papers, Manuscripts Division, Library of Congress.

NOTES, CHAPTER 6

1. Quoted in Edward Waldo Emerson, *Saturday Club,* p. 32.

2. Henry Adams, *The Education of Henry Adams* (New York, 1918; 1931 printing), p. 60. See Ernest Samuels, *The Young Henry Adams* (Cambridge, Mass., 1948), p. 22.

3. Quoted in Edward Waldo Emerson, *Saturday Club,* p. 31.

4. February 22, 1848, College Papers, XV, 333.

5. May 20, 1854, Agassiz Papers, Houghton Library.

6. James Lawrence to William T. Andrews, November 12, 1855, Corporation Records, IX, 371–72.

7. Agassiz to James Walker, June 9, 1859, College Papers, XXVI, 189.

8. "Report of the Committee . . . of the Overseers," Overseers Reports, Instruction Series, Vol. I [January, 1856].

9. *Boston Advertiser,* May 22, 1855.

10. Agassiz to James Walker, December 22, 1854, College Papers, XXI, 401.

11. [October] 20, 1856, copy made by William Gray, July 5, 1883, Agassiz Papers, Museum of Comparative Zoology at Harvard College.

12. Agassiz [obituary of Francis C. Gray], *Proceedings of the American Academy of Arts and Sciences,* III (1852–57), 347–49, quotation from p. 348.

13. William Gray to the President and Fellows of Harvard College, December 20, 1858, College Papers, XXV (December 24, 1858), 334–36.

14. Corporation Records, X (January 17, May 30, 1857), 1–3, 22–25; Josiah Parsons Cooke to James Walker, January 17, 1857, College Papers, XXIV, 29–31.

15. [May, 1857], Joseph Henry Papers, Smithsonian Institution Archives.

16. Rouland to Agassiz, August 19, 1857, Agassiz Papers, Houghton Library; Elizabeth Cary Agassiz to Mrs. Thomas G. Cary, September 19, 1857, in Paton, *Elizabeth Agassiz,* pp. 55–56.

17. Agassiz to Auguste Rouland, September 25, 1857, Agassiz Papers, Houghton Library.

18. Agassiz to Sir Philip Egerton, [September, 1857], Agassiz Papers, Houghton Library.

19. Elizabeth Cary Agassiz to Mrs. Thomas G. Cary, September 19, 1857, in Paton, *Elizabeth Agassiz,* pp. 55–56.

20. *Boston Advertiser,* October 1, 1857; *National Intelligencer,* October 12, 1857.

21. November 17, 1857, Agassiz Papers, Houghton Library.

22. Élie de Beaumont to Agassiz, January 17, 1858, Charles Martins to Agassiz, January 4, 1858, Agassiz Papers, Houghton Library; George Engelmann to Asa Gray, June 1, 1858, Historic Letter File, Gray Herbarium.

23. Agassiz to Rouland, April 26, 1858, Agassiz Papers, Houghton Library.

24. [November, 1858], in E. C. Agassiz, *Agassiz*, II, 553–54.

25. Agassiz to James Walker, April 26, 1858, Agassiz Papers, Houghton Library.

26. Humboldt to Ticknor, May 9, 1858, Ticknor to Agassiz, June 4, 1858, Agassiz Papers, Houghton Library. See *Boston Courier,* June 9, 1858.

27. July 8, 1858, Agassiz Papers, Houghton Library.

28. May 27, 1858, Agassiz Papers, Houghton Library.

29. July 18, 1858, Historic Letter File, Gray Herbarium.

30. Corporation Records, X (June 5, 1858), 67.

31. "Report to the . . . Committee of the Overseers . . . ," Overseers Reports, Instruction Series, Vol. II (December 15, 1858).

32. William Gray to the President and Fellows of Harvard College, December 20, 1858, College Papers, XXV, 334–36, officially received in Corporation Records, X (December 24, 1858), 109–10.

33. *Boston Courier,* January 7, 1859.

34. Jacob Bigelow, W. W. Greenough, and James Lawrence, printed invitation to James Walker, January 18, 1859, College Papers, XXVI [January, 1859], 16.

35. "The Agassiz Museum . . . January 26, 1859," printed circular, College Papers, XXVI [January, 1859], 27.

36. John Clifford, Jacob Bigelow, *et al.,* "Report of Committee Appointed by the . . . Overseers . . . ," Overseers Reports, Instruction Series, Vol. II (January 27, 1859).

37. *Boston Courier,* March 27, 1859, giving a full report of the hearings which were also reported in *Boston Daily Evening Transcript,* March 27, 1859.

38. "Commonwealth of Massachusetts . . . An Act to Increase the School Fund, and to grant aid to the Museum of Comparative Zoology . . . April 2, 1859 . . . ," "Commonwealth of Massachusetts . . . An Act to Incorporate the Trustees of the Museum of Comparative Zoology . . . April 6, 1859," printed in College Papers, XXVI [April, 1859], 101–2.

39. *Report of the Trustees of the Museum . . . 1861* (Boston, 1861), pp. 26–27.

40. February 16, 1859, in Katherine M. Lyell (ed.), *Sir Charles Lyell,* II, 318.

41. April 2, 1859, Agassiz Papers, Houghton Library.

42. Corporation Records, X (June 2, 1859), 135–37; Henry W. Greenough and George Snell, "The Specifications for the Museum . . . ," [July, 1859] printed form in College Papers, XXVI [July, 1859], 204–5.

43. Agassiz, "Rules and Regulations Submitted to the Faculty of the Museum . . . May 4, 1859," College Papers, XXVI [May, 1859], 133–37; "Articles of Agreement, Museum of Comparative Zoology, May 30, 1859," Corporation Records, X (June 2, 25, 1859), 138–45. These documents comprise the original rules for museum government and administration. Agassiz to T. G. Cary, Jr., September 17, 1870, Theodore Lyman Papers, Museum of Comparative Zoology, gives his explanation of the early regulations.

44. Addison Emery Verrill, Private Journal, Harvard University Archives (cited hereafter as "Verrill Journal").

45. July 30, 1859, in Paton, *Elizabeth Agassiz*, p. 60.

46. Verrill Journal, January 26, 1860.

47. "Sketch of a Plan for the Organization of the Museum . . . Submitted to the Faculty . . . by L. Agassiz, Director and Curator . . . ," College Papers, XXVI [November, 1859], 360–68.

48. Verrill Journal, January 26, 1860.

49. Nathaniel Southgate Shaler, *The Autobiography of Nathaniel Southgate Shaler* (Boston, 1909), pp. 93, 103.

50. *Ibid.*, pp. 99–100.

51. This account by Morse appears in Verrill Journal, January 9, 1860. It is also given in Dorothy G. Wayman, *Edward Sylvester Morse* (Cambridge, Mass., 1942), pp. 84–85, in slightly different form.

52. Wayman, *Morse*, p. 141, reprinting Morse's Journal for October 6, [1860].

53. Verrill Journal, July 27, 1860.

54. *Ibid.*, December 29, 1859, April 6, 1860.

55. *Ibid.*, January 12, 24, 1860.

56. Philip P. Carpenter to Alpheus Hyatt, January 29, 1859, Alpheus Hyatt Autograph Collection, Maryland Historical Society.

57. Agassiz to Baird, April 12, December 6, 1860, January 28, 1861, Baird Papers; Edward S. Morse, Journal for September 29, [1860], in Wayman, *Morse*, p. 137.

58. Agassiz to Oswald Heer, December 4, 1860, Agassiz Papers, Houghton Library; Agassiz, "Report to the Committee of Overseers . . . ," [December 28, 1859], Overseers Reports, Professional Series, Vol. II.

59. *Report of the Trustees of the Museum . . . 1861*, pp. 46–48, quotation from p. 48.

60. "Report of the Committee of the Overseers . . . 1860," *ibid.*, p. 12.

61. To George P. Bond, January 6, 1860, Agassiz Papers, Museum of Comparative Zoology.

62. August 6, 1860, Gray Papers.

63. Verrill Journal, November 13, 1860.

64. Agassiz to the Editors of the *Transcript,* November 7, 1860, Manuscript Collections, New York Public Library; *Boston Courier,* November 14, 1860; *Boston Daily Evening Transcript,* November 14, 1860.

65. Agassiz to Spencer F. Baird, December 6, 1860, January 28, 1861, Baird Papers; *Report of the Trustees of the Museum . . . 1861*, pp. 6, 38–39, 41–42.

66. *Contributions,* III (Boston, 1860), 3–301, with twenty-nine plates.

67. Verrill Journal, January 14, 1860.

NOTES, CHAPTER 7

1. March 26, 1860, Agassiz Papers, Museum of Comparative Zoology.

2. Recorded in "Resolutions . . . Board of Overseers . . . Louis Agassiz, December 18, 1873," Agassiz Papers, Houghton Library.

3. November [11], 1859, Agassiz Papers, Houghton Library.

4. Agassiz, marginal notes in Charles Darwin, *On the Origin of Species by*

Means of Natural Selection (London, 1859), pp. 183, 194, on deposit in the rare book case, Museum of Comparative Zoology.

5. To Sir Charles Lyell, June 25, 1859, in Leonard Huxley, *Life and Letters of Thomas Henry Huxley* (2 vols.; London, 1900), I, 174.

6. *Notice sur la géographie des animaux* (Neuchâtel, 1845), pp. 29–31, quotation from p. 31.

7. Gray to John Torrey, January 24, 1847, Gray Papers.

8. Agassiz to Rose Agassiz, December 2, 1846, Agassiz Papers, Houghton Library.

9. George R. Gliddon to Samuel George Morton, January 9, 1848, Manuscript Collections, Library Company of Philadelphia.

10. See, for example, "Some Observations on the Ethnography and Archeology of the American Aborigines," "Value of the Word Species in Zoology," *American Journal of Science*, 2d ser., II, XI (July, 1846, March, 1851), 1–17, 275–76.

11. Mobile, Ala., 1844.

12. Morton, "Hybridity in Animals, Considered in Reference to the Question of the Unity of the Human Species," *American Journal of Science*, 2d ser., III (January, March, 1847), 39–50, 203–11; Nott, *An Essay on the Natural History of Mankind, Viewed in Connection with Negro Slavery* (Mobile, Ala., 1851). See Arthur O. Lovejoy, "The Argument for Organic Evolution before the Origin of Species, 1830–1858," in Bentley Glass *et al.* (eds.), *Forerunners of Darwin, 1745–1859* (Baltimore, 1959), pp. 394–95.

13. "Geographical Distribution of Animals," *Christian Examiner*, XLVIII (March, 1850), 181–204, quotation from p. 188.

14. *Proceedings of the Ameican Association for the Advancement of Science*, III (1850), 106–7, quotation from p. 107.

15. May 4, 1850, Manuscript Collections, Library Company of Philadelphia.

16. "The Diversity of Origin of the Human Races," *Christian Examiner*, XLIX (July, 1850), 110–45, quotation from p. 138.

17. See John Bachman, *A Notice of the "Types of Mankind"* (Charleston, S.C., 1854), p. 13; "Natural History of Man," *United States Magazine and Democratic Review*, XXVI, XXVII (April, July, 1850), 327–45, 41–48.

18. *Christian Examiner*, L (January, 1851), 4.

19. *Ibid.*, pp. 1–17. See the perceptive description of special creationist metaphysics in Lovejoy, "The Argument for Organic Evolution," pp. 412–13.

20. "Sketch of the Natural Provinces of the Animal World and Their Relation to the Different Types of Man," in J. C. Nott and George R. Gliddon, *Types of Mankind* (Philadelphia, 1854), pp. lvii–lxxviii, quotation from p. lxxviii. The *Types* appeared in ten editions between 1854 and 1871.

21. "Prefatory Remarks," in J. C. Nott and Geo. R. Gliddon, *Indigenous Races of the Earth* (Philadelphia, 1857), pp. xiii–xv.

22. Lyell to George Ticknor, January 9, 1860, in Katherine M. Lyell (ed.), *Sir Charles Lyell*, II, 331.

23. *The Geological Evidences of the Antiquity of Man . . .* (London, 1863), pp. 387–88.

24. *Principles of Geology* (11th ed.), II, 481.

25. See the revealing analysis of this problem in Lovejoy, "The Argument for Organic Evolution," pp. 413–14.

26. December 13, 1856, in Jane Loring Gray (ed.), *Gray*, II, 424.

27. *American Journal of Science*, 2d. ser., XVII (March, May, 1854), 241–52, 334–50.

28. To Gray, January 26, 1854, in Leonard Huxley, *Life and Letters of Sir Joseph Dalton Hooker* (2 vols.; London, 1918), I, 475.

29. To Hooker, March 26, [1854], in Francis Darwin (ed.), *Life and Letters of Charles Darwin* (2 vols.; New York, 1896), I, 403.

30. "Statistics of the Flora of the Northern United States," *American Journal of Science*, 2d ser., XXII, XXIII (September, 1856, January, May, 1857), 204–32, 62–84, 369–403.

31. *Contributions*, I, 135.

32. Marcou, *A Geological Map of the United States . . . with Explanatory Text . . .* (Boston, 1853); Dana to Hall, November 28, December 11, 1853, Hall Papers, reporting Hall's reaction to the map and text.

33. Agassiz to Dana, December 15, 1853, Dana Papers; [James Hall], "Notice of a Geological Map of the United States . . . ," *American Journal of Science*, 2d ser., XVII (March, 1854), 199–206.

34. Marcou, *Carte géologique des États-Unis . . .* (Paris, 1855). The map was also published in German (Gotha, 1855) and in German and French journals. William P. Blake, "Review of a Portion of the Geological Map of the United States . . . by Jules Marcou," *American Journal of Science*, 2d ser., XXII (November, 1856), 383–88.

35. Marcou, *Geology of North America . . .* (Zurich, 1858). James D. Dana, "Review of Marcou's Geology of North America," *American Journal of Science*, 2d ser., XXVI (November, 1858), 323–33, quotations from 331, 333.

36. The contents of Agassiz's letter are reported in Wolcott Gibbs to Alexander D. Bache, November 22, 1858, Bache Papers, Dana to Agassiz, June 14, 1859, copy in Dana to Asa Gray, June 14, 1859, Historic Letter File, Gray Herbarium, and "Marcou's Strictures on North American Geologists," *American Journal of Science*, 2d ser., XXVIII (July, 1859), 153.

37. Agassiz, "On Marcou's 'Geology of North America' " and Dana, "Reply to Prof. Agassiz on Marcou's Geology of North America," *American Journal of Science*, 2d ser., XXVII (January, 1859), 134–40.

38. Dana to Agassiz, June 14, 1859, copy in Dana to Asa Gray, June 14, 1859, Historic Letter File, Gray Herbarium.

39. Dana to Gray, June 14, 1859, Historic Letter File, Gray Herbarium.

40. [The Editors], "Marcou's Strictures on North American Geologists," *American Journal of Science*, 2d ser., XXVIII (July, 1859), 153–54, quotation from p. 154.

41. June 1, 1858, in Jane Loring Gray (ed.), *Gray*, II, 445.

42. January 7, 1859, Gray Papers.

43. Gray to John Torrey, January 7, 1859, Gray Papers.

44. *An Essay on Classification* (London, 1859). This was published early in January, 1859. Owen to Agassiz, December 9, 1857, Agassiz Papers, Houghton Library, commenting on the "Essay" in the *Contributions*.

45. *Essay on Classification,* p. 13.

46. *Proceedings of the American Academy of Arts and Sciences,* IV (1857–60), 109–30; "Diagnostic Characters of New Species of Phaenogamous Plants, Collected in Japan . . . ," *Memoirs of the American Academy of Arts and Sciences,* N.S., VI, Part 2 (1859), 377–452.

47. *Proceedings of the American Academy,* IV, 132.

48. *Ibid.*

49. *Ibid.,* pp. 133–34.

50. [February, 1859], Historic Letter File, Gray Herbarium.

51. *Proceedings of the American Academy,* IV, 175.

52. *Ibid.,* p. 177.

53. *Ibid.,* p. 178; Hooker to Gray, [March, 1859], Historic Letter File, Gray Herbarium.

54. Hooker to Gray, [March, 1859], Historic Letter File, Gray Herbarium.

55. *Proceedings of the American Academy,* IV, 195; italics supplied.

56. Lyell to George Ticknor, November 29, 1860, in Katherine M. Lyell (ed.), *Sir Charles Lyell,* II, 340.

57. "Extract . . . of a 'Memoir on the Botany of Japan' . . . ," *American Journal of Science,* 2d ser., XXVIII (September, 1859), 187–200; Durand to Gray, April 16, 1859, Curtis to Gray, May 25, 1859, Historic Letter File, Gray Herbarium.

58. *Essay on Classification,* p. 31.

59. *Ibid.,* p. 179.

60. See Ernst Mayr, "Agassiz, Darwin, and Evolution," *Harvard Library Bulletin,* XIII (Spring, 1959), 169–82, for the most incisive appraisal of the background of Agassiz's natural philosophy. I am indebted to Professor Mayr for futhering my understanding of problems discussed in this section.

61. *Essay on Classification,* p. 84.

62. Joseph Le Conte, *Evolution and Its Relation to Religious Thought* (New York, 1888), pp. 32–49.

63. See *Proceedings of the American Academy,* V (1860–62), 243–47. This is an appraisal of Tiedemann's life and work that was obviously written by Agassiz. Credit is given here to Johann Meckel as recognizing ideas of recapitulation that "were far more satisfactorily demonstrated and illustrated by Tiedemann" (p. 245).

64. *Methods of Study in Natural History,* p. 23. Agassiz first announced his advocacy of this aspect of the recapitulation idea in *Poissons fossiles,* I (1833–43), 81, 102. These passages were written in 1833.

65. *Essay on Classification,* p. 175.

66. *Ibid.,* pp. 266–67, italics supplied.

67. For a complete analysis of the relationship of Von Baer, Darwin, and Agassiz to the recapitulation concept see Jane Oppenheimer, "An Embryological Enigma in the Origin of Species," in Glass *et al.* (eds.), *Forerunners of Darwin,* pp. 292–322.

68. Oppenheimer, "An Embryological Enigma . . . ," in Glass *et al.* (eds.), *Forerunners of Darwin,* pp. 321–22. Agassiz thought Haeckel the worst example of a materialist in biology and disliked him intensely.

69. See G. R. de Beer, *Embryos and Ancestors* (Oxford, 1951), pp. 1–10,

139–42. This should be compared with Lovejoy, "Recent Criticism of the Darwinian Theory of Recapitulation: Its Grounds and Its Initiator," in Glass *et al.* (eds.), *Forerunners of Darwin*, pp. 438–58. See also Mayr, "Agassiz, Darwin, and Evolution," pp. 191–92.

70. *Twelve Lectures on Comparative Embryology*, pp. 27–29. See pp. 26–27, where Agassiz asserts his general agreement with the recapitulation idea.

71. *Essay on Classification*, pp. 69, 116, 177–78.

72. Hooker, "On the Origination and Distribution of Species: Introductory Essay to the Flora of Tasmania," *American Journal of Science*, 2d ser., XXIX (January, May, 1860), 1–25, 306–26, introduction and quotation from Gray in note, p. 1.

73. January 5, 1860, in Francis Darwin (ed.), *Life . . . of Darwin*, II, 63.

74. Verrill Journal, January 14, 1860.

75. *Proceedings of the American Academy*, IV, 360, 362.

76. *Ibid.*, p. 410.

77. *Ibid.*

78. *Ibid.*, pp. 410–15, 428–29.

79. *Ibid.*, pp. 424–26, quotation from p. 424.

80. Gray to Darwin, February 20, 1860, Darwin to Gray, March 8 [1860], Historic Letter File, Gray Herbarium.

81. A.[sa] G.[ray], "Review of Darwin's Theory on the Origin of Species by Means of Natural Selection," *American Journal of Science*, 2d ser., XXIX (March, 1860), 160.

82. *Ibid.*, p. 179.

83. *Ibid.*, p. 180; entire review on pp. 153–84.

84. March 31, 1860, General Historical Letters, Royal Botanical Garden, Kew, England. I am indebted to A. Hunter Dupree for a copy of this letter.

85. *Proceedings of the Boston Society of Natural History*, VII (1859–61), 231.

86. *Ibid.*, pp. 232–35, quotation from 233.

87. *Ibid.*, pp. 241–45.

88. *Ibid.*, p. 271. For other discussions see pp. 246–49, 250–52, 271–75.

89. "Professor Agassiz on the Origin of Species," *American Journal of Science*, 2d ser., XXX (July, 1860), 142–54, quotation from p. 143.

90. *Ibid.*, p. 143. See Mayr, "Agassiz, Darwin, and Evolution," p. 175.

91. "Professor Agassiz on the Origin of Species," p. 154.

92. Hooker to Gray, August 9, 1860, Lesquereux to Gray, July 17, August 25, 1860, Historic Letter File, Gray Herbarium; Darwin to Lyell, August 11, [1860], reporting a letter from Gray in Francis Darwin (ed.), *Life . . . of Darwin*, II, 124.

93. "Discussion between Two Readers of Darwin's Treatise . . . ," *American Journal of Science*, 2d ser., XXX (September, 1860), 226–39.

94. "Darwin on the Origin of Species," "Darwin and His Reviewers," *Atlantic Monthly*, VI (July, August, October, 1860), 109–15, 229–39, 406–25.

95. *Contributions*, Vol. III (Boston, 1860). The review was reprinted on pages 88–99, and 112–13.

96. *Proceedings of the American Academy*, V (1860–62), 72.

97. *Ibid.*, p. 102.

98. June 11, 1861, Lyman Papers.

NOTES, CHAPTER 8

1. December 6, 1860, Museum Letter Books, II, 29–30, Museum of Comparative Zoology at Harvard College (cited hereafter as "Museum Letter Books").

2. Verrill Journal, February 2, 1861.

3. September 9, 1862, Lyman Papers.

4. "Methods of Study in Natural History," *Atlantic Monthly,* IX (January, 1862), 13.

5. *Ibid.,* X (July, 1862), 87.

6. These articles appeared as "Methods of Study in Natural History" in Volume IX (January, February, March, April, May, June, 1862), 1–13, 214–22, 327–37, 446–60, 570–78, 754–62, and Volume X (July, September, November, 1862), 87–98, 325–36, 571–80.

7. *Methods of Study . . .* (Boston, 1863), pp. iii, iv. The volume was published by Ticknor and Fields, publishers of the *Atlantic Monthly.*

8. *The Structure of Animal Life* (New York, 1865, 1874).

9. These articles appeared in Volume XI (March, April, May, June, 1863), 373–82, 460–71, 615–25, 747–56, Volume XII (July, August, September, November, December, 1863), 72–81, 212–24, 333–42, 568–76, 751–67, Volume XIII (January, February, 1864), 56–65, 224–32, and Volume XIV (July, 1864), 86–93.

10. "America and the Old World," *Atlantic Monthly,* XI (March, 1863), 376.

11. *Geological Sketches* (Boston, 1866), including *Atlantic Monthly* articles from March, 1863, through January, 1864.

12. To Agassiz, October 20, 1863, Agassiz Papers, Houghton Library.

13. Typical examples of the use of Agassiz's arguments appear in *Methodist Quarterly Review,* XLIII (October, 1861), 605–25; *Monthly Religious Magazine,* XXVI (December, 1861), 396–97; *Baptist Quarterly,* II (July, 1868), 257–74.

14. Gray to Joseph D. Hooker, May 11, 1863, General Historical Letters, Royal Botanical Garden, Kew, England. I am indebted to A. Hunter Dupree for making a copy of this letter available to me.

15. May 26, 1863, Gray Papers.

16. July 21, 1863, Gray Papers.

17. Darwin to Gray, August 4, [1863], Historic Letter File, Gray Herbarium; Gray to Darwin, November 23, 1863, Gray Papers.

18. November 15, 1863, Dana Papers.

19. To John Gould, March 11, 1860, in Dorothy G. Wayman, *Morse,* p. 226.

20. Journal, October 4, 1861, *ibid.,* p. 181.

21. Journal, December 30, 31, 1861, *ibid.,* p. 183.

22. Alexander Agassiz to Theodore Lyman, January 6, 1862, Lyman Papers.

23. Agassiz to Dr. Henry Wheatland, December 18, 1863, Museum Letter Books, II, 354–55.

24. "Regulations for the Museum of Comparative Zoology, November 5, 1863," printed circular, Agassiz Papers, Museum of Comparative Zoology.

25. Agassiz to Phillips R. Uhler, April 16, 1864, Museum Letter Books, II, 411.

26. Bickmore to Alpheus Hyatt, [November, 1864, March, 1865], Alpheus Hyatt Autograph Collection, Maryland Historical Society.

27. Journal, April 25, 1861, in Wayman, *Morse*, p. 172.

28. *Contributions*, I, xv–xvi.

29. *Contributions*, III (Boston, 1860), vi; IV (Boston, 1862); Agassiz to Clark, March 21, 1863, Clark to Agassiz, June 25, 1863, College Papers, XXX, n.p.

30. Agassiz to Clark, March 21, 1863, Clark to the President and Fellows of Harvard College, April 24, 1863, to Agassiz, June 25, 1863, College Papers, XXX, n.p.

31. Clark to the President and Fellows of Harvard College, April 24, 1863, to Agassiz, June 25, 1863, Hill to Clark, May 12, 1863, Agassiz to Hill, May 14, 1863, College Papers, XXX, n.p.

32. Agassiz to Hill, August 23, 1863, College Papers, XXX, n.p.; Scientific School Records, September 9, 1863.

33. Agassiz [to Thomas Hill], October 10, 1863, College Papers, XXX, n.p.

34. Agassiz to Thomas Hill, September 11, October 12, 1863, Hoar to Thomas Hill, October 16, 1863, College Papers, XXX, n.p.; Clark to Hill, October 28, 1863, Agassiz Papers, Museum of Comparative Zoology.

35. E. R. Hoar and Thomas Hill, "Report . . . November 28, 1863," College Papers, XXX, n.p.; Corporation Records, X (November 28, 1863), 336; Clark, *Mind in Nature* (New York, 1865), pp. liv, 174, 304 n.

36. Scientific School Records, October 7, 1861.

37. *Ibid.*, November 4, 11, 1861; Henry James, *Charles William Eliot* (2 vols.; New York, 1930), I, 94–98.

38. Scientific School Records, November 18, December 2, 1862; Eliot to Hill, November 15, 1862, College Papers, XXIX, n.p.

39. Thomas Hill to Dr. Putnam, November 24, 1863, College Letters, Vol. VI. Cf. Joseph Lovering *et al.* to the President and Fellows of Harvard College, July 23, 1863, College Papers, XXX, n.p.

40. Agassiz to John A. Andrew, December 16, 1862, Museum Letter Books, II, 91.

41. Agassiz to Andrew, December 16, 1862, *ibid.*, pp. 90–92.

42. Agassiz to Andrew, December 22, 23, 25, 1862, *ibid.*, pp. 96–103.

43. Agassiz to Andrew, January 4, 1863, *ibid.*, pp. 104–5.

44. Charles W. Eliot to the President and Fellows of Harvard College, May 19, 1863, in Thomas Hill and George Putnam, "Report of a Committee on the Rumford Professorship, May 30, 1863," College Papers, XXX, n.p.; Corporation Records, X (June 12, 1863), 313; James, *Eliot*, II, 98–112; Scientific School Records, September 9, 1863.

45. George J. Brush to J. D. Whitney, June 30, 1863, in James, *Eliot*, II, 111–12; Agassiz to Thomas Hill, July 25, 1863, Autograph File, Houghton Library, Harvard University; Agassiz to Thomas Hill, January 25, 1863, to E. R. Hoar, February 8, 1863, Museum Letter Books, II, 119–21, 155–56.

46. February 5, 1863, Museum Letter Books, II, 149.

47. Captain Charles H. Davis, *Life of Charles Henry Davis* . . . (Boston and New York, 1899), pp. 283–85; Frederick W. True (ed.), *A History of the First Half Century of the National Academy of Sciences* (Washington, 1913), pp. 1–13.

48. Charles H. Davis [to Harriette Blake Davis], February 24, 1863, in Captain Davis, *Davis*, p. 290; True, *First Half Century*, pp. 11–12; Agassiz to Baird, February 18, 1863, Baird Papers; John Torrey to Asa Gray, March 9, 1863, Historic Letter File, Gray Herbarium.

49. February 5, 1863, Museum Letter Books, II, 150.

50. August 13, 1864, in A. Hunter Dupree, "The Founding of the National Academy of Sciences—a Reinterpretation," *Proceedings of the American Philosophical Society*, CI (October, 1957), 439.

51. February 5, 1863, Museum Letter Books, II, 150.

52. December 4, 1870, *ibid.*, VI, 76.

53. Charles H. Davis [to Harriette Blake Davis], February 24, 27, 1863, in Captain Davis, *Davis*, pp. 290–91; *New York Daily Tribune*, April 23, 1863; *Proceedings of the National Academy of Sciences*, I (1877), 1–2. For a full description of the founding of the Academy see Dupree, *Science in the Federal Government*, pp. 135–41.

54. True, *First Half Century*, pp. 14, 16, 102–5, 109, 202; William Barton Rogers [to Henry Darwin Rogers], April 28, 1863, in Emma Rogers (ed.), *Life and Letters of William Barton Rogers*, II, 161; George Engelmann to Asa Gray, April 4, 1863, Historic Letter File, Gray Herbarium.

55. William Barton Rogers [to Henry Darwin Rogers], April 28, 1863, in Rogers, *Rogers*, II, 162; True, *First Half Century*, pp. 21–22; Peter Lesley, Jr., [to Susan I. Lesley], April 23, 1863, in Mary Lesley Ames, *Peter and Susan Lesley*, I, 419; *Proceedings of the National Academy of Sciences*, I (1877), 29 (sec. 3); Joseph Henry to Asa Gray, April 15, 1863, July 8, 1868, John Torrey to Asa Gray, April 30, 1863, Historic Letter File, Gray Herbarium.

56. George Engelmann to Asa Gray, April 4, [August], 1863, Historic Letter File, Gray Herbarium; Asa Gray to Joseph Henry, April 18, 1863, Joseph Henry Papers, Smithsonian Institution Archives; Gray to Henry, [May, 1863], to George Engelmann, September 2, December 11, 1863, Gray Papers; Joseph Leidy to Joseph Henry, December, 1863, Wolcott Gibbs to Leidy, July 15, 1866, Leidy Papers.

57. May 22, 1864, Museum Letter Books, III, 4–8.

58. April 21, 1863, *ibid.*, II, 200–202.

59. Agassiz to Ticknor, October 14, 1863, Agassiz Papers, Houghton Library; Agassiz to Charles Henry Davis, May 23, 1863, to Alexander D. Bache, May 23, 1863, Museum Letter Books, II, 240–42; "Directions for Collecting Objects of Natural History," [May, 1863], printed circular, Agassiz Papers, Museum of Comparative Zoology.

60. July 25, 1864, Lyman Papers.

61. *Report of the Trustees of the Museum . . . 1862* (Boston, 1863), p. 13.

62. January 20, 1865, Edwin M. Stanton Papers, Manuscripts Division, Library of Congress.

63. November 24, 1864, Museum Letter Books, III, 131–32. See also Gray to Darwin, August 7, 1866, to George Engelmann, March 18, 1865, Gray Papers.

64. Agassiz to Dana, October 15, 1866, Museum Letter Books, IV, 37. Cf. *American Journal of Science,* 2d ser., XLI (May, 1866), 418–22. Agassiz's name does not appear on the *Journal* title page after October, 1866.

65. Agassiz to Henry, August 8, 1864, Museum Letter Books, III, 41–45.

66. Gray to Darwin, November 6, [1866], to Engelmann, November 20, 1866, to J. D. Hooker, February 8, 1886, Gray Papers; Agassiz to Joseph Henry, August 8, 1864, Museum Letter Books, III, 41–45.

67. August 13, 1864, in Dupree, "The Founding of the National Academy," p. 440.

68. Agassiz to Henry, November 15, 1864, Museum Letter Books, III, 125.

69. Agassiz to Rose Agassiz, March 22, 1865, Agassiz Papers, Houghton Library.

70. Allan McLane to Agassiz, March 7, 1865, Agassiz to Nathaniel Thayer, March 14, 1865, Agassiz Papers, Houghton Library; Professor and Mrs. Louis Agassiz, *A Journey to Brazil* (Boston, 1868), p. ix; the expedition adventures are recounted in a series of letters by Elizabeth Agassiz to her family in Paton, *Elizabeth Agassiz,* pp. 71–102.

71. Agassiz to Rose Agassiz, March 22, 1865, Agassiz Papers, Houghton Library.

72. [To Henry James, Sr.], September 12–17, 1865, William James Papers, Houghton Library.

73. James [to Mrs. Henry James, Sr.], March 31, 1865, James Papers.

74. James [to Mrs. Henry James, Sr.], August 23, [1865], James Papers.

75. *Ibid.*

76. James, Brazilian Diary, 1865–66, Houghton Library.

77. James A. Garfield, Diary, December 13, 1867, James A. Garfield Papers, Manuscripts Division, Library of Congress.

78. November 8, 1865, Agassiz Papers, Museum of Comparative Zoology.

79. Joseph Henry to Gray, October 31, 1866, Historic Letter File, Gray Herbarium; Gray to Charles Darwin, November 6, [1866], to George Engelmann. November 20, 1866, Gray Papers.

NOTES, CHAPTER 9

1. Agassiz to Lyman, September 15, 17, 1866, Elizabeth Agassiz to Lyman, September 15, 1866, Lyman Papers.

2. "An Amazonian Picnic," "Physical History of the Amazons," *Atlantic Monthly,* XVII (March, 1866), 313–23, XVIII (July, August, 1866), 49–60. 159–69.

3. *Proceedings of the National Academy of Sciences,* I (1877), 58. The paper was not printed but its contents were reported in Asa Gray to Charles Darwin, August 27, 1866, Gray Papers.

4. To Charles Bunbury, September 3, 1866, in Katherine M. Lyell (ed.), *Sir Charles Lyell,* II, 410.

5. September 10, [1866], Historic Letter File, Gray Herbarium. A full exposition of the glacial discoveries in South America was given by Charles F. Hartt, *Thayer Expedition: Scientific Results of a Journey in Brazil* . . . (Boston, 1870), pp. 45–133. For analyses of Agassiz's South American glacial concepts see James Geikie, *The Great Ice Age*, p. 723, and George P. Merrill, *The First One Hundred Years of American Geology*, p. 430.

6. *A Journey in Brazil* (Boston, 1868), p. 398.

7. To Agassiz, December 26, 1867, Agassiz Papers, Houghton Library.

8. January 3, 1868, Agassiz Papers, Houghton Library.

9. Holmes to Agassiz, January 5, 1868, Agassiz Papers, Houghton Library.

10. To Agassiz, February 23, 1868, Agassiz Papers, Houghton Library.

11. September 17, 1868, Leidy Papers.

12. Agassiz to Andrew D. White, August 29, 1867, Agassiz Papers, Museum of Comparative Zoology; *Autobiography of Andrew Dickson White* (2 vols.; New York, 1905), I, 337; Albert Hazen Wright, "Agassiz and Cornell," in *Pre-Cornell and Early Cornell*, I ("Studies in History," No. 15 [Ithaca, N.Y., 1953]), 6–8, reprinting the inaugural address of Agassiz.

13. Agassiz to Hill, July 30, 1868, Museum Letter Books, IV, 171.

14. Henry James, *Eliot*, I, 186–87, 193–94, 197–98; Samuel Eliot Morison, *Three Centuries of Harvard*, pp. 327–28; Charles William Eliot, "The New Education—Its Organization," *Atlantic Monthly*, XXIII (February, March, 1869), 203–20, 358–67; Wolcott Gibbs, letter of February 19, 1869, in *Atlantic Monthly*, XXIII (April, 1869), 514.

15. James, *Eliot*, I, 193–94, 198.

16. Thomas Hill to William H. Glen, September 11, 1865, College Letters, Vol. VI.

17. *Report of the Trustees of the Museum* . . . *1866* (Boston, 1867), pp. 8–9, italics supplied.

18. *Ibid.*, pp. 10–16, quotation from p. 16.

19. *Ibid.*, *1867* (Boston, 1868), pp. 4–8, 10–11.

20. *Ibid.*, *1868* (Boston, 1869), p. 6.

21. May 29, 1868, Leidy Papers.

22. *Address Delivered on the Centennial Anniversary of the Birth of Alexander von Humboldt* . . . (Boston, 1869), pp. 6–64.

23. Marcou, *Agassiz*, II, 178.

24. December 20, 1870, Museum Letter Books, VI, 124–26.

25. Agassiz to William Claflin, December 26, 1870, *ibid.*, pp. 131–32; [Thomas G. Cary, Jr.], *An Account of the Organization and Progress of the Museum of Comparative Zoology at Harvard College* (Cambridge, Mass., 1871).

26. February 4, 1871, Agassiz Papers, Houghton Library.

27. February 26, 1871, Agassiz Papers, Museum of Comparative Zoology.

28. June 1, 1871, in G. R. Agassiz (ed.), *Alexander Agassiz*, p. 119.

29. March 4, 1872, *ibid.*, pp. 119–21.

30. July 28, 1872, Agassiz Papers, Houghton Library.

31. Agassiz to Benjamin Peirce, December 15, 1871, January 16, 1872, Agassiz Papers, Houghton Library.

32. Elizabeth Cary Agassiz, "The Hassler Glacier in the Straits of Magel-

lan," *Atlantic Monthly,* XXX (October, 1872), 472–82; Agassiz to Benjamin Peirce, July 29, 1872, Agassiz Papers, Houghton Library.

33. August 27, 1872, Lyman Papers.

34. To J. B. Schlesinger, November 6, 1872, Agassiz Papers, Museum of Comparative Zoology.

35. January 21, 1873, Henry Papers.

36. January 28, February 2, 12, 1873, Othniel Charles Marsh Papers, Peabody Museum of Natural History, Yale University. I am indebted to Thomas G. Manning for copies of these letters.

37. Abby P. Godfrey *et al.* to Agassiz, April 25, 1873, Agassiz Papers, Houghton Library; *Report of the Trustees of the Museum . . . 1872* (Boston, 1873), pp. 4–5; G. R. Agassiz (ed.), *Alexander Agassiz,* pp. 84–90.

38. *New York Times,* March 12, 1873; Agassiz to Anderson, March 15, 1873, in *The Organization and Progress of the Anderson School of Natural History* (Boston, 1874), pp. 3–4.

39. Agassiz to Anderson, March 22, 1873, Anderson to Agassiz, March 19, 1873, in *Anderson School,* pp. 7–11.

40. *Ibid.,* pp. 11–14; *Tribune Popular Science* (Boston and New York, 1874), pp. 44–65, reporting all of Agassiz's lectures. Whittier's poem, "The Prayer of Agassiz," is reprinted in Marcou, *Agassiz,* II, 203–6.

41. July 22, 1868, Museum Letter Books, IV, 169–70.

42. August 19, 1868, Agassiz Papers, Houghton Library.

43. Anon., class notes on evolution lectures, March 10, 17, 1871, Agassiz Papers, Museum of Comparative Zoology.

44. November 16, 1873, William Dean Howells Papers, Houghton Library.

45. "Evolution and Permanence of Type," *Atlantic Monthly,* XXXIII (January, 1874), 94.

46. *Ibid.,* p. 95.

47. *Ibid.,* p. 92.

48. *Ibid.,* p. 98.

49. *Ibid.,* p. 101.

50. *The Last Lecture . . . Fitchburg, December 2, 1873* (Boston, 1874).

51. December 4, 1873, Lyman Papers.

52. John Peter Lesley, August 26, 1871, Agassiz Papers, Houghton Library.

53. Quoted in Tharp, *Adventurous Alliance,* p. 241.

54. *Boston Journal,* December 15 through 19, 22, 24, 1873; *Boston Daily Advertiser,* December 25, 1873; *Christian Register,* December 27, 1873; *Boston Courier,* December 15, 19, 1873; G. R. Agassiz, *Alexander Agassiz,* pp. 128–29.

55. James Russell Lowell, "Agassiz," *Atlantic Monthly,* XXXIII (May, 1874), 586–96, quotations from pp. 586–87, 596.

ESSAY ON SOURCES

A COMPLETE LISTING of the primary and secondary sources relating to Agassiz's life would be far too long for inclusion here. The purpose of this essay is to call attention to the nature and value of major source materials, with emphasis on manuscripts, other types of primary sources, and certain secondary publications bearing directly on Agassiz's career. The major secondary sources used in this biography have been cited in the footnotes, with the full title given at the time of initial citation.

THE PUBLICATIONS OF LOUIS AGASSIZ

There is no complete bibliography of Agassiz's writings. For an early compilation see Jules Marcou, *Life, Letters, and Works of Louis Agassiz* (2 vols.; New York, 1896), II, 258–303. This lists 425 books and articles published between 1828 and 1880. The Marcou list is sometimes inaccurate. For a select list of Agassiz's writings, with emphasis on his American career, see Edward Lurie, "Louis Agassiz and American Natural Science" (Ph.D. diss., Department of History, Northwestern University, 1956), pp. 507–20. A list of publications of the period 1828–46, compiled by Agassiz, is in his monumental *Bibliographia zoologiae et geologiae* (4 vols.; London, 1848–54), I, 98–103. This lists eighty-one titles but does not include translations and works of dual authorship. Agassiz's writings of the period 1846–65 will be found under particular headings in Max Meisel, *A Bibliography of American Natural History: The Pioneer Century, 1769–1865* (3 vols.; Brooklyn, N.Y., 1924–29). The Meisel volumes are fully indexed. They are an indispensable guide to the study of natural history and scientific organization in nineteenth-century America.

BIOGRAPHIES AND APPRAISALS OF AGASSIZ

Elizabeth Cary Agassiz (ed.), *Louis Agassiz, His Life and Correspondence* (2 vols.; Boston, 1885), is much more than the usual Victorian "Life and Letters" written by a devoted relative. Elizabeth Agassiz brought to this study of her husband the perception and insight she evidenced in the years of their marriage. Nearly all the letters to and from Agassiz included in these volumes are on deposit in the Agassiz Papers at the Houghton Library, Harvard University. Letters dealing with Agassiz's life in Europe were reprinted almost exactly. Those dealing with his life in America, however, were often only partially reproduced. Jules Marcou's *Life, Letters, and Works of Louis Agassiz* (2 vols.; New York, 1896) was based in large degree on Mrs. Agassiz's *Life*. Where Marcou departed from this account, the results were often strange and misleading. Certain discussions, as for example those regarding the glacial theory, are excellent; others, as in the case of the evolution controversy, are entirely unreliable. Alice Bache Gould's *Louis Agassiz* (Boston, 1901) is a short and perceptive appreciation. James David Teller, "Louis Agassiz, Scientist and Teacher," *Graduate School Studies, The Ohio State University* (Columbus, 1947), is written entirely from published sources. It presents a useful appraisal of Agassiz's impact on American education. Of the great variety of shorter articles and assessments of Agassiz's career and phases of his work that appeared during and after his lifetime, a number deserve particular notice. Asa Gray was by circumstance and insight a perceptive student of Agassiz, and his two appraisals, "Louis Agassiz," *Nation*, XVII (December 18, 1873), 404–5, and "Louis Agassiz," *Andover Review*, 1886, reprinted in Charles Sprague Sargent, *Scientific Papers of Asa Gray* (2 vols.; Boston and New York, 1889), II, 483–90, reflect this understanding. Volume II of the Sargent compilation contains, incidentally, many valuable biographical sketches by Gray of his contemporaries in American and European natural science. William James's beautiful tribute to Agassiz that appeared in *Science*, N.S., V (February 19, 1897), 285–89, and as a pamphlet (Cambridge, Mass., 1897), stands alone. Ernest Favre, "Louis Agassiz, a Biographical Notice," *Annual Report of the Smithsonian Institution, 1878* (Washington, 1879), pp. 237–61, is useful for the Neuchâtel period. The same is true of Arnold Guyot, "Memoir of Louis Agassiz . . . ," in National Academy of Sciences, *Biographical Memoirs*, II (Washington, 1886), 39–73. Lane Cooper, *Louis Agassiz as a Teacher* (Ithaca, N.Y., 1917; rev. ed., 1945), is a valuable contribution. The author selected passages from the writings of Agassiz's students—Shaler, Verrill, Wilder, and Scudder—describing his teaching methods and impact. There is also included a partial list of Agassiz's students and a worthwhile appraisal by Mrs. Malcolm D. Brown. Finally, Tilly Edinger, "Agassiz Lebt," *Natur und Volk*, LXXXII (October, November, 1952), 318–25, 354–61, by a scientist who deeply appreciates Agassiz's life and work, remains the warmest and most understanding brief analysis.

The following special studies of phases of Agassiz's ideas and contributions to science are particularly noteworthy: Walter Baron, "Zu Louis Agassiz's Beurteilung des Darwinismus," *Sudhoffs Archiv . . . ,* XL (1956), 259–77, which stresses the significance of natural theology in Agassiz's conception of nature; Eugene R. Corson, "Agassiz's Essay on Classification Fifty Years After," *Scientific Monthly*, XI (July, 1920), 43–52; Ernst Mayr, "Agassiz, Darwin, and Evolution," *Harvard Library Bulletin*, XIII (Spring, 1959), 165–94, a masterful treatment of the internal logic of Agassiz's special creationism which greatly aided my understanding of the subject; and George Frederick Wright, "Agassiz and the Ice Age," *American Naturalist*, XXXII (March, 1898), 165–71.

MANUSCRIPT SOURCES

These materials have furnished the basic data for my account of Agassiz's life and times. They consist primarily of letters written by or to Agassiz, but manuscripts reflecting the life of his colleagues are also important. Again, a full listing of such data is not possible within the scope of this essay. Particular collections and their depositories are noted individually in the footnotes. For a survey of the depositories that contain material relevant to this biography see Edward Lurie, "Some Manuscript Resources in the History of Nineteenth Century American Natural Science," *Isis*, XLIV (December, 1953), 363–70.

The following are the major collections of Agassiz manuscripts: The primary source for Agassiz's career is the Agassiz Papers on deposit at the Houghton Library, Harvard University. This collection contains over seven hundred letters. There are over two hundred Agassiz letters reflecting all phases and periods of his life, and also letters from such important correspondents as Humboldt, Lyell, Dana, Augustus A. Gould, and Samuel Gridley Howe. The Agassiz Papers on deposit at the Museum of Comparative Zoology consist of about one hundred letters written to Agassiz by such men as Baird, Henry, and Leidy. In addition, important Agassiz correspondence not contained in the official museum letter books is in this collection. The letter folders in the library of the Museum of Comparative Zoology contain correspondence from Agassiz's many European and American colleagues, newspaper clippings, reports on lectures, student notes, and manuscript worksheets used by Agassiz in the preparation of books and articles. A large and important manuscript source for Agassiz's life is the official Letter Books of the Museum of Comparative Zoology. These are six letter-press volumes containing both the official correspondence of the museum and Agassiz's private correspondence during the period 1859–73. In all, there are over 1,700 pages of correspondence in these volumes, representing about 1,300 Agassiz letters. Many of these are duplicated in other collections, notably the Agassiz Papers at the Houghton Library. When Agassiz was away from the museum, the correspondence was carried on by Alexander Agassiz. The Theodore Lyman Papers at the Museum of Comparative Zoology are another basic source, important for the period from 1862 to 1873. They include correspondence between Agassiz and Lyman and between Alexander Agassiz and Lyman.

The following collections are valuable for Agassiz letters and for material relevant to his life: The Augustus Addison Gould Papers, Houghton Library, comprise about one hundred letters, including ten from Agassiz to Gould. The Alexander Dallas Bache Papers, Manuscripts Division, Library of Congress, are important for Lazzaroni activity, and contain twelve letters from Agassiz to Bache. The Spencer Fullerton Baird Papers, Smithsonian Institution Archives, are a most important source for the general history of American natural history and contain over three hundred letters from Agassiz to Baird, as well as correspondence from Dana, Haldeman, Hall, Henry, and Lewis R. Gibbes. The James Dwight Dana Papers, Rare Book Room, Yale University Library, are not a separate collection but a number of folders of letters to Dana from various correspondents, the most notable of which were Charles Darwin and Arnold Guyot. There are twenty Agassiz letters here. The Lewis R. Gibbes Papers, Manuscripts Division, Library of Congress, are a significant body of documents relating to the history of American natural science; they contain over six thousand letters, three of which are from Agassiz. There are letters to Gibbes from J. W. Bailey, Francis S. Holmes, Josiah Clark Nott, and Dana in this collection. The Asa Gray Papers, Gray Herbarium, Harvard University, include autograph letters by Gray and typescripts of original letters on deposit at the New York Botanic Garden and other institutions, many of which are printed in Jane Loring Gray's *Letters*. These are a basic source for the history of nineteenth-century American natural history. The Samuel S. Haldeman Papers at the Academy of Natural Sciences of Philadelphia are a small collection of over 300 letters, including correspondence from John G. Anthony, Baird, and Thaddeus W. Harris. There are fifteen Agassiz letters in this collection. The James Hall Papers at the New York State Museum, Albany, are the largest single source collection in the history of nineteenth-century American natural history known to me. The papers cover the period 1803–1904, are filed in 250 file boxes, and total nearly 20,000 manuscripts. There are seventy-five letters from Agassiz to Hall, as well as correspondence from every major American naturalist.

The Harvard University Archives contain manuscript materials of primary significance for Agassiz's career as it related to the university, the college, the Lawrence Scientific School, and the Museum of Comparative Zoology. The various record collections contain many Agassiz letters to the Corporation, communications to him by this body and the president, and reports relative to scientific education at the university. These collections are the Records of the Corporation and Fellows, Volumes VIII–XI, covering the period 1836–73 and containing the official proceedings of college and university government; the Harvard College Papers, Second Series, Volumes XIV–XXX, covering the period 1846–63 and containing reports of Corporation proceedings, letters to and from the president, and material some of which is also included in the Corporation Records; the Harvard College Letters comprise six volumes for the period 1846 to 1868 and contain letters to and from the president, after 1863 continuing the College Papers material to 1868 (these volumes are designated by the name of the president then in office); Reports to the Overseers, Volumes VIII and IX, Reports to the Overseers, Instruction Series, Volumes I and II, and Reports to the Overseers, Professional Series, Volume II, contain the reports of the various visiting committees of the Overseers for the period 1848–74. These are an important source for Agassiz's relationship to scientific education at the College prior to 1859; after that date, while he still made reports and visiting committees reported on the progress of the museum, the data contained in the official museum reports are of more value. These were issued annually beginning in 1861 and printed, the first report covering 1859 and 1860. The Records of the Faculty and Administrative Board of the Lawrence Scientific School are another important source on deposit at the Archives.

The Historic Letter File, Gray Herbarium, is a collection of primary significance, containing eighty file cases of letters written to Gray by many naturalists of importance in America and many distinguished Europeans. They reflect the stature of Gray in national and international science. Particularly valuable are letters of Dana, Engelmann, Henry, and Torrey. There are no Agassiz letters, but many references to him. The Joseph Leidy Papers, Academy of Natural Sciences of Philadelphia, contain about one hundred folders with letters from such scientists as Baird, Hall, Hayden, Meek, and J. D. Whitney. There are ten Agassiz letters to Leidy. The Manuscript Collections of the Library Company of Philadelphia contain letters of Dana, George R. Gliddon, Haldeman, Morton, and Nott. There are ten letters from Agassiz to various correspondents. The Samuel George Morton Papers, American Philosophical Society, are a small group of manuscripts containing letters to and from Morton principally of importance with regard to the unity-plurality controversy.

The Dom Pedro II Letters, Houghton Library, are a series of sixty letters from Agassiz to Dom Pedro for the period 1866–73 reflecting the success of the Brazil journey. They are photostat copies of originals on deposit at the Imperial Museum, Petrópolis, Brazil. The Records of the United States Coast and Geodetic Survey, Record Group No. 23, United States National Archives and Records Service, include a vast body of data and are quite probably the largest source in the history of a special phase of American science. Nathan Reingold compiled a valuable *Preliminary Inventory* (Washington, 1958) of these documents, which are important not only for the Coast Survey but for the wide range of Bache's organizational activities. There are ten letters from Agassiz to Bache and four from Agassiz to Peirce in this collection.

The William C. Redfield Letter Books, Rare Book Room, Yale University, are a small but important source, comprising three volumes of autograph letters covering the period 1842–57 and including fifteen letters from Agassiz to Redfield. The Benjamin Silliman Papers, Historical Society of Pennsylvania, include some important correspondence of Dana and Silliman relating to the years 1845–54. The Benjamin Silliman Papers, Rare Book Room, Yale University Library, contain five letters from Agassiz to Silliman, Silliman's diary written between 1820 and 1854, and material relating to the history of scientific education at Yale College. The Charles Sumner Papers at Houghton Library contain thirty letters from Agassiz to Sumner. Addison Emery Verrill's Private Journal, on deposit in the Harvard University Archives, relates to the years 1859–64 and is the only manuscript source dealing with student life at the Museum of Comparative Zoology. Like the accounts in Dorothy G. Wayman, *Edward Sylvester Morse: A Biography* (Cambridge, Mass., 1942), and *The Autobiography of Nathaniel Southgate Shaler* (Boston, 1909), it is an invaluable source for Agassiz's life in this period.

In addition to these collections, the Houghton Library of Harvard University contains a number of individual collections with material bearing on Agassiz's life. These include the Charles S. Bowen Autograph Collection, the Ralph Waldo Emerson Memorial Association

Papers, the Oliver Wendell Holmes Papers, the William Dean Howells Papers, the William James Papers (including a diary account of the Brazil voyage), the Henry Wadsworth Longfellow Papers, the Charles Eliot Norton Papers, and the Jared Sparks Papers. Other collections that contain material of relevance are housed at the following depositories: Manuscripts Division, Library of Congress (James A. Garfield Papers, Matthew Fontaine Maury Papers, Charles Wilkins Short Papers, and Ephraim George Squier Papers); Smithsonian Institution Archives (Joseph Henry Papers and Diary of Miss Mary Henry, 1863–1868); Maryland Historical Society (Alpheus Hyatt Autograph Collection and James A. Pearce Papers); American Philosophical Society (Le Conte Family Papers and John Warner Papers); Manuscript Collections of the New York Public Library (containing ten Agassiz letters); Academy of Natural Sciences of Philadelphia (George Ord Papers, and Isaac Lea Papers); United States National Archives and Records Service (Records of the Mexico and Texas Boundary Commission and Records of the Office of Explorations and Surveys).

MAGAZINES AND SERIAL PUBLICATIONS

Among the magazines of general interest, I have relied a good deal on the *Atlantic Monthly*, a periodical that served as the leading popular forum for Agassiz's educational efforts in natural history. A most informative account of the magazine in the Agassiz period is found in James C. Austin's *Fields of the Atlantic Monthly; Letters to an Editor* (San Marino, Calif., 1953). Similar periodicals that were of value are the *Christian Examiner*, the *Massachusetts Quarterly Review*, the *North American Review*, and the *United States Magazine and Democratic Review*.

The following journals of a specialized nature were particularly useful: the *American Journal of Science and Arts* is a mine of information reflecting every phase of the history of nineteenth-century science and the activities of its practitioners. The major scientific emphases of the *Journal* are analyzed in Edward Salisbury Dana *et al.*, *A Century of Science in America* . . . (New Haven, 1918). Covering the period 1818–1918, this is much more than an internal history of the periodical; it is in fact an excellent source for the general history of nineteenth-century American science. The *Christian Register*, *Charleston Medical Journal and Review*, and *Scientific American* were also of value.

In regard to serial publications, the *Proceedings of the American Academy of Arts and Sciences* is a basic source for Agassiz's career in the United States. The same is true of the *Proceedings of the American Association for the Advancement of Science* and the *Proceedings of*

the Boston Society of Natural History. As noted, the Annual Report
of the Trustees of the Museum of Comparative Zoology is of first importance. Each report included a statement by Agassiz on the progress
of the museum. Other serial publications reflecting the wide scope of
Agassiz's interests are the Proceedings of the National Academy of
Sciences (Washington, 1877), dealing with the years 1863–76, and the
Annual Report of the Secretary to the Board of Regents of the Smithsonian Institution, Volumes I–XXVIII (1846–73).

SECONDARY SOURCES

The material for an understanding of Agassiz's life is naturally divided into two major areas, the European and American phases. Certain publications, of course, span both aspects of his life and work.
For the formative period of Agassiz's life, the chief sources are his
European publications, the correspondence recorded in Mrs. Agassiz's
Life, and the documents on deposit at the Houghton Library. There
are certain general and special studies which deserve particular men-
tion for their usefulness in understanding Agassiz's life in Europe and
the sources and character of his ideas and contributions to science.
These are Frank Dawson Adams, The Birth and Development of the
Geological Sciences (Baltimore, 1938; Dover Edition, 1954); Charles
Coulston Gillispie, Genesis and Geology: A Study in the Relations of
Scientific Thought, Natural Theology, and Social Opinion in Great
Britain, 1790–1850 (Cambridge, Mass., 1951; Harper Edition, 1959), a
book that is fundamental to any understanding of the pre-Darwinian
history of biology; Bentley Glass, Owsei Temkin, and William Straus,
Jr. (eds.), Forerunners of Darwin, 1745–1859 (Baltimore, 1959), a collection of essays by specialists which is an invaluable contribution to
knowledge of the intellectual and internal history of biology in a
crucial period in its development and which includes a revised and
enlarged version of Arthur O. Lovejoy's "The Argument for Organic
Evolution before the Origin of Species, 1830–1858"; Arthur O. Lovejoy, The Great Chain of Being (Cambridge, Mass., 1936, 1957 printing), a volume that must rank as a prime source for the study of the
intellectual antecedents and relationships of biological thought; Arthur
William Meyer, Human Generation: Conclusions of Burdach, Döllinger, and von Baer (Stanford, Calif., 1956); Joseph Needham, A History of Embryology (Cambridge, 1934); Erik Nordenskiöld, The History of Biology, translated by Leonard Bucknall Eyre (New York,
1928); Emmanuel Rádl, The History of Biological Theories (London,
1930); Charles Singer, A History of Biology (2d ed., rev.; New York,
1950); Karl Alfred von Zittel, History of Geology and Paleontology,
translated by Maria M. Ogilvie-Gordon (London, 1901).

The following biographies, many of which are really primary sources because of the letters they contain, are of particular value: Pierre Florens, *Cuvier* . . . (Paris, 1845); Sarah Lee Bowditch, *Memoirs of Baron Cuvier* (New York, 1833); Francis Darwin (ed.), *The Life and Letters of Charles Darwin* (2 vols.; New York, 1888; 1896 ed.), and *More Letters of Charles Darwin* (2 vols.; New York, 1903); Helmut de Terra, *Humboldt: The Life and Times of Alexander von Humboldt* (New York, 1955); Leonard Huxley, *Life and Letters of Sir Joseph Dalton Hooker* (2 vols.; London, 1918), and *The Life and Letters of Thomas Henry Huxley* (2 vols.; London, 1900); Katherine M. Lyell (ed.), *Life, Letters, and Journals of Sir Charles Lyell, Bart.* (2 vols.; London, 1881); Alexander Ecker, *Lorenz Oken* . . . , translated by Alfred Tulk (London, 1883).

Concerning the study of science in America, the following books are of particular importance: Whitfield J. Bell, Jr., *Early American Science: Needs and Opportunities for Study* (Williamsburg, Va., 1955), a work that is a bibliography, appraisal, and guide to further research from the early seventeenth century to 1815; I. Bernard Cohen, *Some Early Tools of American Science* . . . (Cambridge, Mass., 1950), a book that is indispensable for a knowledge of early American science and particularly valuable for science in New England and at Harvard College; Edward Salisbury Dana *et al.*, *A Century of Science in America*, noted above; A. Hunter Dupree, *Science in the Federal Government* . . . (Cambridge, Mass., 1957), a book that describes and analyzes a vital aspect of the social and cultural history of science in America; Thomas Cary Johnson, *Scientific Interests in the Old South* (New York, 1936); Dirk J. Struik, *Yankee Science in the Making* (Boston, 1948), an important contribution that traces the development of science and technology in this region to 1865.

The history of American natural history has not been written. There is some valuable special literature that should be noted. George Brown Goode, *The Beginnings of Natural History in America* (Washington, 1886); John C. Greene, "The American Debate on the Negro's Place in Nature," *Journal of the History of Ideas*, XV (June, 1954), 384–96; Bert James Loewenberg, "The Reaction of American Scientists to Darwin," *American Historical Review*, XXXVIII (July, 1933), 687–701; "The Controversy over Evolution in New England, 1859–1873," *New England Quarterly*, VIII (June, 1935), 232–57; "Darwinism Comes to America," *Mississippi Valley Historical Review*, XXVIII (December, 1941), 339–68; George P. Merrill, *The First One Hundred Years of American Geology* (New Haven, 1924), with important material on the state geological surveys; Sidney Ratner, "Evolution and the Rise of the Scientific Spirit in America," *Philosophy of Science*, III (January, 1936), 104–22; William Martin Smallwood, *Natural History and the American Mind* (New York, 1951), an interesting effort to evaluate cultural and intellectual influences upon biological thought from the eighteenth century to 1846; R. T. Young, *Biology in America* (Boston, 1923), a work that is unfortunately sketchy and incomplete in many aspects. In addition to these titles, the E. S. Dana volume cited above contains special studies on the development of zoology, vertebrate paleontology, botany, and geology. The Meisel bibliography cited above also provides valuable material on the institutional aspects of the history of biology in America.

The following biographies of American scientists are of particular value and relevance to the study of Agassiz's life. G. R. Agassiz (ed.), *Letters and Recollections of Alexander Agassiz* (Boston, 1913); Merle M. Odgers, *Alexander Dallas Bache* . . . (Philadelphia, 1947); William Healey Dall, *Spencer Fullerton Baird* (Philadelphia, 1915); Daniel C. Gilman, *The Life of James Dwight Dana* (New York, 1899); Captain Charles H. Davis, *Life of Charles H. Davis, Rear Admiral* (Boston, 1899); Jane Loring Gray (ed.), *Letters of Asa Gray* (2 vols.; Boston, 1893); John M. Clarke, *James Hall of Albany* (Albany, 1921); Thomas Coulson, *Joseph Henry: His Life and Work* (Princeton, N.J., 1950); Mary Lesley Ames, *Life and Letters of Peter and Susan Lesley* (2 vols.; New York, 1909); Charles Schuchert and Clara M. Levene, *O. C. Marsh, Pioneer in Paleontology* (New Haven, 1940); Emma Rogers (ed.), *Life and Letters of William Barton Rogers* (2 vols.; Boston, 1896); George P. Fisher, *The Life of Benjamin Silliman* (2 vols.; New York, 1866), John F. Fulton and Elizabeth H. Thomson, *Benjamin Silliman, 1779–1864: Pathfinder in American Science* (New York, 1947). In addition to these volumes, there is much pertinent information in the *Biographical Memoirs* of the National Academy of Sciences (Washington, 1877——) and in two volumes of biographical sketches and studies: W. J. Youmans (ed.), *Pioneers of Science in America* (New York, 1896), and David Starr Jordan (ed.), *Leading American Men of Science* (New York, 1910).

RECENT SOURCES

THE FOLLOWING BOOKS and articles, published since 1960, have revealed new dimensions of Agassiz's career. In 1962 I edited the *Essay on Classification* (Cambridge, Mass.: Harvard University Press), with bibliographic sources corrected and modernized. In 1974 I published *Nature and the American Mind: Louis Agassiz and the Culture of Science* (New York: Neale Watson Academic Publications), a study of Agassiz and the Penikese experience. Stephen J. Gould, *Ontogeny and Phylogeny* (Cambridge, Mass.: Harvard University Press, 1977), contains a brilliant analysis of the recapitulation concept and its significance for Agassiz's thought. Mary Pickard Winsor, "Louis Agassiz and the Species Question," in William Coleman and Camille Limoges (eds.), *Studies in the History of Biology*, vol. 3 (Baltimore: Johns Hopkins University Press, 1979), is similarly outstanding for her contribution to illuminating Agassiz's position on evolution. S. M. Andrews, *The Discovery of Fossil Fishes in Scotland up to 1845* (Edinburgh: Royal Scottish Museum, 1982), is a singular work of historical and scientific detection, placing Agassiz's investigation of fossil fishes in new perspective by underscoring his indebtedness to others and the imperfection of some results compared to public reputation. A similar fundamental reappraisal has been accomplished by Jane Oppenheimer in "Louis Agassiz as an Early Embryologist in America," in Randolph S. Klein (ed.), *Science and Society in Early America* (Philadelphia: American Philosophical Society, 1986). Two books assess the work of two of Agassiz's most notable students: Lester D. Stephens, *Joseph LeConte: Gentle Prophet of Evolution* (Baton Rouge: Louisiana State University Press, 1982), and David N. Livingstone, *Nathaniel Southgate Shaler and the Culture of American Science* (Tuscaloosa: University of Alabama Press, 1987). Robert V. Bruce, *The Launching of Modern American Science, 1846–1876* (New York: Alfred A. Knopf, 1987), provides fulsome detail on the culture Agassiz entered in 1846 and on the Scientific Lazzaroni.

437

INDEX